A History of Astronomy

from 1890 to the Present

David Leverington

A HISTORY OF ASTRONOMY

from 1890 to the Present

Springer
London Berlin Heidelberg New York
Paris Tokyo Hong Kong
Barcelona Budapest

Front cover: The Great Vienna Telescope made by Howard Grubb of Dublin in 1880 with a 27 inch (69 cm) diameter objective. For five years it was the largest refractor in the world. The picture of the Hubble Space Telescope is reproduced by courtesy of NASA.

ISBN 3-540-19915-2 Springer-Verlag Berlin Heidelberg New York

British Library Cataloguing in Publication Data
Leverington, David
History of Astronomy: From 1890 to the Present.
I. Title
520.904
ISBN 3-540-19915-2

Library of Congress Cataloging-in-Publication Data
Leverington, David, 1941–
A history of astronomy from 1890 to the present / David Leverington
p. cm.
Includes bibliographical references (p.) and index.
ISBN 3-540-19915-2 (pbk. : alk. paper)
1. Astronomy—History—20th century. I. Title.
QB22.L48 1995 95-12034
520′.9′04—dc20

Printed in Great Britain
2nd Printing with corrections, 1996

Typeset by Wilmaset, Birkenhead, Wirral
Printed by The Alden Press Ltd, Osney Mead, Oxford, UK
34/3830-54321 Printed on acid-free paper

Contents

Preface

The history of astronomy is, like most history, a multidimensional story, and when writing about a specific period, the author has to decide how to handle all the developments of earlier times in order to set the scene. I have done this by starting most chapters of the book with a summary of astronomical knowledge at the beginning of our chosen period, together with a brief review of how such knowledge had been gained. This story is not only interesting in itself, but it will also assist those readers that would appreciate a brief reminder of some of the basic elements of astronomy.

It is also necessary to decide when to start our history. Should it be the year 1900 or 1890, or should it be linked to some key development or investigation, e.g. the discovery of the electron by J. J. Thomson in 1897, or the discovery of spectroscopic binary stars by Pickering and Vogel (independently) in 1889, or maybe the year 1890 in which Thomas Edison tried unsuccessfully to detect radio waves from the Sun and Johannes Rydberg published his formula for atomic spectra?

I have, in fact, decided to start this history at about 1890, as it was the year of publication of the Draper Memorial Catalogue of stellar spectra which, together with its updates, provided essential data for the understanding of stellar spectra until well into the twentieth century. This date also gives a clear hundred years up to the present.

As astronomy is such an enormous subject, I have described progress in each of the main subject areas of the Solar System, the Stars and the Galaxies sequentially, rather than try to paint the developing picture in all these areas together. Then follow parts on the development of Instruments, Facilities and Techniques mainly in the guise of Telescopes, Radio Astronomy and Space Research.

It is not practical, in one volume, to describe all the developments in astronomy over the last one hundred years, and so some selection of material is inevitable. My aim is to give an outline of developments, and enable those who wish to investigate the subject further, to do so by consulting the books listed at the end.

This book is written for the reader with a basic understanding of astronomy, but if I have failed to convey the story clearly enough in some areas, I would be grateful to receive any suggestions for improvement.

I would like to express my special thanks and appreciation to Alan Cooper, Roger Emery, Mike Inglis and Stuart Clark who had the kindness and patience to read the text and suggest modifications of both fact and style, to make the book more accurate and readable. If there are any errors of fact, misinterpretation or misrepresentation remaining, however, they are entirely mine and I apologise in advance.

Finally, in any project of this nature, the pressure does not stop with the writer, and this book would not have been completed except for the patience and encouragement of my wife Chris. She deserves an extra vote of thanks, for putting up with me and piles of books and papers all over the house, during the time that it has taken to turn this book into a reality.

Introduction

Astronomy is the oldest and most fundamental science. It attempts to explain not only what the Universe is today and how it works, but also how it started (if indeed there was a starting point), how it evolved to the present day and how it will develop in the future.

Astronomy is also about a Universe that can be seen, free of charge, by anyone who cares to glance at the night sky, weather and light pollution permitting. Can anyone looking at a really dark sky, girdled by the Milky Way, not be impressed by what he sees, and wonder how we on our small insignificant planet fit into all of this?

For many centuries man has observed the heavens and has tried to explain what he sees. To us, with the benefit of hindsight, a great many of the explanations are obvious, but in the past, many excellent astronomers have held ideas which seemed logical to them, but now seem strange to us. For example, William Herschel, one of the greatest astronomers of all time, believed, at the end of the eighteenth century, that the surface of the Sun was dark underneath its bright atmosphere, and that it could be inhabited by living beings. It was only in the mid nineteenth century that it was realised that the Sun was gaseous throughout, and it required the advent of atomic physics in the twentieth century to explain how the Sun generates its heat.

There is no reason why many of our own theories should not be inaccurate, or even wrong, of course, and it is interesting to speculate how our knowledge of today may appear in time to come. Will the Big Bang theory have been abandoned, and will black holes have been shown to be just a figment of our imagination? Most astronomers think not, but who knows?

The twentieth century, for so long synonymous with progress and modern thinking, has but a short way to go before it becomes the last century. It seems an appropriate time, therefore, to review what has happened in the science of astronomy over the last hundred years or so. A period in which travel and communications, for example, have been revolutionised by scientific discoveries and technological developments which appear to be occurring at an ever increasing rate.

We tend to think of the period in which we live as completely different from any other because of this fast rate of progress. Yet this same feeling is prevalent in many astronomy books written over the last hundred years, and for many years before then. In 1881, for example, Edmund Ledger wrote (Reference 1) "The progress of astronomical science during the last five-and-twenty or thirty years has been so rapid as almost to approach the marvellous." We are tempted to wonder what the situation will be 100 years from now. Looking backwards in time may help us to get some perspective. Who knows what radical new theories are just around the corner, and what theories that we think secure today will turn out to be wrong in the future?

A hundred years ago, Newtonian physics ruled supreme. Planck's Quantum Theory, Einstein's Theory of Relativity and Heisenberg's Uncertainty Principle did not exist. The photographic plate and spectroscopy were still relatively modern developments; radio astronomy and space research were both some decades in the future. What did people know, or rather not know, about the Universe and how has this knowledge developed since then? That is the story of this book.

1 The Sun

Early Work*

Sunspots, Prominences and the Disturbed Sun

As long ago as the early seventeenth century, the famous Italian astronomer Galileo Galilei had discovered that the Sun was rotating on its axis by observing the motion of sunspots across the Sun's disc. In 1859 Richard Carrington, an English amateur astronomer and son of a wealthy brewer, found that the surface of the Sun was rotating faster at the equator than at middle latitudes. This was correctly interpreted by the Jesuit priest and astronomer, Angelo Secchi, as showing that the Sun was gaseous.

Close observation of the surface of the Sun showed that, not only was it speckled with sunspots (Figure 1.1), but it had a generally granulated structure which gave the appearance of bright clouds over a darker substrate. Sunspots were thought, by some astronomers, to be nothing more than holes in the bright photosphere through which this darker substrate could be seen. The problem was that the lower layers of the Sun ought to be hotter and lighter than the surface, rather than darker. Hervé Faye, of the École Polytechnique, sidestepped this problem by proposing that a sunspot is a whirlpool, sucking into it cooler and hence darker gases, mainly hydrogen, from the outer regions of the Sun. There did not appear to be any vorticular motion around the sunspots, however, so Secchi suggested in 1872 that matter was ejected from the Sun at the edges of the spot. This matter then cooled and fell back into the centre of the spot, producing its dark central region.

Large loops and filaments of rapidly moving gas, were seen at the edge of the Sun during total solar eclipses. These prominences, as they were called,

---

*Sections headed "Early Work" outline the state of knowledge in about 1890.

Figure 1.1 *Drawing of the Sun made in 1870 by Tacchini at Palermo showing sunspots and bright markings, or faculae. (From* Memoirs of the Italian Spectroscopical Society, *Vol. vi.)*

had first been seen in 1733, but it was not until 1860 that they were proved to be connected with the Sun, rather than the Moon, when it was observed that they did not follow the Moon as it crossed the solar disc during an eclipse. But what were the prominences?

P. Jules Janssen, a French astronomer working at Meudon Observatory near Paris, went to India to observe the total solar eclipse of 18th August 1868, where he was able to examine the spectrum of a large prominence visible at totality. He and other eclipse watchers found that the spectrum consisted of a number of bright hydrogen lines (see Pages 290–291), showing for the first time that prominences consisted of hot hydrogen gas. They also found, in the same spectrum, a bright yellow line that was initially taken to be the D line of sodium, but which was afterwards found to be of a slightly different wavelength. Janssen decided to see how long after totality he could see these bright prominence lines, finding, much to his surprise, that they could still be seen when the eclipse was completely over.

Independently, 2 months later, J. Norman Lockyer in England also found that he could see the bright line spectrum of prominences in full sunshine, something that he had long suspected should be possible and, by a strange coincidence, the notification of his discovery reached the French Academy of Sciences on the same day as that of Janssen. Lockyer also saw the yellow line in the prominence spectrum, suggesting that it was due to a new element which he called helium, after *helios*, the Greek word for the Sun. Not all astronomers accepted that this yellow line was due to a new element, however, and in 1890 the origin of the line was still unresolved.

The most spectacular (or eruptive) prominences were seen to be connected with sunspots, whereas the more cloud-like prominences were associated with faculae, or very bright markings, on the surface of the Sun (they can be seen near the edge of the Sun in Figure 1.1). The eruptive prominences contained hydrogen, helium and various metallic elements, of which iron, titanium, calcium, barium, strontium, sodium and magnesium were the most evident, whereas the cloud-like prominences, associated with faculae, consisted almost entirely of hydrogen, helium and calcium. Sunspots tended to occur in zones from about 10° to 40° latitude on either side of the equator and were rarely seen outside these zones, whereas faculae were seen at any latitude.

Over a period of 6 months in 1852, Edward Sabine in England, Rudolf Wolf in Switzerland and Alfred Gautier in France independently concluded that there was a strong correlation between the occurrence of large sunspots and disturbances in the Earth's magnetic field (and the visibility of the aurora borealis). The actual mechanism for the link between the Sun and these disturbances was unknown, although the existence of such a link indicated that sunspots may be connected with electric disturbances on the Sun.

In the late eighteenth century, the great English astronomer William Herschel had suggested that the heat output of the Sun should be increased when there were many sunspots, and he attempted to find a correlation between sunspots and the Earth's weather. In the absence of good meteorological data, he tried to correlate sunspot numbers with the price of wheat, which should depend on the weather, but with no success. In 1843 Heinrich Schwabe of Dessau in Germany discovered that the number of sunspots on the Sun varied with a period of about 10 years. Rudolf Wolf then showed, by looking through historical records, that the average period between maxima was about 11.1 years, but with a range varying from 7 to 17 years. The reason for this sunspot cycle was unknown, and no-one was able to detect any associated variation in the solar heat output. There continued to be much speculation about a possible link between sunspot numbers and the Earth's weather, although there was little evidence of an 11 year weather cycle. Although no such weather cycle could be found, E. Walter Maunder at Greenwich did find that sunspots had been virtually non-existent between 1645 and 1715, coinciding with a period of relatively cold winters in the Earth's northern hemisphere. No other correlations of sunspots with the weather were known.

Carrington found that the solar latitudes at which sunspots were seen gradually changed over the 11 year sunspot cycle, starting at about 30° to 40° latitude immediately after sunspot minimum, and gradually reducing as the cycle progressed. Sunspots were at about 15° latitude when the cycle went through its maximum, and at about 6° latitude when the minimum of the cycle was reached. After the minimum, sunspots then started to appear at mid latitudes once more.

The spectrum of the Sun as a whole was known to consist of a bright continuous spectrum, crossed by numerous dark lines called Fraunhofer absorption lines, after the German physicist Joseph Fraunhofer who first investigated them in the early nineteenth century (see Page 290). Gustav Kirchhoff, working in Heidelberg, published the first analysis of these Fraunhofer absorption lines in the solar spectrum in 1861. Then, 5 years later, Lockyer discovered that these absorption lines were often broader for light coming from sunspots than for normal sunlight, because the material in sunspots is in more rapid motion than for the remainder of the Sun. There were also very sharp bright lines in the middle of the hydrogen and calcium absorption lines for light coming from sunspot areas. These bright lines were generally attributed to high layers of hot gas above sunspots.

The Quiet Sun

Charles Young, who was professor of Astronomy at Princeton, New Jersey, discovered a most remarkable effect when he observed the limb of the Sun through his spectroscope during the total solar eclipse of December 1870. At the moment when the Moon cut off light coming from the visible disc of the Sun, the previous bright continuous solar spectrum with dark absorption lines was replaced instantaneously by a bright line emission spectrum. The bright lines were in apparently the same places as the previous dark ones. This "flash spectrum", as it was known, was seen for only a second or two, disappearing when the Moon covered the thin "reversing" layer that generated these bright lines just above the visible surface of the Sun. Clearly the hot gas of the reversing layer was emitting a bright line spectrum in its own right but, as it was cooler than the surface of the Sun, it also absorbed these same wavelengths from the light emitted by the surface, causing the Fraunhofer absorption lines in the normal solar spectrum.

So the lowest level that can be seen on the Sun was known to be the compressed gas of the hot photosphere which gives the continuous part of the solar spectrum. Above that is the cooler reversing layer, which produces most of the Fraunhofer absorption lines, then comes the chromosphere and prominences and, finally, the corona and the zodiacal light, which are discussed below.

By 1890 it was known that hydrogen, calcium, magnesium, carbon, sodium, iron, titanium and various other metals, making a total of 34 elements, were present in gaseous form in the reversing layer, and that hydrogen, helium and calcium were present in the chromosphere. The proportions of the elements in the various layers of the Sun were unknown, although it was assumed that the layers of the Sun were cooler the further from the centre they were, with the corona being the coolest part of all.

The bright solar corona, which was seen to surround the Sun during total solar eclipses, was known to extend for millions of kilometres from the Sun, but in the words of Professor Young (Reference 2), little was known about it other than it was "an inconceivably attenuated cloud of gas, fog, and dust, surrounding the Sun, formed and shaped by solar forces". During the total solar eclipse of 1868 the corona was seen to have a faint continuous spectrum, indicating that it was scattering sunlight, and this was confirmed by Janssen in 1871 when he found faint Fraunhofer absorption lines in its continuous spectrum. In the meantime, during the total solar eclipse of 7th August 1869, the Americans Charles Young and William Harkness had independently discovered a bright green spectral line in the corona, indicating that the corona was also self-luminous. The bright green line was thought to be due to iron.

The presence of the heavy element iron in the corona provided a major problem, as astronomers generally believed that the heavy elements like iron were only in the lower reversing layer of the solar atmosphere, with the cooler, more tenuous higher layers containing only the lighter elements. In 1876 Young showed that the line of iron in the laboratory was a doublet, while the coronal line was single, casting doubt on the cause of the coronal line, although in 1890 most astronomers still thought that the line was due to iron.

The two total solar eclipses of 1889 proved that the solar corona is more uniform in shape at times of solar maximum than at times of solar minimum. Sometimes, in the latter case, the corona could be seen to extend up to 15 million km from the Sun, in the plane of the Sun's equator.

The zodiacal light is seen just after sunset, or just before sunrise, as a faint band of light which is brightest near to the horizon. Nineteenth century astronomers concluded that it was aligned along the ecliptic, and that it extended from the Sun to beyond the Earth's orbit. Giovanni Cassini, the director of the Paris Observatory, had suggested that the zodiacal light was due to a very large cloud of dust-like particles, but others, including the French mathematician Pierre Laplace, had thought that it was due to a very tenuous gas. Neither theory was proven.

The Colour and Temperature of the Sun

One of the biggest problems facing astronomers in 1890 was the lack of a coherent theory of radiation to enable them to understand the Sun and stars. (Wien's black body radiation law, for example, relating the wavelength of peak energy output to temperature, was not discovered until 1893).

In 1879 it was found experimentally by the Austrian physicist Joseph Stefan that the total radiation energy output for a perfect emitter (or "black body") is proportional to the fourth power of its absolute temperature. This was a major step forward in the theory of heat, and it limited the wild guesses of the surface temperature of the Sun produced in previous decades. In order to produce an accurate estimate of the surface temperature of the Sun, however, it was necessary to know how much energy was absorbed by both the solar and Earth's atmospheres.

The Sun was known to appear brighter at the centre than the edge, and this was attributed to the absorption of the Sun's atmosphere. Hermann Vogel in Germany, and Samuel Langley and Edward Pickering in America independently estimated this absorption from measurements made at various wavelengths at a number of points across the solar disc. These indicated that the Sun's atmosphere absorbed significantly more blue light than red, leading to the belief that the surface of the Sun was both brighter and bluer than it appeared to be.

The French astronomer Jules Violle studied the absorption of the Earth's atmosphere in 1875, comparing sea level measurements of the Sun with those taken from the summit of Mont Blanc. A few years later, Langley analysed similar measurements made from the summit of Mount Whitney in the United States. These studies indicated a preferential absorption of blue light by the Earth's atmosphere, leading to the conclusion that the Sun would look distinctly blue, and be three or four times brighter, if the solar and Earth's atmospheres were removed. The surface temperature of the Sun was then estimated to be about 10,000 K*.

The Generation of Heat

The biggest question about the Sun that exercised the minds of nineteenth century astronomers was, how does the Sun generate its heat? It was known that the Sun could not generate enough heat by mere combustion, as this would have allowed it to exist for only a few thousand years. So how could the Sun have produced heat and light for at least the age of the Earth, which was thought to be measured in hundreds of millions of years?

The generally accepted theory was first proposed by the Scottish engineer John Waterston, who suggested that the Sun was generating heat by its gravitational contraction. His idea was developed by Hermann von Helmholtz in 1854, who calculated that the reduction in diameter required to produce its present heat output was about 75 m per year, but even this mechanism could only have kept the Sun producing heat at the required level for something like 25 million years. Unfortunately, this did not seem to be long enough, as geological analysis of the surface of the Earth indicated that the Earth had been receiving heat from the Sun for much longer than that.

So the 1890 astronomers had a problem. Were Helmholtz's calculations correct, or was there a completely different mechanism which had not been discovered that could have provided the Sun's heat for a much longer time? Was the Earth really much older than 25 million years? The answers to these questions were uncertain in 1890, although most geologists were convinced that the answer to the last question was "yes".

The Temperature of the Sun and its Generation of Energy

The temperature of the Sun had been estimated by 1890, using the Stefan–Boltzmann law of radiation, as about 10,000 K. In 1893 Wien showed that

---

*Temperatures are generally quoted in kelvin, abbreviated to K. They are degrees above absolute zero, which is −273°C.

the wavelength of the maximum energy radiated from a perfect black body is inversely proportional to its temperature, and this enabled a much more accurate estimate of the solar temperature to be made of 6,000 K.

The American engineer James Homer Lane and the German physicists Arthur Ritter and Robert Emden, like many other researchers, had assumed that heat was transported from the interior to the exterior of the Sun by convection, but in 1894 the British astronomer R. A. Sampson suggested that the primary mechanism in the Sun's atmosphere was radiation. Karl Schwarzschild, the director of Göttingen Observatory, then used this concept to explain the limb darkening of the Sun. Arthur Eddington (Figure 1.2) working at Cambridge University extended the concept of radiative equilibrium to the internal structure of the Sun and stars (see Page 136) and deduced, in 1926, that the Sun's central temperature was a startling 39 million K.

Cecilia Payne, in her doctoral thesis at Harvard, made the revolutionary proposal that hydrogen and helium were the main constituents of the atmospheres of the Sun and stars. Her suggestion was not immediately accepted and she met with some resistance, until Albrecht Unsöld, a young German astrophysicist working at the Mount Wilson Observatory, persuaded Henry Norris Russell of Princeton University that hydrogen was present in the solar atmosphere in enormous quantities. Russell, whose views were widely respected, published his definitive paper on the subject in 1929, concluding that hydrogen was the main constituent of the solar atmosphere. Three years later, Eddington concluded that hydrogen was the main constituent in the Sun as a whole. This required modifications to

Figure 1.2 *Pioneering astronomers of the 1920s taken during an International Astronomical outing to Plymouth, Mass. in 1932. left to right: A. S. Eddington, J. S. Plaskett, W. S. Adams, J. H. Oort, H. N. Russell, H. Shapley, W. K. Miller, F. W. Dyson, F. Slocum and B. Lindblad. (Courtesy* Sky and Telescope *Magazine.)*

Eddington's estimate of the central temperature of the Sun, which he reduced to 19 million K in 1935.

In the nineteenth century the best theory of energy production in the Sun was the Waterston–Helmholtz contraction theory described above, but this could only explain a solar lifetime of about 25 million years or so, which even at that time appeared to be too short.

Eddington suggested two alternative mechanisms for energy generation in the Sun in 1920, based on Ernest Rutherford's and Francis Aston's recent research into atomic structure at Cambridge University. Energy could be produced either by the mutual annihilation of protons and electrons, or when hydrogen atoms fuse to make helium atoms and atoms of higher mass. The Sun could go on shining using either process for billions* of years. Then, in 1938, Hans Bethe in America and Carl von Weizsäcker in Germany independently proposed a fusion theory that was so convincing that the alternative mass annihilation theory rapidly fell out of favour. They suggested that solar energy is produced by hydrogen nuclei being transformed into helium nuclei, with carbon as a catalyst (see Page 143), and Bethe estimated that the central temperature of the Sun would be 18.5 million K, assuming a composition by weight of 35% hydrogen. This was virtually identical with Eddington's estimate of 19 million K, which was based on gas dynamics and was independent of the energy-production process. Bethe also estimated that the Sun would continue to produce energy for another 12 billion years.

Further work has shown that these estimates of the core temperature of the Sun were somewhat too high, and a more likely temperature of 15 million K was established. This in turn meant that the proton–proton cycle proposed by Charles Critchfield (see Page 142) is dominant in the Sun, rather than the carbon cycle assumed by Bethe and von Weizsäcker.

In 1931 the distinguished Austrian physicist Wolfgang Pauli had proposed the existence of a massless particle, called a neutrino, to explain the conservation of energy in a nuclear process called beta decay, but it was not until the 1950s that the first concrete evidence was found for the existence of the neutrino. At this time the possible nuclear processes in the Sun were also being analysed in more detail, and it was shown that there were three different proton–proton (PP) reactions contributing to the Sun's energy. Although most of this energy is emitted by the Sun in the form of electromagnetic radiation (like light), a small amount is carried away by neutrinos, so in the early 1960s Raymond Davis of the Brookhaven National Laboratory decided to try to measure these neutrinos to check on the theory of energy generation. But how could these neutrinos be detected?

Theory showed that if neutrinos with a minimum energy of 0.81 MeV† react with the isotope chlorine 37 they will produce argon 37. Unfortun-

---

*I have used the American definition of billion as 1,000 million in this book, with apologies to British traditionalists, as this is now the general practice in astronomy.
†Million electron volts. 1 MeV $\equiv 1.6 \times 10^{-13}$ J.

ately, of the three PP reactions in the Sun, only the so-called PP III reaction produced neutrinos with energies significantly above this 0.81 MeV threshold, and this reaction was expected to be the rarest of the three in the Sun. Nevertheless, in 1967 the Brookhaven National Laboratory commissioned a neutrino detector which consisted of an 85,000 gallon (380,000 litre) tank of dry-cleaning fluid, perchloroethylene, to detect these PP III neutrinos, by measuring the amount of argon produced. The tank was large as the number of neutrinos expected to be detected was very small (only about one every three days even with this tank), and it was placed deep underground in the Homestake gold mine in South Dakota to reduce spurious signals from energetic cosmic rays.

Early results produced by Davis showed a neutrino flux that was too low but, as it was close to the detection limit of the system, improvements were made in 1970 to allow more accurate results to be produced. These confirmed that the number of neutrinos was too low by at least a factor of two. One possibility was that there was something wrong with Davis' equipment, but his results were confirmed in the late 1980s by Kamiokande II, which was a different type of detector located in the Japanese Alps.

Even today the discrepancy between the observed and expected flux of neutrinos coming from the Sun has not been satisfactorily explained. It is expected that the cause of the problem will be some subtle twist in the theory of neutrinos or in the theory of energy generation and transfer through the Sun. But it would not be the first time in history that an apparently small nagging inconsistency between theory and experiment has resulted in the major modification, or even abandonment, of a major theory which is convincing in so many other respects – in this case the theory of energy generation in the Sun and stars.

The Corona

Young and Harkness had discovered a bright green emission line in the spectrum of the Sun's corona in 1869, which was thought to be due to iron. In 1898, however, it was shown that the wavelength of this coronal line was 530.3 nm,* and not 531.7 nm as previously measured. No known terrestrial line fitted this new wavelength, and so a previously unknown element called coronium was proposed as the cause. Early in the twentieth century, this was becoming a more and more unlikely possibility, as the gaps in the periodic table of the elements were filled.

Ira Bowen, of the California Institute of Technology (Caltech), had solved a similar problem with emission lines in nebulae in 1927, when he found that they were caused by ionised oxygen and nitrogen which made what are

*Nanometres. 1 nm = 10^{-9} m or 10 Å.

called "forbidden transitions".* Similar explanations failed with the coronal line which, by now, had increased to 19 different lines. None of these lines could be explained.

In the early twentieth century, astronomers assumed that the Sun was hottest in the centre, becoming cooler as the Sun and its atmosphere was traversed all the way out to the corona. Mitchell's work on the flash spectrum of the Sun in the early part of this century showed, however, that the solar atmosphere was more ionised the further it was from the surface. Using Bohr's theory of the atom (see Page 295), this seemed to indicate that the higher layers of the atmosphere were at a higher temperature. The Bengali physicist Megh Saha showed in 1920, however, that high ionisation can also be explained by low gas pressure and, as the gas pressure in the corona was known to be low, the concept of the temperature of the Sun's atmosphere decreasing with increasing height returned. The corona, which surrounds the atmosphere, was still thought to be cool, but not for long.

Walter Grotrian, of the Potsdam Astrophysical Observatory, concluded in 1934 that the coronal temperature must be an astonishing 350,000 K, in order to explain its spectrum. Then, 3 years later, Grotrian read a report by the Swedish physicist Bengt Edlén describing the emission spectra of iron atoms stripped of nine or ten of their electrons by high voltage sparks. These spectra could explain two of the coronal lines, so he wrote to Edlén explaining his idea, suggesting that Edlén looked into the matter further. Edlén sent a polite reply, but didn't investigate Grotrian's suggestion.

Grotrian decided, in 1939, to go into print with his ideas, and this persuaded Edlén to check through his unpublished spark spectra, where he found that calcium atoms, stripped of 11 or 12 of their electrons, appeared to produce two more of the coronal lines. After further work, he finally found that 13 times ionised iron atoms produced the intense 530.3 nm coronal line.

So, in 1941, Edlén concluded that the coronal lines are produced by highly ionised iron, calcium and nickel atoms, in a corona with a temperature of at least 2 million K, but how such a temperature could be produced when the photosphere had a temperature of only 6,000 K was a mystery. In 1963 Herbert Friedman and his team at the American Naval Research Laboratory recorded the X-ray spectrum of highly ionised iron in the solar corona confirming Edlén's conclusion.

Observations of the corona in the nineteenth century had showed that it

---

*When an atom is excited, one of its electrons goes from a low energy state to a high energy state. It usually stays in the latter for a very short period of time (typically about 10^{-8} s), before it spontaneously reverts to a lower energy state, releasing energy. Some atoms have some energy states with very long lifetimes (the so-called metastable states), and, in laboratory conditions, these atoms lose energy by collision with other atoms, before they have had time to spontaneously lose energy. In the very tenuous gas of a nebula, however, the collision frequencies are very low, and such atoms lose energy spontaneously by so-called "forbidden transitions" from the metastable states, thus producing emission lines not observable in laboratory conditions.

appeared to emit a continuous spectrum with Fraunhofer absorption lines, in addition to the emission line spectrum discussed above. It was thought that this continuous spectrum was due to sunlight scattered by particles in the corona, but it could have been produced by sunlight being scattered in the Earth's atmosphere. It was not until the early twentieth century that the coronal origin was proven.

Sunspots and the Disturbed Sun

Charles Young discovered in 1892 that, at very high dispersions, many absorption lines in the sunspot spectrum appeared to have a sharp bright line in their centre. Shortly afterwards, the Dutch physicist Pieter Zeeman showed, in the laboratory, that spectral lines can be split into two when they originate in the presence of a magnetic field, with each of the lines oppositely polarised.

In 1908, George Ellery Hale and Walter Adams found, at the Mount Wilson Solar Observatory, that photographs of the Sun taken in light of the 656.3 nm hydrogen line showed patterns around sunspots that looked like iron filings in a magnetic field. Hale wondered if the patterns could be caused by magnetic fields associated with sunspots. If so, maybe the Young effect, mentioned above, could be caused by Zeeman splitting of the absorption lines. To check on this, Hale examined sunspot spectra, after the light had been passed through a polariser, and found that the two components of the spectral line pairs had opposite polarisations, confirming that they were caused by Zeeman splitting in a magnetic field. By comparing the splitting of the spectral lines with those produced in the laboratory, he was also able to estimate the magnitude of the magnetic fields associated with the sunspots as about 3,000 gauss, or 10,000 times that of the Earth's magnetic field measured at the surface of the Earth.

Hale and his colleagues at Mount Wilson started monitoring the polarities of sunspots, hoping to find a pattern, but they found spots of both polarities on both sides of the equator. They then noticed that the spots generally occurred in pairs, with one spot of the pair almost always having a different polarity to the other. In addition, the sunspots that were in the lead of the pair as they moved across the disc, the so-called preceding or p spots, were all found to have the same polarity in one hemisphere, but the opposite polarity in the other hemisphere. This pattern was well established by 1912, when the observers at Mount Wilson noticed that the polarities of the p spots had reversed in both hemispheres at the solar minimum that year.

Hale had used the new 60 ft (20 m) high solar tower telescope to discover the magnetic fields in sunspots and, on the strength of that work, he had obtained funding from the Carnegie Institution for an even larger tower telescope (see Page 267). This new telescope came into operation just after the polarity reversal had been discovered. Armed with these two magnificent instruments, Hale and his staff continued to monitor sunspot polarities, expecting them to reverse again, possibly at the solar maximum expected in 1917. No change was observed at that maximum, and so the next solar minimum, expected in 1922, was awaited with great interest. On 24th June 1922, a single spot of the new cycle was found by Ferdinand Ellerman, one of Hale's colleagues, to have a reversed polarity. (Single spots normally have the polarity of a lead spot). The evidence was not conclusive, but by the following year the polarity reversal had become clear, leading Hale to suggest that the solar cycle should be considered as a 22 year cycle, rather than an 11 year cycle, as it would take 22 years for the spots to revert to their original polarities.

In 1908, Alfred Fowler of London University and Hale independently proved (see Page 22) that sunspots are cooler than the surrounding Sun. In the same year, John Evershed at the Kodaikanal Solar Observatory in India found that gas flows outwards from sunspots. Then, five years later, C. Edward St John, one of Hale's colleagues, found that the gas moves back inwards towards the spots at higher levels in the atmosphere, thus confirming Secchi's theory of the nineteenth century.

Hale developed Secchi's theory in the early part of the twentieth century to explain how sunspots occur in pairs, with one spot having an opposite polarity to the other. He suggested that the two members of a pair are at the top of the two sides of a U-shaped tube, which is joined together under the surface of the Sun. Gas spirals inside this U-shaped tube so, as seen from above, the gas rotation is in opposite senses at either end of the tube, thus producing spots of opposite polarities. Hale's theory was further developed by the Norwegian meteorologist Vilhelm Bjerknes who suggested, in 1926, that all the spots in one hemisphere are part of a single tubular vortex that runs under the surface of the Sun. From place to place, this tube breaks through the surface, producing one spot, and travels a short distance before diving below the surface once again, producing a second spot. This theory explained why the preceding spots in one hemisphere have the same polarity as each other.

Bjerknes also pictured the tubular vortex as gradually moving towards the equator as the sunspot cycle progresses, thus explaining the observed change in the preferred latitude for spots through the cycle. It was also suggested that there were two tubular vortices in each hemisphere, one being near the surface, as outlined above causing sunspots, and one being much further down, so that it does not break through the surface. While the upper vortex moves towards the equator during the sunspot cycle, the lower one moves from the equator towards higher latitudes. When the upper one reaches the equatorial region it falls deep into the Sun, while the lower one, which has now reached high latitudes, rises to the surface and starts a new solar cycle.

Sabine, Wolf and Gautier had found a correlation, in the mid-nineteenth century, between sunspots and disturbances in the Earth's magnetic field (and visibility of the aurora). Maunder then showed in 1913 that the larger magnetic storms on Earth start about 30 hours after a large sunspot crosses the centre of the solar disc. Smaller storms did not seem to be generally associated with sunspots, however, but Chree and Stagg showed in 1927 that they had a tendency, not shown by the greater storms, to have a recurrence frequency of 27 days, which is the Sun's synodic period of rotation. The German geophysicist Julius Bartels called the invisible source on the Sun of these smaller magnetic storms, M regions. Both the larger magnetic storms (called flare storms) and the smaller storms (called M storms) were assumed to be caused by particles ejected by the Sun that had a travel time of about a day.

In the mid-1920s, as short-wave communications became important, a number of radio engineers also noticed brief, sudden fade-outs in radio reception. As the sunspot cycle was near to its maximum, some engineers wondered if these fade-outs were solar-related, and then E. Hans Mögel, a young radio engineer in Berlin, showed that they only occurred on the daylight side of the Earth. In 1935 near the next solar maximum J. Howard Dellinger, of the American Bureau of Standards, investigated four major radio fade-outs that had occurred that year. He checked on solar activity with the Mount Wilson solar observatory and, much to his delight, discovered that major solar flares had occurred just before two of the fade-outs. Unfortunately, astronomers at the Mount Wilson Observatory were not observing the Sun at the time of the other two fade-outs.

Over the next two years, Dellinger investigated over 100 radio fade-outs, but found that, although there was a correlation with solar flares, that correlation was by no means perfect. He suggested that some unknown solar phenomenon was generating invisible radiation that penetrated the Earth's ionosphere which is responsible for reflecting short-wave radio signals. This invisible radiation then ionised atoms at lower levels in the atmosphere, causing the radio signals to be absorbed, rather than transmitted.

In 1942 James Hey, a physicist working in the British Army Research Group, found that the Sun emitted radio waves which interfered with radar signals and, 3 years later in Australia, Joe Pawsey, Ruby Payne-Scott and Lindsay McCready found that the radio emissions from the Sun increased with greater sunspot numbers.

Horace Babcock, of the Mount Wilson Observatory, designed and built a solar magnetograph in 1952 to measure weak magnetic fields on the Sun, at the request of his father Harold Babcock, a retired solar physicist. Their magnetograph measured the magnitude of the Zeeman effect across the Sun, and with it they were able to detect magnetically disturbed regions of the Sun, both before and after the appearance of sunspot groups. In 1954 they were also able to measure the general magnetic field of the quiet Sun as about 1 gauss (or 10^{-4} tesla).

The Babcocks noticed that there were both bipolar and unipolar regions on the Sun. In the bipolar regions the magnetic flux leaving the Sun was about equal to that entering it, and in the unipolar regions the magnetic flux was either leaving or entering the Sun. They found that the bipolar regions could last for as long as 9 months, with sunspots forming and dispersing within them, whereas the unipolar regions did not appear to be connected with sunspots at all.

They suggested that ions and electrons leaving the Sun in bipolar regions would follow the field lines above those regions, and would collide over the Sun, generating radio noise and forming visible prominences. Ions and electrons (called corpuscular radiation) leaving the Sun in unipolar regions, on the other hand, would stream away from the Sun, and some of them would reach the Earth, causing radio interference. Corpuscular radiation would also leave the Sun near the poles and follow the Sun's general field lines, but how far out in space these field lines went was unknown.

Horace Babcock returned to his work on magnetic stars in 1955, while Harold continued to observe the Sun, discovering the reversal of its general field 2 years later, during solar maximum.

In 1963 Conway Snyder and Marcia Neugebauer of NASA's Jet Propulsion Laboratory (JPL) showed, using data from the American Mariner 2 spacecraft, that there was an excellent correlation between the velocity of the high-speed ions of the solar wind and the level of geomagnetic activity on Earth, thus proving that magnetic storms on Earth are caused by high-speed ions from the Sun. These ions were the invisible ionising radiation postulated by Dellinger, and the corpuscular radiation deduced by Harold and Horace Babcock. Try as they might, however, Snyder and Neugebauer were unable to pinpoint the sources on the Sun of these high-speed streams, the areas that Bartels called M regions, and which the Babcocks had identified as unipolar regions.

The breakthrough in the search for the solar M regions came in 1973 with the launch of the American solar observatory, Skylab, when Allen Krieger, Adrienne Timothy, and Edmond Roelof compared X-ray intensities of the

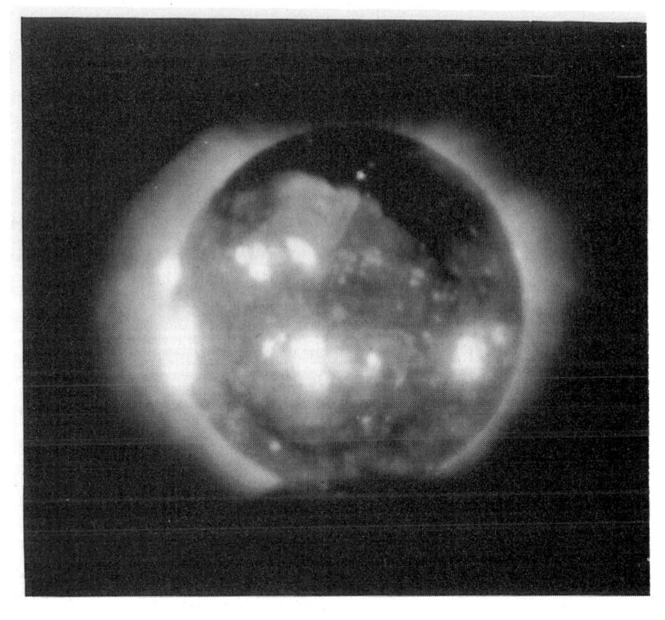

Figure 1.3 *The Sun in X-rays taken by Skylab in 1973. The dark region near the top of the disc is a coronal hole. (Courtesy* NASA/Smithsonian Astrophysical Laboratory.*)*

Sun with the velocity of the solar wind. They found that a large coronal hole, which was dark in X-rays, correlated with a fast solar wind, and further work by other researchers, showed a good correlation between coronal holes and a fast solar wind. These results demonstrated that the Babcocks' idea was basically correct, that fast corpuscular radiation emanates from unipolar regions where the magnetic field lines run freely into space. The M regions are coronal holes (see Figure 1.3).

The Quiet Sun and the Interplanetary Plasma

Carrington had observed the rate at which sunspots crossed the solar disc, and had discovered, in the mid-nineteenth century, that the surface of the Sun rotates more quickly at the equator than at middle latitudes. In addition, the Doppler shift of the spectral lines at both edges of the Sun had been measured by Young in 1876, and this technique helped to determine the rotation rate of the Sun at high latitudes, which are usually devoid of sunspots.

Faculae, unlike sunspots, occur at all latitudes. So Maunder decided in 1924 to look at the historical records of faculae in order to improve estimates of the rotation period of the Sun at high latitudes. Doppler-shift measurements were also perfected independently in the early twentieth century by St John, Adams and Evershed, showing that the photosphere, reversing layer and lower chromosphere rotate around the Sun at different velocities.

The famous German astronomer Johannes Kepler had suggested, in 1619, that comets' tails point away from the Sun because material coming from the head of the comet is pushed away by particles of solar radiation, but in the nineteenth century this theory was no longer considered plausible, because light was then thought to be a wave motion, not a stream of particles. The generally accepted theory in the nineteenth century, was that this effect is due to electrical repulsion between the Sun and the cometary material.

In the twentieth century, light was shown to be both a wave motion and a particle stream (a stream of photons), and light was observed to exert pressure on objects in its path. Thus Kepler's explanation of the direction of comets' tails became popular once more.

As the twentieth century progressed, solar physicists were beginning to think that a weak continuous flux of charged particles was evaporating from the Sun, causing the persistent faint aurorae at the Earth's poles, and constant minor fluctuations of the Earth's magnetic field. They suggested that this continuous flux of charged particles was channelled towards the Earth's poles by its magnetic field.

Ludwig Biermann, of the Max Planck Institute for Physics in Göttingen, suggested, in 1951, that comets' tails point away from the Sun because they are affected by the same continuous stream of solar ions and electrons that cause minor magnetic disturbances and faint aurora on the Earth. Biermann showed, theoretically, that radiation pressure could not exert enough force to produce comets' tails, but charged particles could do so if their velocities were between 500 and 1,000 km/s, and their density between 100 and 1,000/ cm^3 at the distance of the Earth, resulting in an average plasma flux of about 4×10^{10} ions/cm^2/s.

In the nineteenth century, the corona had been seen to extend up to 15 million km from the Sun, and the zodiacal light appeared to extend even beyond the Earth's orbit. Alfred Behr and Heinrich Siedentopf of Göttingen's Astronomical Institute calculated, in 1953, that the density of electrons in the zodiacal light near the Earth was about 600/cm^3, which was consistent with Biermann's calculations for the solar plasma near the Earth. Two years later the English geophysicist Sydney Chapman calculated that a very hot corona, consisting mainly of electrons and protons, should exist beyond the Earth's orbit. Chapman's extended solar corona was basically static, whereas that of Biermann was clearly dynamic. They couldn't both be right; either the particles emitted from the Sun in Biermann's model would drive away Chapman's extended corona, or Chapman's corona would distort Biermann's solar wind. Eugene Parker of the University of Chicago's Enrico Fermi Institute tried to link the two theories and, when he added dynamic terms into Chapman's equations, he found that the results were consistent with Biermann's model.

Over the next few years, Parker developed these concepts, but there was a basic problem because, even when he chose Biermann's lowest density estimate of 100 ions/cm^3, and Edlén's lowest coronal temperature of 2 million K, he could find no satisfactory energy source from the Sun to keep the solar wind blowing. His theory showed, essentially, that the corona temperature could not be maintained at 2 million K, with such a strong and relatively dense wind blowing through it, given the likely energy input to it from the Sun. He suggested that the energy input could be increased by hydromagnetic waves propagating through the photosphere, but this suggestion was dismissed by other solar physicists.

The only way to help to resolve this dilemma was to measure the solar plasma in situ, which became a distinct possibility with the advent of the space age, starting with the launch of Sputnik 1 on 4th October 1957.

The first spacecraft to measure the plasma, away from the influence of the Earth's magnetic field, was the Soviet Union's Lunik 2, en route to the Moon, in September 1959. The results indicated that the extended corona, at the distance of the Earth from the Sun, contained high-speed ions, thus supporting Biermann's theory of a dynamic environment. The measured ion (or plasma) flux of 2×10^8 ions/cm^2/s appeared to be two orders of magnitude lower than Biermann's prediction, however, because the measured plasma density was only 1 or 2 ions/cm^3, compared with Biermann's prediction of 100 ions/cm^3, minimum (see the Table below).

The American Explorer 10, launched in March 1961, was the first spacecraft to measure the solar wind in detail. Unfortunately it did not quite escape from the influence of the Earth's magnetic field (i.e. from the magnetosphere), and so its results were a little suspect. Nevertheless, at the Massachusetts Institute of Technology (MIT) Herbert Bridge was able to estimate a plasma flux of 4×10^8 ions/cm²/s, an average ion (probably proton) velocity of 300 km/s, and an average density of about 10 ions/cm³. Explorer 10 was also able to measure the direction of flow of the plasma, which the earlier Lunik 2 could not, and it was found to be away from the Sun, as expected.

In summary, the Lunik and Explorer results compared with Biermann's prediction as follows:

	Biermann's prediction (1951)	**Lunik 2 (1959)**	**Explorer 10 (1961)**
Plasma velocity (km/s)	500–1,000	~ 1,000	120–660
Plasma density (ions/cm³)	100–1,000	1–2	6–20
Plasma flux (ions/cm²/s)	4×10^{10}	2×10^8	4×10^8
Flow of plasma	Away from Sun	Not measured	Away from Sun

The plasma fluxes measured by Lunik 2 and Explorer 10 were very similar to each other, but the plasma density measured by Explorer was appreciable higher than that measured by Lunik. Was this because Explorer had not escaped completely from the Earth's magnetosphere? Only a new measurement campaign would tell.

Mariner 2, launched in August 1962, had a more sensitive ion detector on board, designed by Conway Snyder and Marcia Neugebauer of JPL. It generated over 3 months' worth of scientific data, covering virtually the whole of the spacecraft's trajectory from the Earth to Venus, which it flew by in December 1962. The plasma velocity varied between 400 and 700 km/s, but occasionally it reached the maximum detectable speed of 1250 km/s, with flux and density measurements similar to those of Explorer 10. The solar wind was observed to flow continuously from the Sun and, on a number of occasions, it showed sudden increases in both density and velocity. The ions were generally found to be protons, but some helium nuclei were also detected.

Explorer 18 (otherwise called IMP 1) was launched in November 1963 into a highly eccentric orbit, with an apogee of almost 200,000 km. The major axis of the orbit was originally oriented at about 30° to the Sun–Earth line, with the apogee on the sunward side of the Earth. The spacecraft thus spent about 75% of each orbit in interplanetary space, outside of the Earth's magnetosphere and the bow shock wave where the solar wind is deflected by the Earth's magnetic field. As the mission proceeded, however, the angle between the major axis of the spacecraft's orbit, and the Sun–Earth line, gradually increased, because of the Earth's orbital motion around the Sun, and so the spacecraft spent less and less time in the undisturbed interplanetary plasma. That time reduced to zero in mid February 1964 but, in the meantime, the Sun had rotated through almost three revolutions.

In 1958, Parker had examined the way that ions of the solar wind would "carry" the Sun's magnetic lines of force into space to create the interplanetary magnetic field. If the ions left the Sun radially from its equatorial region at about 400 km/s, for example, they would take about 4.5 days to reach the position of the Earth's orbit, and, during this time, the Sun would have rotated through about 600. Parker suggested, therefore, that the Sun's magnetic lines of force would be carried into space in a spiral pattern, similar to the pattern produced by a lawn sprinkler, and, at the position of the Earth's orbit, the lines of force of the resulting interplanetary field would make an angle of about 450 to the Sun–Earth line.

Norman Ness of the Goddard Space Flight Center and John Wilcox of the University of California analysed the results of the Explorer 18 magnetometers, and found in 1964 that, although the field lines did make an angle of about 450 to the Sun–Earth line, there were a number of cases when the direction of the field suddenly changed by 1800. The boundaries at which the field changed direction were in the same position, relative to the Sun's surface, for each of the (almost) three revolutions of the Sun that had been observed, except when a storm on 2nd December suddenly caused the field direction to change from sunward to anti-sunward.

Ronald Rosenberg and Paul Coleman of the University of California, after analysing the interplanetary magnetic field data collected by ten spacecraft, concluded, in 1969, that the field polarity of the Sun was generally different in its two hemispheres. A number of research groups analysed the constant stream of satellite data being sent to Earth over the next few years and, by 1975, there was a broad consensus on the general shape of the solar wind's three-dimensional structure. The neutral sheet, that separated the opposite field polarities, was approximately in the Sun's equatorial plane, but it was crinkled, or warped, so that, as the Sun rotated, alternately positive and negative field polarities were detected by spacecraft in Earth orbit. This three-dimensional model was given strong support, in 1976, when Edwin Smith reported that the sector pattern, between alternate field polarities, disappeared when Pioneer 11 was above 15°N in heliographic co-ordinates.

Around the solar maximum of 1979, when the Sun's general field changed polarity, there was a confusing situation which lasted for about 3 years, in total. The neutral sheet changed a good deal during this period, and secondary sheets were often present.

The Solar Constant

William Herschel had suggested, in the late eighteenth century, that the heat output of the Sun (the so-called "solar constant") should be higher when there are sunspots, with a consequential effect on the Earth's weather. No variations in solar heat output had been detected in the nineteenth century, however, and the only link with the Earth's weather was the mini Ice Age, between 1645 and 1715, which correlated with a period during which there was an almost complete absence of sunspots (in the Maunder Minimum, as it is called).

Charles Greeley Abbott, who was to become the director of the Smithsonian Astrophysical Observatory, attempted to measure variations in the heat output of the Sun starting in 1902, but he found that atmospheric effects were drowning any solar variations. In 1911, therefore, he started to obtain data from dry mountainous areas in America and Africa. At first, he thought that he had detected variations of 3% in the monthly mean values, with daily variations of about 10%, but as his researches continued, he kept re-analysing his data, producing differing results. So his data were regarded as questionable by many astronomers.

Abbott was hampered by both the variability of the Earth's atmosphere and the poor accuracy of the radiometers available at the time. Little progress was made in measuring variations in the solar constant until it became possible, with the advent of the space age, to make continuous measurements of the Sun from above the Earth's atmosphere.

In the early 1960s, temperature readings of the surface of several spacecraft were found to be inconsistent with the pre-launch predictions based on ground simulation tests. The differences could be due to errors in the solar constant used, inaccurate radiometers in the simulation chamber, poor thermal modelling, or in-orbit changes in the thermal properties of spacecraft materials. In 1964, JPL, which was carrying out investigations into the problem, gave a contract to the Eppley Laboratory of Newport, Rhode Island, to measure the solar constant, to see if the correct value had been used in the ground tests.

Andrew Drummond, the chief scientist at the Eppley Laboratory, decided to build a 12-channel radiometer, using thermopiles as detectors and, in 1967, it was flown at an altitude of 77 km in the X-15 rocket aircraft. The measured value of the solar constant, of 1361 W/m^2 $\pm$ 1%, was about 2.5% lower than the previously accepted figure, which partially solved the problem.

Unfortunately, the Eppley thermopile radiometer flown on the Nimbus 6 spacecraft in 1975 produced unreliable results, so Nimbus 7, that was launched in October 1978, included an Eppley *cavity* radiometer (see Page 305). This was to prove much more reliable, and, between November 1978 and May 1979, it measured an average solar constant of 1376 W/m^2, which was very similar to the X-15 result of the previous decade. More importantly, John Hickey, of the Eppley laboratory, found reductions of 0.4% in the solar constant in both August and November 1979, which correlated with periods of intense solar activity. These Nimbus 7 results were consistent with the observation that sunspots were cooler than undisturbed areas of the Sun's surface.

Meanwhile, John Eddy had been researching the historical data linking sunspots to climatic change on the Earth while he was working as a part-time lecturer at the University of Colorado. He found, in 1975, that the production of carbon 14 by cosmic rays was much higher than usual during the Maunder Minimum, according to the isotopic abundance of carbon 14 measured in tree rings. This was consistent with the known effect that cosmic rays are more prevalent during periods of low solar activity.

Eddy also analysed data on naked-eye sunspots, aurorae at medium latitudes, and carbon 14 abundances, for earlier centuries, and found evidence for a minimum in solar activity in the twelfth and thirteenth centuries, like the Maunder Minimum, which he called the Spörer Minimum. The consistency of the data between these different parameters then gave him confidence in using carbon 14 data for even earlier periods. He was able to identify a number of periods of maxima and minima in solar activity over the last 5,000 years, and show, at least for the northern hemisphere, that periods of prolonged minima were associated with severe winters. This effect could, conceivably, be caused by disturbances of the Earth's magnetosphere or ozone layer, by variations in the solar wind or ultraviolet emissions, respectively, but Eddy preferred the more direct theory, that these climate changes were linked to a change in the heat emitted by the Sun.

Unfortunately, Eddy could find no 11 year cycle in measurements of the solar constant made between 1908 and 1955 at the Smithsonian Observatory. This led him to conclude, in 1976, that the correlations between sunspots (or solar activity), the solar heat output, and the Earth's weather, were only present in the longer term (i.e. over decades or centuries), rather than in the shorter term (daily, monthly or yearly). He also concluded that the solar constant only varied over the longer term, and not over the 11 year solar cycle.

Peter Foukal, a solar physicist at the Harvard–Smithsonian Center for Astrophysics, suggested, in 1975, that sunspots may reduce the solar flux, while faculae may increase it, and he tried to use data on sunspot and faculae variabilities to try to correlate them with solar flux variations. Foukal, Pamela Mack and Jorge Vernazza analysed the old Smithsonian data and radiometer data from Mariners 6 and 7, to see if they could detect any variations in the solar constant that could be correlated with sunspots or faculae. The Mariner data showed no correlation of solar flux with sunspots or faculae, whereas the Smithsonian data indicated that, while increased sunspot areas reduced the solar flux, increased facular areas increased it, exactly as Foukal had suggested.

Foukal, Mack and Vernazza suggested that the reason that variations in solar flux were detected on Earth (using Smithsonian data), rather than by satellites in Earth orbit, was because changes in the Sun's ultraviolet radiation was causing the Earth's atmospheric transmission to change. This reasoning was rejected by Foukal and Vernazza 2 years later, however, because they could not conceive of a mechanism in the Earth's atmosphere that would react quickly enough to such changes in the Sun's output. They concluded, instead, that they had measured a real short-term correlation between intense magnetic fields on the Sun, which are associated with sunspots and faculae, and the radiant energy emitted by the Sun. A similar correlation between high solar sunspot activity, and reductions in the Sun's irradiance, was found, independently, in the same year by John Hickey using Nimbus 7 data (see above). Foukal and Vernazza published their results in 1979, and Hickey published his in 1980.

The American Solar Maximum Mission spacecraft, that was launched in February 1980, had, in its payload, a high precision cavity radiometer designed by a group led by Richard Willson at JPL. Using this radiometer, Richard Willson and his collaborator Hugh Hudson, a solar physicist at the University of California, found a reduction of 0.15% in the solar irradiance on 9th April 1980, compared with the average solar flux measurement of 1368.3 W/m^2 for the first five months of operation. They also found that a second reduction of 0.08% occurred on 28th May and that both of these reductions correlated with the passage of large sunspot groups across the central solar meridian. So both the Nimbus 7 and Solar Max spacecraft had shown that reduced levels of solar flux correlated with high sunspot activity over timescales of a few days. This was, of course, completely contrary to Eddy's conclusions following his analysis of historical data, that cold winters, and low levels of solar flux, correlated with low solar activity.

There were many small variations in the solar irradiance measurements made by Solar Max, in addition to the large reductions which correlated with large sunspots. Some of the reductions in solar flux were found to be due to small sunspots, and some of the increases were due to faculae. Initially, there were many disagreements between the various groups analysing the data, but, by 1986, a general consensus had been reached that, on average, increases in solar irradiance due to faculae, balanced reductions due to sunspots.

It had been known since 1960 that the surface of the Sun oscillates with a period of about 5 minutes, but, in 1973, Robert Dicke of Princeton University discovered that the whole Sun is oscillating or vibrating with this period. Martin Woodard and Hugh Hudson of the University of California then decided to examine the Solar Max irradiance data in 1980, to see if variations with a similar period could be detected in it. They were soon successful, and announced their discovery of solar irradiance oscillations, with a period of 5 minutes, in the following year.

The Solar Max spacecraft lost its pointing control in December 1980, and for the next $3\frac{1}{2}$ years the irradiance data were much more difficult to interpret, with a consequent loss in accuracy. In April 1984, however, NASA organised a successful repair mission. There was considerable interest to compare the high accuracy solar irradiance data for 1984, near the solar minimum, with that of 1980, near the solar maximum. Willson, Hudson, Fröhlich and Brusa reported, in 1985, that there had been a reduction of 0.02% per year in irradiance for the first 5 years of operation of the Solar Max radiometer and, in 1988, they reported that the reduction in irradiance had levelled out, and it was increasing now that the sunspot cycle had passed though its minimum. Shortly afterwards, similar results were published by Hickey and R. A. Kerr for Nimbus 7. Thus on this longer timescale of a solar cycle, low solar activity was correlated with low solar irradiance. (Eddy had been almost correct, therefore, although he had thought that such correlations only occurred over timescales of tens of years, at least, and not over the 11 year solar cycle.)

As the presence of sunspots had been shown to be correlated with reduced solar irradiance, their progressive reduction in numbers towards sunspot minimum would cause the solar irradiance to increase, not decrease as had been observed. Faculae could play a role in this reduction of solar radiance towards solar minimum, however, as their effect was known to be contrary to that of sunspots. Foukal and Lean investigated this, and in 1988, they showed that the reduced solar irradiance, between 1980 and 1984, correlated with a reduced solar faculae index that represented the output from both faculae associated with the disturbed parts of the Sun, and the bright facular grains all over the Sun. So faculae seem to be the dominant source of solar irradiance variations over periods as long as a solar cycle. Maybe the solar heat output was lower during the Maunder Minimum when sunspots, and possibly faculae, were virtually non-existent.

The Solar Spectrum

Lockyer, Janssen and others had, in 1868, seen a bright yellow line in the spectrum of prominences, and Lockyer had suggested that this was caused by a new element called helium. This explanation was disputed by some astronomers in the late nineteenth century, but the matter was resolved when helium was discovered on Earth in 1895.

Henry Rowland at Baltimore, using his new concave diffraction grating, published a catalogue in 1896 of 20,000 Fraunhofer lines in the normal solar spectrum, and this enabled him to identify 36 terrestially-known elements in the Sun, which was increased to 51 by St John in 1928. The spectrum of the Sun is basically that of an ordinary G-type star of the Harvard classification system (see Page 123).

The spectral sensitivity of Rowland's photographic plates limited his work to the range of from 295 nm to 733 nm, but his work was the standard against which all other work was judged for over 50 years. In the late 1940s, the range was extended in the infrared up to 1,350 nm by Babcock and Moore, although results above 900 nm are affected by absorption of the Earth's atmosphere. Infrared detectors allowed further expansion of the detection range up to 24 microns (24,000 nm) in 1952, but the use of these detectors only came into their own with the development of high altitude observatories, sounding rockets and satellites (see Chapters 13 and 16) that operated above most or all of the absorbing atmosphere.

Norman Lockyer suggested that sunspot spectra indicated that sunspots were cooler than the surrounding photosphere. Extensive work by Adams, King and Gale supported Lockyer's suggestion and then, in 1908, Fowler and Hale independently showed that various molecules (including titanium oxide) were present in sunspots. These molecules, which can only exist at lower temperatures, were the first unambiguous evidence of the lower temperatures of sunspots. Finally, Charlotte Moore of Princeton, who worked at the Mount Wilson Solar Observatory, published the first extensive catalogue of sunspot spectra in 1933, showing that the high temperature lines in normal sunlight were weaker in the sunspot spectrum, whereas the low temperature lines were stronger, thus proving that sunspots were regions of lower temperature.

2 The Moon

Early Work

The Moon's Orbit

The Greek astronomer Hipparchos had understood the basic geometry of the Moon's orbit around the Earth over 2,000 years ago. He realised that the Moon's orbit was not circular, and that the plane of the Moon's orbit around the Earth made an angle to the ecliptic (the plane of the Earth's orbit around the Sun).

The apogee of the Moon's orbit is the point at which it is furthest from the Earth, and the perigee is where it is nearest. Hipparchos realised that the line of apsides (the line joining the apogee and the perigee of the Moon's orbit) was not fixed in space, but progressed with a period of 8.85 years, and that the line of nodes (the line of intersection of the plane of the Moon's orbit with that of the Earth) was regressing (going backwards) with a period of 18.6 years.

Over the next 2,000 years, the detailed movements of the Moon were subject to great scrutiny and analysis, so that by the late nineteenth century they were quite well understood. Unfortunately, however, there are so many disturbing influences to the Moon's motion, and that of the Earth from which it was observed, that there were still small unexplained residuals.

The Moon revolves around the Earth as the Earth revolves around the Sun, and so the Moon's motion in the solar system is basically governed by the gravitational pull of both the Earth and Sun. The Earth's orbit around the Sun is, however, perturbed by the Moon and planets, so the planets have a small (or second order) effect on the Moon, which is currently causing the eccentricity of the Earth's orbit to reduce and the Moon to slowly accelerate in its orbit around the Earth (see Page 48). The effect is very small but, in 1853, it was shown by the gifted Cambridge mathematician and astronomer John Couch Adams to advance the sidereal position of the Moon in its orbit by about 5.7 arcsec* in 100 years. This effect is countered, to some extent, by the tides in the seas and oceans of the Earth, which slow the Moon down. Although the planetary effect has been the larger over the last few thousand years, it is a periodic effect; sometimes, as at present, accelerating the Moon in its orbit, but sometimes slowing it down. So, in the longer term, the tidal effect, which consistently slows the Moon down, causing its orbit to get gradually larger, is the more important.

It is often said that the Moon revolves around centre of the Earth, but it does, in fact, revolve around the centre of mass of the Earth–Moon system, which is about 5,000 km from the centre of the Earth. Accurate observations of the Moon's movement enabled the position of this centre of mass to be determined and this, in turn, enabled the mass of the Moon to be calculated relative to that of the Earth. This and other methods in the nineteenth century gave a very accurate estimate of the density of the Moon of about 62% that of the Earth.

The Lunar Surface

Astronomers generally thought that the craters on the Moon had been formed millions of years ago by volcanic eruptions when the Moon still possessed a large molten core, although in 1824 Franz von Paula Gruithuisen of Munich had suggested that the craters had been produced by the impact of meteorites. In the nineteenth century there was no way of knowing which theory was correct.

Did the Moon possess an atmosphere? If it did, it was known to be very tenuous. Arthur Berry (Reference 2), for example, thought in 1898 that the Moon had an atmosphere, although its density was less than 1% of that of the Earth's, whereas Anton Pannekoek (Reference 3), quoting observational data from 1834, could see no evidence for an atmosphere and suggested that the maximum density could be only 0.05% of that of the Earth at best. Berry further suggested that, if a tenuous atmosphere did exist, it could be of considerable depth. (It is interesting to note that Robert Ball, in Reference 4, made a similar suggestion for possible large volume/low density atmospheres on the minor planets or asteroids; see Page 105. Clearly the implications of the concept of escape velocity on planetary atmospheres, outlined by the Irish physicist Dr Johnstone Stoney in 1870, had not been fully appreciated.)

The question of whether there was life on the Moon or planets was one that fascinated many people in the nineteenth century; astronomers and the general public alike. By 1890, most astronomers had concluded that there was no life on the Moon because it had no water and lacked a significant

---

*1 arcsec ≡ 1/60 arcmin ≡ 1/3600 degree.

atmosphere. It was considered quite likely, however, that there would be life elsewhere in the solar system, albeit very different from that on Earth.

The Origin of the Moon

George Darwin's theory of the origin of the Moon, which he proposed in 1879, was generally accepted in the late nineteenth century. George, the second son of the famous biologist Charles Darwin, suggested that when the proto-Earth was still molten, it had gradually contracted as it cooled. Its rotation rate had increased as it contracted (according to the principle of conservation of angular momentum), and when its rate had increased to about 3 to 5 hours per revolution, it had broken into two unequal parts, one being the Earth and the smaller one the Moon. Darwin realised that the rotation of the proto-Earth would not have been sufficient on its own to cause fracture, and suggested that the break-up had been facilitated by a resonance* coupling between the tidal forces acting on the proto-Earth from the Sun, and the natural oscillation frequency of the molten body.

After break-up of the proto-Earth, both the Earth's rotation on its axis and the Moon's rotation around the Earth had slowed down because of tidal forces, as both bodies were still molten, causing the Moon's orbit to become gradually larger. Tidal forces of the Earth on the less-massive Moon also locked the axial spin rate of the Moon to that of its orbital rotation period around the Earth, thus ensuring that the Moon always had its same side facing the Earth. The gravitational attraction of the Earth also caused the Moon to have a small Earth-facing bulge, which remained facing the Earth, even after the Moon had almost completely solidified.

Originally, when the Moon had just separated from the Earth, the time taken for the Moon to orbit the Earth (i.e. the month) was the same as that taken for the Earth to rotate on its axis (i.e. the day) at about 3 to 5 of our current hours. Both the month and day have been getting longer since then, but the ratio of the month to the day has increased from the initial value of 1 to about 27 today. This ratio was thought to have gone through a maximum value of about 29 some time ago, and so it is 27 now and reducing. If the Earth–Moon system could be isolated from all other outside influences, particularly that of the Sun, the ratio would probably continue reducing until it was 1 again, but instead of the day and month both being equal to 3 to 5 hours, they would probably both be equal to about 1,400 hours (about 58 of our current days). In this case the Earth, as well as the Moon, would have its same side permanently facing the other body.

Darwin had shown theoretically that the proto-Earth could have been unstable at a speed of about 3 to 5 hours per revolution, but it was left to the great French mathematician Henri Poincaré a few years later to show that, in such a case, the body would have broken up into two unequal parts.

---

*Resonance occurs when the frequency of application of periodic forces acting on a body exactly match the natural oscillation frequency of that body.

In the late nineteenth century it was recognised that there was a possible problem with Darwin's theory, as there would have been a tendency for the Earth to break up the Moon by tidal forces when it was still very close. This break-up would not happen immediately after separation, however, and the Moon could possibly have moved away from the danger area before it had had time to break up. Whether this was possible or not was unclear.

If the Earth and Moon had once been part of the same body, the internal temperature of the Moon today would be much lower than that of the Earth, as the Moon is much smaller and would, therefore, have cooled down faster. There were no active volcanoes seen on the Moon, but it was not known whether the centre of the Moon was still molten or whether it had solidified. There had been a number of sightings of small changes on the surface of the Moon but they were hotly disputed. The only change, having some level of acceptance, being the disappearance of the 10 km crater Linné which was found by J. F. Julius Schmidt, the director of the Athens Observatory, to have been replaced by a small white spot in 1866. Some observers agreed with Schmidt, but Johann Mädler, the great lunar expert, and others could find no change in Linné's appearance. Whether Linné had disappeared or not, it was clear that the surface of the Moon was very stable.

The Surface

The question as to whether there is life on the Moon was still unresolved until well into the twentieth century. William Pickering, the brother of Edward Pickering who was director of the Harvard College Observatory, suggested in 1903 that there may be plants, snow and river beds on the Moon. Then in 1921 he suggested that there may be a low form of vegetation that completes its life cycle in the 14 days of sunlight, being fed by carbon dioxide released from rock fissures. In 1969 the first astronauts to set foot on the Moon spent some time in quarantine, on their return to Earth, in case they had been contaminated with elementary organisms. No such organisms were found, even as fossils, and so there has probably never been life of any sort on the Moon.

In 1924 Bernard Lyot, of the Meudon Astrophysical Observatory near Paris, deduced that the Moon was covered in volcanic ash, by measuring the polarisation of light scattered from the lunar surface. This conclusion was consistent with the 190 K fall in temperature observed for the surface of the Moon during the 1927 lunar eclipse by S. B. Nicholson and E. Pettit at the Mount Wilson Observatory. In the 1950s Thomas Gold of Cornell University and others had concluded that the lunar maria are covered in volcanic ash or dust up to a few metres deep, but other estimates put the thickness at no more than a few centimetres. If Gold was correct this would cause a problem with possible spacecraft landings.

In the late nineteenth and early twentieth century it was still not clear whether the lunar craters had been formed by the impact of solid bodies or by volcanic eruptions. A number of astronomers favoured the impact theory that Gruithuisen had outlined in 1824, and every now and again it had been resurrected in various forms over the next hundred years or so, although most astronomers favoured the volcanic theory.

One of the problems, that the adherents of the impact theory had to solve, was why most of the craters are circular, in spite of the fact that the impacts should have occurred at all angles of incidence to the surface. Grove K. Gilbert, the renowned geologist, suggested that the impacting bodies were small natural Earth satellites, which would have had only small velocities relative to the Moon, and so would have fallen onto the Moon almost vertically. After the First World War, however, it was realised that the shape of lunar craters resembled shell craters, and that, as craters are formed by the shock wave of the impact or explosion, a non-vertical impact can still produce a circular crater. Later calculations showed that the shock wave produced by a 1 km meteorite impacting the Moon at a velocity of 30 km/s could, for example, have produced the 100 km diameter crater Copernicus. Spacecraft probes have now shown that most of the lunar craters were formed by meteoritic impact, although some are probably of volcanic origin.

American lunar spacecraft showed in the 1960s that the crater Linné, which was thought to have possibly disappeared in the nineteenth century, still exists (see Figure 2.1). During the twentieth century, however, reports of other small changes on the Moon have continued. In particular, Dinsmore Alter took some photographs that appeared to show a temporary local obscuration in the large walled plane of Alphonsus. On 3rd November 1958, Nikolei Kozyrev, of the Crimean Astrophysical Observatory, photographed the spectrum of a bright red patch that appeared briefly in the same area, and, on 30th October 1963, Greenacre and Barr of the Lowell

Figure 2.1 *The crater Linné on the Moon photographed by an Apollo spacecraft. (Courtesy* National Space Science Data Center, World Data Center-A for Rockets and Satellites, NASA; Principal Investigator, F. J. Doyle; the Apollo program.*)*

Observatory saw a disturbance in the Aristarchus area. Thus the surface of the Moon does not appear to be completely dead.

The Soviet Lunik 1 spacecraft started man's exploration of the Moon in January 1959, showing that the Moon had virtually no overall magnetic field and, later that year, Lunik 3 took the first photograph of the far side of the Moon. It appeared to be broadly similar to the visible side, although there are no large maria. Seven years later the Soviets soft-landed Lunik 9 near the edge of the Oceanus Procellarum on the Earth-facing side and showed that, at least at the landing site, the surface dust was very thin, contrary to Gold's prediction of deep dust.

The American Surveyor 5 spacecraft soft-landed on the Moon in 1967 and undertook a chemical analysis of the soil, confirming that it was made of basalt (i.e. volcanic rock) as most astronomers had predicted. Rock samples returned from Tranquillity Base by Apollo 11 were also basalt which appeared, from their microscopic structure, to have been ejected from depths ranging from 150 to 400 km.

The Apollo astronauts set up seismometers on the Moon to measure moonquakes, and a number of their lunar modules were deliberately crashed onto the surface. The Moon was found to vibrate for almost an hour after such deliberate impacts.

The Apollo seismometers and thermometers indicated that the Moon still has a liquid core of about 800 km diameter, with a temperature estimated at over 1,000 K. Above that is a partially molten asthenosphere, followed by a 1,000 km iron-rich mantle, capped by a crust about 70 km thick. The surface regolith is of loose rock only a few metres thick. Unlike the Earth, the Moon has only one lithospheric plate, so there has never been any plate tectonics.* The altimeters in the orbiting Apollo command modules showed that the surface of the Moon is several kilometres nearer to the centre of the Moon on the near (i.e. Earth-facing) side than on the far side, indicating that the material in the nearest hemisphere is slightly more dense than in the furthest hemisphere.

The isotope potassium 40 decays to argon 40 with a half-life† of 1.28 billion years, and so the measurement of the proportions of potassium and argon in crystals within basalt enables its age to be determined. It was found from the Apollo samples that lunar basalts from the "seas" or "mare" have ages ranging from 3.1 to 3.9 billion years, while the rocks from the highland regions have ages ranging from 3.5 to 4.4 billion years, the latter being almost the same as the estimated age of the Sun and Earth.

Apollo was not able to sample material from all over the Moon, and so these age ranges are minimum ranges. The ages of the youngest maria can

---

*The surface of the Earth consists of a number of plates that are colliding in some places, and separating in others. This process of movement is called plate tectonics.

†Radioactive decay is a statistical process, in that not all of the atoms of the same element decay in the same time. The half-life is the time in which 50% of the atoms decay.

be estimated by comparing their cratering densities with those of mare of known age, and doing this indicates that the youngest mare are about 2.6 billion years old.

Rock samples, and X-ray fluorescence measurements taken from orbiting Apollo command modules, showed that the lunar highlands mainly consist of a type of rock called anorthosite, which is produced when minerals crystallise out of a gradually cooling liquid. Gravitational measurements from other orbiting spacecraft showed that the rocks underneath the lunar highlands are relatively light, indicating that the whole of the 70 km lunar crust is made of anorthositic rock.

The anorthositic rock of the crust could only have been formed from a molten crust, so this proves that the Moon must have been originally molten. Before the Apollo missions, some astronomers believed that the Moon had formed as a relatively cool body, which had since warmed up by the radioactive decay of uranium, thorium, and potassium 40, but this has now been shown to be incorrect. The Moon was hot when it formed, and it has since cooled.

The ALSEP experiments left on the Moon by the American Apollo astronauts showed that the very tenuous lunar atmosphere consisted mostly of helium, which appeared to come from the solar wind, and argon, which appeared to come from the Moon as its density correlated with seismic events. Then, in 1987, Drew Potter and Tom Morgan of NASA found minute amounts of sodium and potassium in the lunar atmosphere. Four years later Mike Mendillo, Alan Stern, Jeff Baumgardner and Brian Flynn of Boston University found that the Moon has a very tenuous sodium tail, similar to that of a comet, pointing away from the Sun. Some process on the surface of the Moon must be producing sodium to replenish the sodium in the "tail", the most likely cause being the vaporisation of small amounts of the lunar surface by the continuos bombardment of micrometeorites.

The Origin and Subsequent History of the Moon

George Darwin's theory of the origin of the Moon, as being spun off from a rapidly rotating, molten Earth was shown, in the early part of the twentieth century, to be untenable. In particular, Forest R. Moulton of the University of Chicago pointed out, in 1909, that the viscosity of the proto-Earth would have been high enough to stop it from breaking up, given the known angular momentum of the Earth–Moon system. If, on the other hand, in some unknown way the spin rate had been high enough to achieve separation, so much angular momentum would have been transferred from the Earth to the Moon, that the Moon would have escaped completely from the Earth's gravitational pull, and would have become another planet of the Sun. In other words, either the angular momentum of the Earth–Moon system was the same as it is today, in which case the separation could not have taken place, or the momentum was high enough to allow separation, in which case the Moon would have escaped from the Earth's gravitational attraction.

Gerstenkorn suggested in 1955 that the Moon had been captured by the Earth, and in 1972/73 Alastair Cameron and Fremlin independently suggested that the Moon had originally been orbiting the Sun inside the orbit of Mercury. Fremlin proposed that a resonant interaction between Mercury and the Moon had caused the orbit of Mercury to become smaller and more eccentric, while the Moon had been ejected into an Earth-crossing orbit where it was captured. Analysis of lunar rocks brought back by the Apollo astronauts showed, however, that the Earth and the Moon have similar relative amounts of oxygen isotopes, indicating that these bodies were formed in the same part of the solar system.

In 1975, William Hartmann and Donald Davis proposed that the Moon had been formed following the impact of a large body with the Earth. A similar theory had been proposed 30 years earlier by the geologist Reginald Daly of Harvard University, but his suggestion had been ignored at the time. Hartmann and Davis suggested that the cores of the Earth and the impacting body had merged, but a cloud of mantle debris had been ejected by the collision, and this debris had subsequently aggregated to form the Moon. This theory solved the angular momentum problem of spontaneous fracture, and was consistent with contemporary theories of the origin of the solar system. It also explained why the Moon has a small core, and why it has proportionally less volatile elements than the Earth (as they were lost by the mantle debris that formed the Moon at the high temperatures generated by the collision). Theories of the origin of the Moon are still speculative, however, although the Apollo missions have enabled us to understand its subsequent history more clearly.

In 1988, the geochemists Richard Carlson and Gunter Lugmair dated one of the oldest samples of lunar crust brought back by the Apollo astronauts, showing that the crust solidified 4.44 ± 0.02 billion years ago, during a period of intense meteoritic bombardment. This bombardment appears to have gradually reduced in intensity about 3.9 billion years ago, when the last major lunar basin, Orientale, was formed. Over the period from 3.9 to 3.1 billion years ago, basaltic lavas from the molten region below the anorthositic crust poured onto the surface to fill the low-lying areas. Early on in this period the crust was not thick enough to support the extra weight and fractured, and these fracture features can still be seen today. Surface activity virtually ceased about 2.5 billion years ago, and the Moon's surface has been modified since then only by the occasional meteoritic impacts.

3 The Origin of the Solar System

Early Theories

Theories of the origin of the solar system that were developed in the eighteenth and nineteenth centuries had to explain, among other things, that:

(i) The planets' orbits are nearly circular and lie in approximately the same plane as each other (see Figure 3.1).
(ii) The Sun's equator is in a very similar plane to that of the planets' orbits.
(iii) The Sun revolves on its axis, and the planets revolve in their orbits, all in the same direction.
(iv) The inner planets are basically small and rocky, whereas the outer ones are much larger and less dense.

In addition, if the theory could explain the distances of the planets from the Sun, the existence of planetary moons, comets, meteorites, and so on, so much the better.

In 1755 the young German philosopher Immanuel Kant had suggested that the solar system had formed out of a nebulous mass of gas, which had developed into a flat rotating disc as it contracted. This disc rotated faster and faster, as it cooled and contracted, and, eventually, it started throwing off masses of gas as it became unstable. These masses cooled to form the planets, and the core of the disc condensed to form the Sun.

In Kant's theory, the nebula only started to rotate as it contracted, which contravened the law of conservation of angular momentum. In 1796, however, Laplace independently published a similar but more detailed theory, in which the original nebula was rotating before it started cooling and contracting. He proposed that, when the angular velocity became too fast for the nebula to remain stable, the matter near the outer edge separated and formed a ring, which eventually coalesced to form the outermost

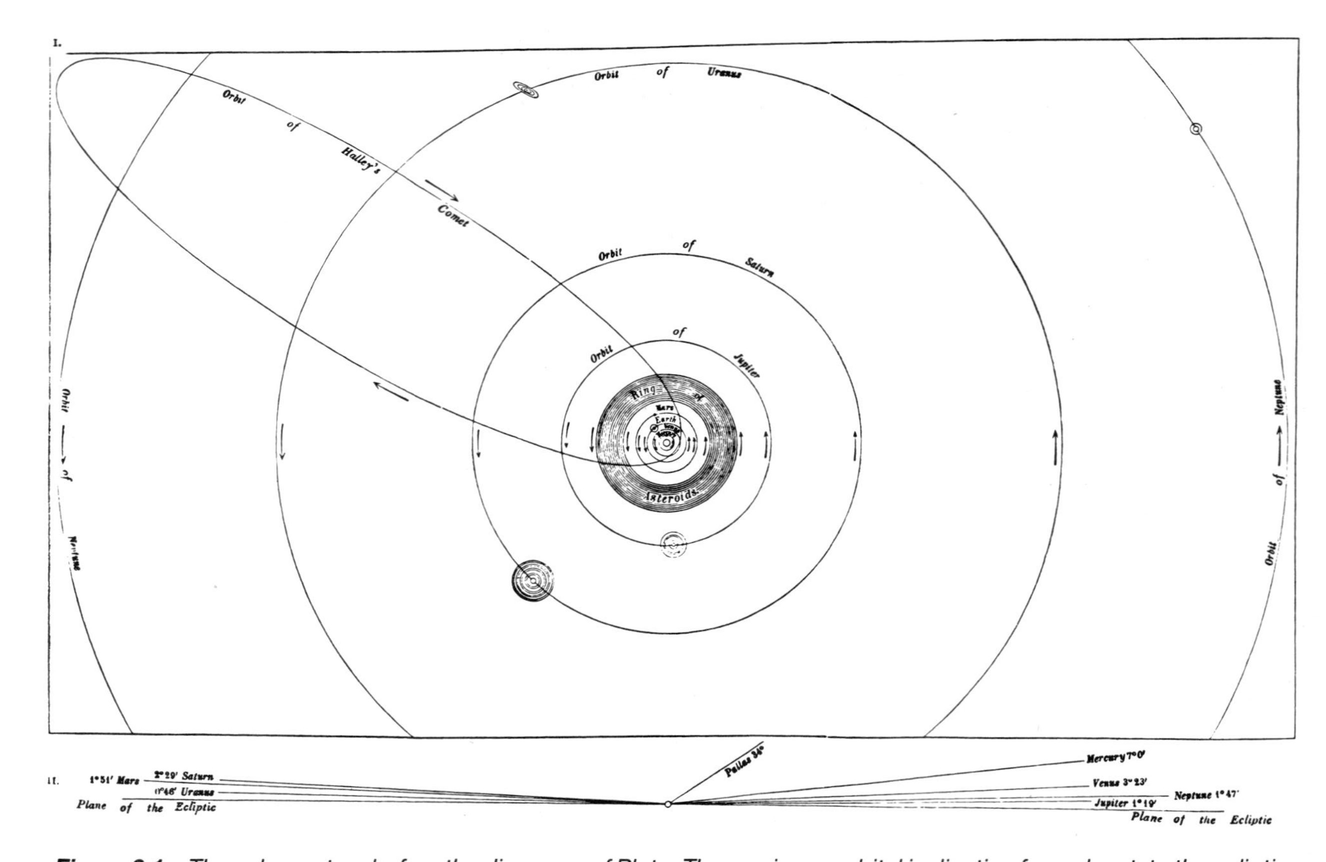

Figure 3.1 *The solar system before the discovery of Pluto. The maximum orbital inclination for a planet, to the ecliptic, is the 7° of Mercury. (From* The Heavens, *by Amédée Guillemin, 1871, Plate I.)*

planet. This process continued, with the nebula throwing off ring after ring as it contracted and speeded up, thus producing rings that eventually formed more planets. Laplace suggested that the planetary moons formed in an analogous way, from condensing rings of material that were thrown off as the protoplanets contracted and speeded up. Saturn's ring was the remainder of the original nebula that did not condense to form a moon as it was too close to the planet. Herschel's detailed observations of nebulae near the end of the eighteenth century, showed that some of them looked like the primitive nebula required by Laplace's theory, so giving it added credibility. Helmholtz's nineteenth century theory of the heat generation of the Sun (see Page 6) described the production of heat by the further contraction of Laplace's solar nebula after the planets had formed as separate entities.

Laplace's theory could not explain the fact that Uranus spun on its side or the apparent retrograde spin of Neptune (see Page 89), but it was suggested that these spin directions could have been changed by some other event after the planets had been formed. There were also other problems with Laplace's theory, but it was the best theory available in 1890, and it was thought to be broadly correct.

Laplace had suggested that the Earth's Moon, and the other planetary moons, had been formed from the condensing solar system nebula, but Darwin had proposed that the Moon had been born when the proto-Earth had split into two unequal bodies. Darwin's theory could not be extended to explain the origin of any of the other moons of the solar system, however, as they were too small compared with their primary body. Likewise, calculation of the various tidal effects between the planets and the Sun showed that the planets could not have been spun off from the early Sun.

Mercury and Venus were so close to the Sun that it was believed, in the late nineteenth century, that their axial rotation rates were the same as their orbital rotation rates, i.e. they were synchronous with one face permanently facing the Sun. It was thought that their original axial rotation rates had slowed down very quickly after formation, because of tides set up in these still-molten planets by the Sun, thus not allowing them to throw off moons like the Earth and other planets. So Mercury and Venus were the only two planets without any moons.

If the Sun and planets had been formed by the contraction of a gaseous nebula then, presumably, stars had also been formed in the same way. In that case, most stars should also possess planets, and some of these planets could also support some form of life. Nineteenth century astronomers doubted, however, that we would ever be able to see planets around stars as they were too far away.

Collisions and Close Encounters

Towards the end of the nineteenth century doubts began to be expressed about Laplace's nebula contraction theory. The angular momentum of the primeval nebula would have been very similar to that of the current solar system, because of the principle of the conservation of angular momentum. Babinet had shown in 1861 that, on this basis, the primeval nebula could not have had enough angular momentum to cause it to spin off material.

The English writer Richard Proctor suggested, in 1870, that the solar system had been formed when another star had had a glancing collision with the Sun. If the planets had been formed in this way, how had the planetary satellites formed? As the planetary satellite systems appeared to mimic the solar system, it was expected that they had been formed in the same way, and yet the probability that the Earth, Jupiter, Saturn, Uranus and Neptune had all received glancing blows was infinitesimally small for the theory to be tenable.

In 1898 the Cambridge mathematician W. F. Sedgwick suggested that the star didn't need to hit the Sun. If a star had passed close by the Sun, it could have drawn out a large amount of material by tidal attraction. The English astronomer Sir James Jeans independently put forward a similar idea in 1901, and worked out the mathematics in 1916. He showed how a long tongue of gas pulled out of the Sun would break up into individual gas clouds. The small clouds would dissipate, because of their low gravitational attraction, but the larger ones would condense into a series of gaseous spheres. Calculations showed that these spheres would have masses comparable with those of the planets of the solar system. While the gaseous planets were cooling, however, their orbits around the Sun would still be highly eccentric, and the Sun would, by tidal attraction, cause material to be pulled out of them. This would condense to form planetary satellites.

The smaller of the planetary spheres would cool to a liquid state more rapidly than the larger ones. The more liquid a sphere, the more difficult it is to break it up, and so the smaller planets resisted the Sun more successfully than the larger ones, hence Mercury and Venus have no satellites, and Jupiter and Saturn have the most.

The newly formed planets would have to plough their way through all sorts of gas, dust and condensates left over from the original event, and this would cause the planetary orbits to become more circular, explaining why most planetary orbits have a very low eccentricity.

The current planets are larger, the further they are away from the Sun, until Jupiter is reached, and then they become smaller with increasing distance. Jeans thought that this was too much of a coincidence and explained it by proposing that the tongue of gas pulled out of the Sun was thickest in its middle than at either end. The encounter with the star took place in a plane a few degrees from the equatorial plane of the Sun, which also explains why the planetary orbits are not quite in that equatorial plane.

Jeans pointed out that such close encounters would be relatively unusual, even though, in the early universe, stars were much closer together than they are now, so planets would be present around only a very small proportion of stars. This was quite a change from the nineteenth century view, where planets were thought to be present around the majority of stars.

Thomas Chamberlin and Forest Moulton of the University of Chicago proposed an alternative star encounter theory in 1905. They suggested that the Sun originally had much larger prominences than now, and these were enormously amplified by a passing star, causing the Sun to eject a great number of small clouds of gas that formed into planetesimals which, in turn, cooled and coalesced to form the present planets. Unfortunately, Jeans showed that the small clouds of gas would have dissipated before they had had time to cool to form the planetesimals. It may appear that Jeans' theory had the same problem in explaining the formation of the planetary satellites, but the gas at this stage would be much cooler than when it was ejected from the Sun in Chamberlin's and Moulton's theory, and so would be much less likely to dissipate.

The rapid explosion of theories in the early part of the twentieth century rapidly fizzled out, as every new theory was found to have serious drawbacks. For example, the American Lyman Spitzer analysed Jeans' close encounter theory in 1939, and showed that the filament of gas drawn out of the Sun would not have condensed, but would have formed a permanent gaseous nebula surrounding the Sun.

Condensing Nebulae Re-examined

Kant's theory of 1755 had been dismissed mainly because a static nebula cannot start to rotate as it contracts, as this would contravene the law of conservation of angular momentum. In the twentieth century, however, it became clear that nebulae, typical of those where stars form, are turbulent with streams of gas moving at velocities of the order of 10 km/s. The Irish mathematician William McCrea showed that, if a dense nebula with streams of gas had condensed to form the solar system, it would have had more than enough angular momentum to produce the solar system that we currently see. The key point is that the gas molecules in these nebulae are not moving completely at random, but are moving in streams, and so the nebula has a net angular momentum, even though it does not appear to be rotating. Given that there was enough angular momentum in such a nebula, how could the nebula condense to form planets?

Carl von Weizsäcker suggested, in 1943, that cells of circulating convection currents, or vortices, formed in the solar nebula after the Sun had condensed, and it was these vortices that caused small chunks of material to condense into planetesimals, in the regions bordered by vortices rotating in opposite ways. The planetesimals grew to form the planets by accretion. The planetary satellites were formed from nebulae surrounding the planets, in an analogous way. Unfortunately, Kuiper showed that the vortices would not have been stable enough to allow planetesimal condensations to take place.

Gerard Kuiper examined the statistics of a condensing nebula, and concluded that any condensing nebula must originally have had (after the Sun had condensed) about 100 times more mass than that currently residing in the planets, otherwise it would have dispersed rather than condensed. So only 1% of the nebula condensed to form the planets, the remainder being lost to the solar system.

The planetesimal theory for the formation of the Earth was resurrected by the distinguished chemist Harold Urey in the 1950s, to explain the depletion of the heavy inert gases on the Earth compared with the Sun. He reasoned that, if the Earth had formed directly from the same nebula as the Sun, the Earth's gravity would also have been sufficient to retain the heavy inert gases, but if the solar nebula had produced only small bodies, the planetesimals, their gravity would have been insufficient to retain these gases.

In parallel with Urey's work, the Soviet theorist Viktor Safronov investigated how planetesimals could aggregate to produce planets. Safronov showed how planets could form with almost circular orbits, provided they were formed by numerous collisions between small bodies. Further work in the 1970s by a number of theoreticians showed that asteroid-sized bodies could be produced by accretion in about 100,000 years, the Earth could have reached about half of its present mass in 1 million years, and be close to its present mass in 20 to 50 million years.

Safronov also found that when a planetary-sized body forms, a few secondary bodies, intermediate in size between asteroids and planets, would have grown nearby. He suggested, for example, that the collision between one of these secondary bodies and Uranus could explain why Uranus is spinning on its side, and in 1975 Hartmann and Davis suggested that a collision between a secondary body and the Earth could have formed the Moon (see Page 30). In the following year, Alastair Cameron of the Harvard–Smithsonian Center for Astrophysics and William Ward of JPL independently came to the same conclusion about the origin of the Moon, and calculated that the secondary body probably was about the size of Mars.

The theories of planetesimal accretion and subsequent collision are by no means generally accepted at present, but they have explained a great deal that previous theories could not, and so appear plausible. More than that, it would be unwise to speculate.

4 The Terrestrial Planets

Mercury

Early Work

Mercury, the innermost planet, had been known since ancient times and its phases had been seen with early telescopes. Unfortunately, Mercury is a difficult object to observe, as it is always too low down in the sky when the Sun is below the horizon, and its image is consequently disturbed by the Earth's atmosphere. As a result, it had proved impossible to see any but the vaguest surface markings with even the best telescopes available in the nineteenth century.

Johann Schröter, an amateur astronomer at Lilienthal near Bremen, thought he saw surface markings on Mercury in 1800, which led him to estimate an axial rotation period of about 24 hours. In about 1880, Giovanni Schiaparelli, who was director of the Milan Observatory, hit on the idea of studying the planet in daylight when it was well above the horizon. He concluded in 1882 that its axial rotation period was 88 days, which is the same as Mercury's period of rotation about the Sun. If Schiaparelli was right, it would mean that Mercury kept the same face turned permanently towards the Sun, like the Moon does to the Earth. This was not considered unreasonable at the time, because Mercury was so close to the Sun that tidal friction in its originally molten crust could well have locked its axial rotation to its orbital period. Percival Lowell confirmed this 88 day rotation period observationally a few years later from his observatory at Flagstaff, Arizona, but other astronomers disagreed with the result, leaving its true axial rotation period in doubt.

Although the diameter of Mercury was accurately known, its mass was uncertain because the only objects whose orbits it perturbed were occasional comets as they approached the Sun. The perturbation measurements were difficult and produced densities for Mercury ranging from 3.8 g/cm^3* to 6.85 g/cm^3. These were of the same order of magnitude as the density of the Earth, so Mercury was clearly a rocky, rather than a gaseous planet, but what its surface was like nobody knew.

In 1859 Le Verrier, the French astronomer who had predicted the position of Neptune, showed that the longitude of the perihelion of Mercury's orbit was increasing at 565 arcsec per century, instead of the 527 arcsec predicted using Newton's laws. The difference was much greater than observational error. One possible cause could be the zodiacal light (see Page 5) but, if it was causing this effect, the zodiacal light must† be symmetrical about Mercury's orbit, which is inclined at about 7° to the ecliptic. Unfortunately, the zodiacal light did not seem to have such an inclination. A more likely possibility was thought to be the perturbation of Mercury by an undiscovered planet, that was given the name of Vulcan, closer to the Sun than Mercury.

One of the triumphs of nineteenth century astronomy had been the discovery of Neptune from predictions made using unexplained deviations in the orbit of Uranus (see Pages 87–89). If this other new planet Vulcan did exist inside the orbit of Mercury, it would be very difficult to find because of the brightness of the Sun. There were two possible ways of seeing it; either by observing the Sun to try to see Vulcan crossing the solar disc from time to time, or by looking for the planet close to the Sun during a total solar eclipse. In fact, Schwabe had been looking for Vulcan crossing the solar disc when he had discovered the sunspot cycle in 1843 (see Page 3), but he had seen no such planet.

A number of observers had reported seeing something resembling a small planet crossing the Sun's disc, but in each case the object had never been seen again. The most famous observation of such a transit was made on 26th March 1859 by Edward Lescarbault, a country doctor and amateur astronomer. He was made a member of the Legion of Honour as a result, although Emmanuel Liais, a French astronomer who was observing the Sun at the same time, saw nothing and declared Lescarbault to be mistaken.

Similarly, a number of astronomers had reported seeing a bright object close to the Sun during total solar eclipses, but all the observations were doubtful. So, although no-one had seen Vulcan for certain, the unexplained deviations in Mercury's orbit and the numerous tentative sightings of a possible planet left a niggling doubt that such a planet may exist.

There was considerable uncertainty in the late nineteenth century as to whether Mercury had an atmosphere or not. Careful observation of the planet during its transit of the Sun in both 1868 and 1878 had yielded contradictory results, with some observers reporting that they had seen the planet surrounded by a halo, although the majority had not seen such an

---

*The density of water on Earth is 1 g/cm^3. 1 $g/cm^3 \equiv 10^3$ kg/m^3.

†As the movement of the node of Mercury's orbit, out of the current plane of its orbit, could be explained by known effects.

effect. The halo could have been an optical illusion, but if it was real, it would have been clear evidence that Mercury had an atmosphere.

The lack of clear markings on Mercury indicated that the surface was being obscured by clouds. On the other hand, Mercury had a noticeably lower reflectivity (or albedo) than Venus, so if clouds were being observed, they were very dull. Spectroscopic measurements indicated that Mercury may have water vapour in its atmosphere, but there were doubts about the measurements, because they had had to be made through the Earth's atmosphere near the horizon, and it was difficult to separate the spectral lines due to the Earth's atmosphere from those due to Mercury.

The general consensus near the end of the nineteenth century was that Mercury probably did have an atmosphere with water vapour as a possible constituent. It was thought unlikely, however, that the planet could support life, with or without clouds, as it would be far too hot.

Axial Rotation Period

A number of observers in the late nineteenth century had thought that the axial rotation period of Mercury was 88 days, or synchronous with its orbital period around the Sun. In 1929 Eugène Antoniadi, the Greek astronomer who spent most of his working life in France, confirmed this synchronous rotation period after 5 years of intensive observations with the 33 inch (85 cm) Meudon refractor, during which he also saw evidence of occasional dust clouds.

The synchronous rotation period was generally accepted by astronomers and was usually stated as a fact until, in 1962, W. E. Howard and his colleagues at Michigan found that the night side seemed to be warmer than it should be if it were permanently in shadow. Then, in 1965, R. Dyce and G. Pettengill measured the Doppler shift of radar signals sent from Earth using the Arecibo radio telescope in Puerto Rico. They found that Mercury's rotation period was not 88 days, but 58.65 days, or exactly 2/3 of the planet's rotation period around the Sun, so Mercury rotates exactly $1\frac{1}{2}$ times on its axis per Mercurian year. This means that, at perihelion, when Mercury is at the closest point in its orbit to the Sun, there are only two positions on Mercury's equator that could be directly facing the Sun. These sub-solar points, which are on opposite sides of the planet, are called the hot poles of Mercury, alternating as sub-solar points between successive perihelia.

Peter Goldreich explained the linkage between Mercury's day and its year by a process of spin–orbit coupling. Mercury's orbit is highly eccentric, with a perihelion of 46 million km and an aphelion of 70 million km, and so the gravitational attraction of the Sun at perihelion is over twice that at aphelion (gravitational attraction being proportional to the square of the distance). Goldreich suggested that Mercury originally had had a fast rotation and, as a result, an equatorial bulge had been formed. As the Sun's gravitational pull is much higher at Mercury's perihelion, compared with aphelion, it slowed the planet's axial rotation, so that the long axis of the bulge pointed at the Sun at every perihelion. The situation stabilised with alternate ends of the long axis pointed at the Sun at successive perihelia.

The Precession of Mercury's Perihelion

Le Verrier had shown in 1859 that the longitude of the perihelion of Mercury's orbit was precessing at the rate of 38 arcsec per century more than could be explained using Newton's celestial mechanics. As the observations became more accurate at the end of the nineteenth century, the deviation was shown by Simon Newcomb to be 41 $\pm$ 2 arcsec per century. Newcomb, who was head of numerical analysis for the "American Nautical Almanac", suggested that the effect could be explained if Newton's gravitational inverse square law had the inverse power increased from 2 to 2.0000001574, but this was shown to be wrong by careful observations of the Moon's orbit. Various other proposals were put forward to explain the discrepancy but none were satisfactory, and the error was one of the largest problems unexplained at the end of the nineteenth century.

Albert Einstein finally solved the problem in November 1915, using his new general theory of relativity, which predicted the magnitude of the effect to be 43 arcsec per century, matching the observed results within observational error. Einstein was ecstatic at the success of his theory in solving this crucial problem.

In 1959 radar was first used to measure the distance of the planets. Over the next decade or so distances in space were measured to unprecedented accuracies, and by 1976 the anomalous part of Mercury's perihelion precession was observed to be 43.11 $\pm$ 0.21 arcsec per century which was, within error, the same as the 42.98 predicted by general relativity. So general relativity did not need any modifications to predict the precession of Mercury's perihelion exactly.

The Surface

Gustav Müller and Paul Kempf's photometric measurements of Mercury (see Page 299) at the Potsdam Astrophysical Observatory, indicated in 1893 that there is no atmosphere on Mercury, and that its surface is darker and rougher than the Moon. Then, in 1924, Lyot deduced, by measuring the polarisation of light reflected from its surface, that Mercury was covered in volcanic ash. Three years later, Donald Menzel, who was then at the Lick Observatory in California, deduced a temperature of 670 K from radiometric measurements of the sunlit part of the planet.

So far, the American Mariner 10 spacecraft is the only one to have visited Mercury. Launched in November 1973, it intercepted Mercury three times, in March 1974, September 1974 and March 1975. Mariner showed a surface that looked very much like the Moon, with a large number of impact craters, although the extensive lunar-like seas were noticeably absent. The nearest equivalent to a lunar sea was the 1,300 km diameter Caloris basin.

The rim of the Caloris basin is defined by the Caloris Mountains which are several kilometres high and up to 50 km across, and the floor of the basin seems to be composed of lava which has been modified to produce a pattern of ridges and grooves. Robert Strom, Newell Trask and J. E. Guest suggested in 1975 that the ridges were produced by compression when the floor of the basin subsided. The grooves, which generally form polygonal patterns, are up to 10 km wide and 500 m (1,700 ft) deep. They seem to be younger than the ridges as some grooves have modified the ridges, but never the other way round.

An area called the "weird terrain" by the Mariner 10 imaging team was found at about 30°S, 20°W on Mercury, consisting of hills and mountains up to 1,800 m (6,000 ft) high, arranged in a strange pattern (see Figure 4.1). This area is at the antipodes of the Caloris basin, which led P. H. Schultz and D. E. Gault to suggest that the impact that had produced the Caloris basin had formed these strangely-shaped hills and valleys. Schultz and Gault showed in 1975 that this was quite possible as the seismic energy from the Caloris event would have been focused by the large planetary core.

Scarps were also found to be common on Mercury, providing evidence for the contraction of the whole planet by about 1 to 2 km radius. This could

Figure 4.1 *Mariner 10 images showing the weird terrain on the opposite side of Mercury to the Caloris basin. (Courtesy* National Space Science Data Center, World Data Center-A for Rockets and Satellites, NASA.*)*

have been due either to cooling, or to the planet taking on a spherical shape as its spin rate reduced, or both.

Astronomers were now in for a big surprise when, in August 1991, radar studies of the planet by Duane Muhleman, Martin Slade and Brian Butler of Caltech indicated that there may be water ice at Mercury's north pole. This very unexpected observation, for a planet that is so close to the Sun, could be explained by the fact that the axis of Mercury is perpendicular to the plane of its orbit around the Sun, and so the Sun, as seen from the poles, is permanently on the horizon. The floors of craters near the poles are, therefore, in permanent shadow, and always very cold. David Page estimated in 1992 that the temperature could be as low as 60 K, well below the temperature of 112 K required to retain water ice on Mercury for billions of years.

The Interior

Mercury is a much smaller planet than the Earth and, because of this, it would have cooled much faster than the Earth after its initial formation. It was generally thought, therefore, that Mercury had an almost solid core. This was consistent with the Mariner discovery that the surface was a single lithospheric plate which, although broken, showed no lateral movement of the pieces.

The magnetic field on Earth was thought to be due to a dynamo effect in its large liquid core. Mercury, on the other hand, is small, with a slow axial rotation and a solid core, and so it was expected that there would be no magnetic field. It was something of a surprise, therefore, when Mariner discovered, in 1974, that Mercury had a magnetic field which, although smaller than that of the Earth, was still appreciable. Later analysis showed that Mercury had an exceptional amount of iron in its core which explains the source of the field, although how the iron became magnetised is unknown.

Venus

Early Work

Venus in the telescope looks like a much brighter and larger version of Mercury, with similar phases and very little visible detail. The albedo (or reflectivity) of Venus is very high at over 0.50 (or 50%), and it was assumed that this was due to the planet being covered by a blanket of clouds.

Evidence for an atmosphere on Venus had first been seen by the Russian chemist Mikhail Lomonosov during Venus' transit of the Sun in 1761, and the atmosphere had been clearly seen by the Scottish astronomer Ralf Copeland and others in the nineteenth century as a thin luminous ring, when the planet was almost in front of the Sun. Spectroscopic evidence had led the respected spectroscopists William Huggins and Hermann Vogel to conclude, independently, that the atmosphere on Venus included small amounts of oxygen and water vapour, although the measurements were difficult, because the light had to pass through the Earth's atmosphere, which added its own absorption bands to any originating on Venus. As far as life was concerned, Ball (Reference 4) suggested in 1897 that "If water be present on the surface of Venus and if oxygen be a constituent of its atmosphere, we might expect to find in that planet a luxuriant tropical life, of a kind analogous in some respects to life on Earth."

It was very difficult to estimate an axial rotation period for Venus because of the lack of any clear features, but Giovanni Cassini and others in the seventeenth, eighteenth and nineteenth centuries observed vague markings which led them to suggest a rotational period of about 24 hours. In 1890, however, Schiaparelli surprised everyone when he used indistinct surface markings to deduce a period of 224.7 days, which, like his rotational period for Mercury, is equal to its period of rotation around the Sun. The evidence for either rotational rate for Venus was not convincing to nineteenth century astronomers, although the idea of a rotation rate locked to its orbital rate had attractions, as it could have been explained by tidal friction in the originally molten planet.

Cassini reported seeing a moon of Venus in the seventeenth century, and other observers reported similar sightings in the eighteenth and nineteenth centuries. In 1887 the Belgian astronomer Paul Stroobant showed that these observations could generally be attributed to stars or to internal telescope reflections, although the possibility that Venus may have a small moon, like the moons of Mars, had not been completely dismissed.

When the phase of Venus was new, an "ashen light" was sometimes seen to cover the dark disc. Many theories existed as to its cause, but the best-regarded were put forward by J. Lamp, who suggested that it was something analogous to our aurora, and by Vogel, who thought that it was due to an extensive twilight caused by scattering in the Venusian atmosphere.

Axial Rotation Period

In the nineteenth century, axial rotation rates of 24 hours and 224.7 days (i.e. synchronous rotation) had been suggested for Venus. Vesto Slipher of the University of Indiana studied spectrograms made in 1903, and finding no Doppler shift due to its rotation, concluded that the rotation period of the visible cloud layer could be no faster than about three weeks. Radiometric measurements of the cloud-tops of Venus in 1926 (see Page 305) indicated that the clouds were moving much faster than the synchronous rotation period of 224.7 days (i.e. one Venusian year).

No substantial new evidence on the rotation period was forthcoming until, in 1957, Charles Boyer, a French amateur astronomer living in the Congo, took a series of photographs in ultraviolet light which was known to emphasise cloud details. Boyer used only a 10 inch (25 cm) reflector, but he found a distinctive V-shaped marking (with the V on its side), which reappeared every 4 days. When this was announced, Henri Camichel found the same period on photographs taken in 1953, and so the rotation period of Venus' clouds, at least, appeared to be 4 days.

Radical new data were produced in 1962 when the Doppler shift of radar signals bounced off Venus indicated a retrograde (i.e. East-to-West) rotation of the surface, with a period of about 240 days. This was confirmed in 1965, using the new Arecibo radio telescope, when a more accurate period of 243 days was measured. Both the 243 day planetary rotation, and the 4 day atmospheric rotation, were retrograde.

The Atmosphere

Percival Lowell astonished the astronomical world when he reported to the Boston Scientific Society in October 1896 that he had seen clear linear markings on Venus. It was well known at the time that Venus had an extensive atmosphere, so Lowell concluded that the atmosphere must be transparent, as he was convinced that he had seen surface markings. Astronomers generally dismissed his findings as spurious, as Venus appeared in even the largest telescopes to be virtually featureless. Lowell, in his defence, quoted Godfrey Sykes' explanation of the ashen light, as being due to the reflection by surface ice of the light received from the Earth and other heavenly bodies. This, Lowell pointed out, also implied that the atmosphere was transparent, but the astronomical community was unconvinced.

In 1929 Lyot deduced, by measuring the polarisation of the light scattered by Venus, that it was covered in a layer of haze that lay above the opaque cloud layer. At about the same time, radiometric measurements indicated cloud-top temperatures of 330 K and 250 K, for the day and night sides, respectively.

Spectroscopic analysis of the atmosphere of Venus in the late nineteenth century had indicated the presence of oxygen and water vapour, but by 1932 this analysis had been discounted (see Page 295) as Walter S. Adams and Theodore Dunham, Jr., with their painstaking work at Mount Wilson, could only find evidence for carbon dioxide. Subsequent research indicated that Venus had a substantial amount of carbon dioxide in its atmosphere, which would generate a greenhouse effect much greater than on Earth. Calculations indicated surface temperatures of 350 to 400 K, which is about the boiling point of water on Earth (i.e. at an atmospheric pressure of 1 bar).

Rupert Wildt at Göttingen suggested in the 1930s that, if the atmosphere of Venus has very low amounts of oxygen, there would be no ozone layer to shield the lower parts of the atmosphere from the Sun's ultraviolet light. Under these conditions, ultraviolet light from the Sun would cause carbon dioxide and water vapour (if it existed) to recombine to form oxygen, that would be absorbed by the rocks, and formaldehyde (CH_2O). Wildt thought, therefore, that the clouds on Venus could consist of formaldehyde and water droplets, but he could find no observational evidence for presence of formaldehyde.

In 1954, Donald Menzel and Fred Whipple of Harvard College Observatory gave renewed credibility to the view popular in the nineteenth century, that there were oceans on the surface, and that the clouds consisted of water droplets. They envisaged Venus as being in a state similar to that of the Earth millions of years ago, but Venus would be hotter as it is closer to the Sun. But how hot was it? If it was hotter than the boiling point of water, the surface must be completely dry and desert-like, but if it was cooler Menzel and Whipple may be correct.

Analysis of the radio emissions from Venus by Mayer of the US Naval Research Laboratory, in 1956, indicated a surface temperature of at least 600 K. Although a high temperature had been expected, this value seemed unreasonably high, and so when the American Mariner 2 spacecraft was launched to Venus in 1962, it was equipped with a small microwave radiometer. This measured a surface temperature of nearly 700 K, so the surface must be a desert, as liquid water could not exist at that temperature. Mariner also tried to measure the magnetic field of Venus, but, unlike in the case of Mercury, could find no evidence of one.

In December 1970, the Soviet Union's Venera 7 spacecraft measured the surface atmospheric pressure on Venus as being an incredible 90 times that on Earth, equivalent to the pressure 1,000 m below the surface of the Earth's oceans. At first these results were treated with caution, but later Soviet spacecraft confirmed them.

Astronomers made another surprising discovery in 1972, when they found that the clouds on Venus were composed of sulphuric acid droplets. A few years later, the probes from the American Pioneer-Venus 2 spacecraft showed that these clouds were in a layer 48 to 58 km above the surface, much higher than the rain clouds on Earth, which are generally below about 10 km in altitude.

The 4 day rotation period of Venus' clouds, deduced by Boyer, was confirmed in 1974 by Mariner 10, as it flew by Venus on it way to Mercury. This meant that the winds at the cloud level in the atmosphere are blowing at 400 km/h relative to a very slow moving surface, and the clouds all appear to be blowing in the same direction. This is very different from the situation on Earth as, although the jet stream at mid latitudes is blowing at roughly the same velocity as on Venus, the surface of the Earth is rotating rapidly, and there are balancing reverse-flow winds over the tropics.

The Surface

The clouds on Venus are so thick that the surface cannot be seen through them, and so the only way to map the surface was to use radar. The first such maps were made using radar beams transmitted from Earth, but the results were relatively crude and difficult to interpret, although the regions known as Alpha and Beta Regiones were discovered.

The first direct views of the surface, from probes landed on Venus in 1975 by the Soviet Union's Venera 9 and 10 spacecraft, showed a dry rock-strewn surface as far as their horizons. Maps of the surface improved dramatically in 1978, when the American Pioneer-Venus 1 spacecraft went into orbit around Venus, equipped with a radar altimeter. It showed huge rolling plains stretching right around the planet, some lowland areas and two highland regions now called Ishtar and Aphrodite. The peaks of the highest mountains, Maxwell Montes, in the eastern part of Ishtar, were found to be 12,000 m (39,000 ft) above the general surface level, so they are appreciably higher than the Himalayas. Aphrodite, which is larger than Ishtar, has a vast rift valley at its eastern end nearly 3,000 m (10,000 ft) deep, 2,200 km long, and 280 km wide. Two shield volcanoes, called Rhea Mons and Theia Mons, which are much larger than any found on Earth, were also found isolated from the two upland areas. Six years later, a great many impact craters and small volcanoes were found by the Soviet Union's Venera 15 and 16 orbiters.

The American Magellan spacecraft entered orbit around Venus in August 1990 and, over the next two years, it completed a detailed radar mapping of the surface at an unprecedented resolution of 150 m. It found that the surface is mostly volcanic, with large lava-flooded plains, and thousands of volcanoes. There are also signs of tectonic activity which has caused, for example, multiple faulting and deep fractures in the Devana Chasma. There are a number of rift valleys, some of which have been partly flooded by molten lava, and a number of impact craters (see Figure 4.2).

The density of impact craters on the surface has enabled the ages of various areas to be estimated. Parts of the surface have no impact craters at all, implying that these areas have an age of no more than a few tens of millions of years, although, at present, no active volcanoes have been found. Active volcanoes would be difficult to detect on Venus, however, because the high atmospheric pressure would prevent material being thrown into the atmosphere. Instead, the lava would tend to ooze directly onto the surface.

The Earth

Early Work

• *Precession, Nutation and Other Oscillatory Motion*

The Dutch astronomer Christiaan Huygens had suggested as long ago as 1683 that the Earth had an equatorial bulge (or polar flattening). Although the Earth is not spherical, at least its shape is virtually constant now, which is more than can be said for the direction of its spin axis or the shape and orientation of its orbit.

The Earth's spin axis, which is inclined at about $23\frac{1}{2}°$ to the ecliptic, is not fixed in space, but describes a circle around the perpendicular to the ecliptic with a period of about 26,000 years. This precession, which affects the

Figure 4.2 *Three large impact craters on Venus imaged by the Magellan synthetic aperture radar in 1990. (Courtesy* National Space Science Data Center, World Data Center-A for Rockets and Satellites, NASA; Experiment Principal Investigator, Dr. Gordon H. Pettengill, The Magellan Project.*)*

position of the celestial pole (the position in the sky about which the stars appear to rotate) was known to Hipparchos over 2,000 years ago. It was eventually explained by Newton in the seventeenth century as being due to the gravitational effect of the Moon and Sun on the Earth's equatorial bulge.

Not only is the Earth's spin axis precessing in space with a period of about 26,000 years but, superimposed on this, there is a nutation, or wobble, of 9 arcsec (half amplitude) with a period of 18.6 years. This nutation had been discovered by James Bradley in the eighteenth century, when he was professor of astronomy at Oxford. He deduced that it was caused by the regression in the line of nodes of the Moon's orbit (see Page 23) that had the same period.

The position of the perihelion of the Earth's orbit, where the Earth is closest to the Sun, was also known to vary. The precession of the Earth's spin axis in space, and this precession of the perihelion of the Earth's orbit, meant that the aphelion (where the Earth is furthest from the Sun) would alternate from midwinter's day in the southern hemisphere to midwinter's day in the northern hemisphere and back about every 21,000 years. So, it was concluded that both hemispheres could expect to see cold or warm winters alternating with a frequency of 10,500 years.

As the eccentricity of the Earth's orbit was much greater about 200,000 years ago than it is today, the magnitude of the cold and warm winter effect would have been much greater then than it is today. It was thought that these effects could explain the Ice Ages, which would have a frequency of about 21,000 years in either hemisphere, and which would have been much more severe about 200,000 years ago compared with today. Unfortunately, estimates of the times of the actual ice ages did not match these periods.

In 1885 Seth C. Chandler, working at Cambridge, Massachusetts, and Friedrich Küstner at Berlin independently discovered a most remarkable effect. The Earth's spin axis was moving by about 0.4 arcsec with respect to the surface of the Earth itself, resulting in a periodic change in the latitude of Cambridge, Mass. and of Berlin by 0.4 arcsec over a period of one year. A similar effect had been measured earlier by observers at Washington, but they could not bring themselves to suggest that the movement was real. The gifted Swiss mathematician Leonhard Euler had, in fact, predicted such an effect as long ago as 1765 by considering the dynamics of a rigid spinning Earth, but his prediction had been largely forgotten.

• *The Secular Acceleration of the Moon and Residual Effects*

Edmond Halley, who is best known for his work on comets, had found that the Moon's position in the sky as seen in the seventeenth century was about 2° in advance of where it should be, when comparing it with ancient eclipse records. This so-called secular acceleration of the Moon could be explained if either the speed of the Moon in its orbit was gradually increasing, and/or the speed of rotation of the Earth on its axis was gradually slowing down (i.e. if the length of the day was increasing).

In the late eighteenth century, Laplace showed that the Moon's rate of rotation around the Earth was increasing slowly, because the gravitational effect of the Sun on the Moon was being changed as the eccentricity of the Earth's orbit was currently being reduced. The effect was very small, giving a change in the position of the Moon in the sky by 10.2 arcsec in 100 years, but this was very close to the actual deviation of 12.2 arcsec.

In the early nineteenth century, various astronomers tried to refine Laplace's theory, and in 1853 John Couch Adams showed that some of the terms that Laplace had omitted from his calculations, thinking that they were unimportant, were, in fact, significant. Including them reduced the effect of the increase in the Moon's orbital rotation to 5.7 arcsec in 100 years, or about half of that observed. In 1866 Charles Delaunay, of the Paris Observatory, showed that the other half of the effect was probably due to the Earth's axial rotation being slowed down by tidal friction, by about

1.8×10^{-8} seconds per day. But when Simon Newcomb analysed all the lunar data in 1878, he found that adding the Laplace–Adams and the Delaunay effects together tended to overestimate the observed effect, although it was accepted that Delaunay's calculation of the tidal effects was difficult and could be in error.

Near the end of the eighteenth century Laplace had found evidence of an oscillation in the position of the Moon with a period of some hundreds of years, that was superimposed on all other effects. In 1846 Peter Hansen of Gotha in Germany suggested that this was caused by perturbations of Venus on the Moon, producing a maximum deviation of 21 arcsec in the Moon's position in the sky over a period of 240 years. Delaunay also calculated the magnitude of this effect and disputed Hansen's value, which Hansen was then forced to admit was empirical. To make matters worse, the Moon then deviated from Hansen's empirical curve over the next few decades, and by 1880 the Moon was lagging 10 arcsec behind its predicted position.

Newcomb decided to take the bull by the horns to get to the bottom of the problem, analysing all the ancient eclipse observations, and all the star occultations by the Moon since the invention of the telescope. This led him to conclude, in 1878, that the observations could be explained by adding an empirical term of 17 arcsec, with a period of 273 years, which was very similar in magnitude to Hansen's empirical value. The cause, however, was still a mystery. Newcomb speculated that it could either be because the Moon's motion is influenced by something other than the gravitational effects of the Sun, Earth and planets, or because the Earth's day is varying in a periodic manner (in addition to it gradual slowing down due to the tides).

• *The Structure and Age of the Earth*

It was thought that the Earth had originally condensed from the same nebula as the remainder of the solar system. The proto-Earth had then cooled to a fluid state and, as it had continued cooling, had contracted and speeded up its rotation to such an extent that it had broken in two unequal parts (see Page 25); one being the Earth and the other the Moon. Eventually a crust formed on the surface of the still-cooling Earth, but because the Earth was spinning, the crust set with a slight equatorial bulge, which resulted today in the Earth's equatorial diameter being 44 km larger than the polar diameter. It was known in the nineteenth century that the Earth still had a liquid core, but its dimensions were unknown, as was the internal structure of the planet. It was clear, however, that the solid skin could not be too thin, otherwise it would be constantly being distorted and fractured by the tidal action of the Sun and Moon.

Estimates of the age of the Earth varied from 100 million years (by the distinguished English physicist William Thomson, later called Lord Kelvin) to 400 million years (by George Darwin). Unfortunately, even the smallest of these estimates was substantially more than the calculated 25 million year age of the Sun (see Page 6), and this discrepancy was still unresolved.

Dynamics

Chandler and Küstner had discovered in 1885 that the axis of the Earth was moving very slightly with respect to the surface of the Earth, over a period of about a year. In 1891 Chandler showed, following more detailed measurements and analysis, that the pole moves in a circle with a period of 427 days and radius 10 m (equivalent to 0.3 arcsec). Euler had shown in the eighteenth century that, if the Earth is rigid, the period should be 306 days. The difference between this value and that observed was attributed by Newcomb to the elasticity of the Earth and the momentum of the oceans, modified by a 12 month period caused by seasonal changes in the ice and air masses of the two hemispheres.

Delaunay had suggested, in 1866, that tidal friction was slowing down the Earth's rotation in a progressive fashion, which explained about half of the secular acceleration of the Moon, but in 1909 Thomas Chamberlin of Chicago concluded that tidal friction would not be enough to cause the observed magnitude of the effect. Geoffrey Taylor analysed the ocean geometry in more detail in 1919, and found that shallow inland seas were much more important than the oceans in slowing down the Earth. In particular, in the following year Jeffreys showed that the Bering Sea is responsible for over half the effect.

Newcomb had shown, in 1878, that the Moon's position appeared to deviate by a small amount in a cyclical fashion, with a period of 273 years, but the cause was unknown. Newcomb had thought that it could be due to variations in the Earth's rotation rate, but he dropped the idea in 1903. If the effect was due to variations in the Earth's rotation rate, it should be evident in the apparent movement of the Sun and planets, as well as that of the Moon and, in 1914, the Anglo-American astronomer Ernest W. Brown showed that there was such an effect in the apparent motions of the Sun, Mercury and Venus. This was later confirmed by Harold Spencer Jones at Greenwich, and by Willem de Sitter at Leiden who noticed a similar effect for the satellites of Jupiter. The Earth was showing random irregularities in its axial rotation rate of from +0.0034 s/revolution in 1870, to −0.0045 s/revolution in 1910. Clearly irregular deviations could not be predicted in future, and so the dream of centuries that it would eventually be possible to predict all the observed motions of the Moon was clearly not feasible. To make matters even more complicated, in 1949 the Belgian N. Stoyko found a seasonal variation in the Earth's rotation period, superimposed on all other variations, which was attributed in the following year to seasonal changes in the distribution of the atmosphere. Evidently the Earth is a most inaccurate clock.

The Surface

Emil Wiechert suggested in 1897 that the Earth consisted of a dense metallic core, mostly of iron, with a density of 7.8 g/cm^3, surrounded by a lighter rocky layer with a density of 3.3 g/cm^3. In 1934 the English geophysicist

Harold Jeffreys proposed that the other terrestrial planets and the Moon had a similar structure, but with the dense metallic core being smaller as the densities decreased from Earth (5.5 g/cm^3) to Mercury (5.4 g/cm^3) to Venus (5.3 g/cm^3) to Mars (3.9 g/cm^3) to the Moon (3.3 g/cm^3), so that in the case of the Moon the core was non-existent. That was why the Moon had a density equal to that of just the light rocky element.

The French Nobel laureates Marie and Pierre Curie discovered, in 1898, that radium exhibited strong radioactivity, causing it to gradually change to an isotope of lead. This enabled the age of rocks to be determined by measuring how much of this isotope of lead and of radium they contained. The measurements in the early twentieth century showed ages of up to 1,500 million years, which were considerably in excess of the 100 to 400 million years previously estimated for the age of the Earth. By 1989 rocks with an age of 3.96 billion years had been found in Canada's Northwest Territories.

The American spacecraft Vanguard 1, launched in March 1958, showed that the shape of the Earth is not the symmetrical ellipsoid of revolution assumed by Huygens, but is pear-shaped, with the south pole nearer to the centre than is the north pole. Vanguards 2 and 3, which were launched during the next 18 months, also showed that the Earth's equator is not circular, but elliptical. Subsequent spacecraft have shown that the Earth has humps and hollows all over it, compared with a mean ellipsoid of revolution of obliquity 1:298.26, the maximum deviation being a deep hollow of about 100 m depth just off the southern tip of India.

The dinosaurs seemed to have become suddenly extinct at the end of the Cretaceous period about 65 million years ago, and this was attributed by the Russian Iosif Shklovskii in 1957 to a supernova explosion that had taken place within about 30 light years of Earth. High concentrations of iridium, and of other elements normally found in meteorites, were then found in clay samples in many different parts of the world, in the boundary layer between Cretaceous and Tertiary rock sediments, laid down at the end of the Cretaceous period. Most of the iridium that was originally present in the molten Earth was thought to have sunk into its core, and so in 1979 Luis and Walter Alvarez of the University of California suggested that an asteroid, meteorite or comet had hit the Earth about 65 million years ago, causing some of this iridium from the Earth's core to be vaporised and redistributed around the Earth. The dust cloud produced by the impact would have blotted out the Sun and have cooled the Earth, causing the extinction of much plant and animal life, including the dinosaurs. But could a large impact crater of the correct age be identified?

Originally it was thought that the impact had occurred near Manson, Iowa, where the remnants of a large crater were found but, in 1992, it was found that a 180 km diameter crater, that lies buried more than 1 km beneath the Yucatan Peninsula in Mexico, has the correct age and size and is probably the impact site. The object that formed this crater was estimated to be a small asteroid, or large cometary nucleus, of about 10 km diameter.

The Atmosphere and the Radiation Belts

The physicists Julius Elster, Hans Geitel and Charles Wilson undertook a number of experiments at the turn of the century to try to understand static electricity and the ionisation of dry air. This led experimenters to compare the ionisation at various locations, over rocks, glaciers and deep lakes, because they suspected that the ionisation was caused by radioactive rocks on the Earth's surface. Wulf took his measuring instrument (an electroscope) up to the top of the Eiffel Tower in Paris, expecting to see the ionisation level reduce, which it did, but only to a small extent.

The Austrian physicist Victor Hess made ionisation measurements from a hot air balloon in 1912, expecting these to decrease with increasing altitude but, much to his surprise, the ionisation increased with altitude above 2 km. The source of the ionisation could not be the Earth, but where else could it come from? The obvious source was the Sun, but the ionisation was the same at night as during the day and, as the variation in the ionisation over a day was not repeatable from day to day, the source could not be a single object in space either. The rays causing the ionisation appeared to be coming from all over the cosmos, causing Robert Millikan to call these rays "cosmic rays".

Over the years, as first balloons, then sounding rockets, and finally spacecraft analysed cosmic rays it became clear that the early measurements had been of secondary cosmic rays, rather than the primary rays that hit the top of the Earth's atmosphere. These primary rays, which produced secondary rays during their collision with air molecules, were found to consist of highly ionised atoms of various elements travelling at speeds close to that of light. The relative abundances of the elements in the primary rays were similar to those found in meteorites and elsewhere in the universe, except that there was much more lithium, beryllium and boron than expected. Analysis showed that these elements could not have been produced in the Big Bang and, although they may be produced in the interior of stars, they would be rapidly destroyed by stellar temperatures. So where did they come from?

Walter Baade and Fritz Zwicky working in California showed, in the 1940s, that supernovae occur often enough in the Milky Way to produce the observed number of cosmic rays. Supernovae were also thought to eject carbon, nitrogen and oxygen nuclei in large numbers, and so this solved the problem of the high abundances of lithium, beryllium and boron, as they could be produced by the collision of cosmic rays of carbon, nitrogen and oxygen nuclei with the hydrogen and helium of interstellar space.

Explorer 1, the first American spacecraft, was launched on 31st January 1958 into a highly eccentric orbit of 360 × 2540 km around the Earth. Prior to the launch, it had been assumed that radiation levels, connected with elementary particles, would continue to increase with altitude (like cosmic rays) as the effect of the Earth's atmosphere was reduced. Much to everyone's surprise, however, James Van Allen's Geiger counters showed that the radiation levels reached an intense maximum at 900 km altitude, and then reduced with greater altitude. More data were produced by Explorer 2, launched on 26th March, and this enabled Van Allen to announce the discovery of large radiation belts, now called the Van Allen belts, surrounding the Earth and constrained by the Earth's magnetic field.

Mars

Early Work

A number of maps of Mars had been produced in the nineteenth century showing relatively consistent markings on its surface, which were generally thought to indicate land and water. Tracking these markings enabled the axial rotation period of Mars to be estimated to within a tenth of a second.

The most famous, or maybe one should say the most "infamous", maps of Mars produced before 1890 were drawn by Schiaparelli during the oppositions of 1877 and 1879. These maps (see Figure 4.3, for example) show a number of intersecting parallel lines, which he called "canali", or "channels" in English, crossing the surface. Unfortunately, the Italian word "canali" was mistranslated as "canals", giving the impression that the linear features had been constructed by intelligent beings. Although, in the twentieth century, a great deal of scorn has been heaped on Schiaparelli's observations, Perrotin and Thollon also observed linear markings from Nice in 1886, and Arthur Berry (Reference 2) wrote in 1898 "These remarkable observations have been to a great extent confirmed by other observers, but remain unexplained". Agnes Clerke (Reference 5) also did not think that the linear markings were illusory, although it is also fair to add that Ledger (Reference 1) was uncertain as to their accuracy.

Anyone who has observed a planet through a telescope will confirm how difficult it is to be certain of its surface details, as turbulence in the Earth's

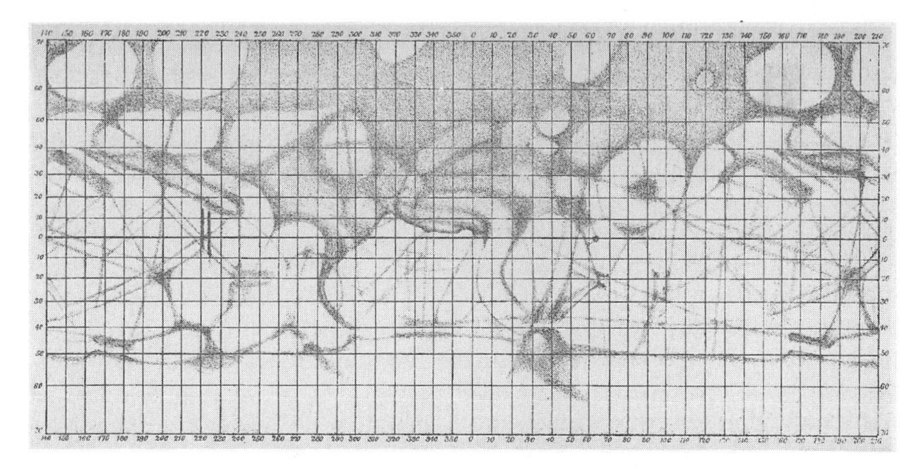

Figure 4.3 *One of Schiaparelli's maps of Mars showing the linear features which he called "canali". (From* The Story of the Heavens, *by Robert Ball, 1897, Plate XVIII.)*

atmosphere can change the image over fractions of a second. In addition, the brain has to be taught how to interpret images that the eye sees, as any observational astronomer or military observer will confirm (often referred to as "training the eye"). Finally, long periods of time at the eyepiece, often in low temperatures, produce fatigue and stress that also affect what one sees. It is not really surprising, therefore, for people to claim to have seen details in an image which are not really there. Were Schiaparelli's observations real? Only time and better telescopes would tell.

Mars was known to be reddish in colour, with large, dark bluish-green areas. There were white areas at both poles that had been observed to expand and contract with the Martian seasons. John Herschel, the son of the famous astronomer William Herschel, concluded in the nineteenth century that the reddish colour was due to the colour of the soil, although some astronomers thought that it could be due to vegetation. The dark bluish-green areas, which cover about half of the surface, were generally considered to be water.

The white polar areas were thought to be accumulations of ice and snow, which was consistent with the observations of water vapour in the Martian atmosphere. The polar ice-caps were seen to be much smaller (in relative area) than those on the Earth, indicating that there may not be much water on Mars, but this was inconsistent with the theory that the bluish-green areas were seas. The polar ice caps were also seen to shrink very rapidly in the Martian summer, giving the impression that Mars is not as cold as the Earth, which seemed strange considering Mars is further from the Sun, and that it also appeared to have a more tenuous atmosphere, which would allow heat to escape. No really satisfactory explanation was available in the nineteenth century to explain these apparent inconsistencies.

In 1860, Emmanuel Liais had suggested that the dark areas could be vegetation. Schiaparelli agreed with this, pointing out that they did not show the Sun's reflection, which they should do if they were large areas of water, but these ideas were generally discounted.

It was observed that there were far fewer clouds on Mars than on Earth, and that its atmosphere appeared noticeably less dense, with a possible density of about 15% of that on Earth. It was thought, however, that there may be life of some kind, as there seemed to be plenty of water and the temperatures did not appear to be as low as those on Earth (at least at the poles). In addition both oxygen and water vapour had been detected in the atmosphere.

The nineteenth century was very much a time when evolutionary theories began to hold sway. Not only did Charles Darwin propose his evolutionary theory for the Origin of the Species, but theories were developed by Laplace and others for the evolution of the solar system, and measurements indicated that the solar system was still changing today. It was also hypothesised by some astronomers that the suitability of various planets for life went through optima at different times. Thus, although the Earth had good conditions for life now, Mars was thought to have probably passed its optimum, whereas the optima for Venus and Mercury were still to come when they had cooled down because of the gradual cooling of the Sun.

The opposition of Mars in 1877 led to the discovery of its two moons. Asaph Hall, using the largest refracting telescope then in existence, the 26 inch at the Washington Naval Observatory, discovered the outer of the two moons (later called Deimos) on 11th August, and found the second (Phobos) 6 nights later. It was impossible to see the discs of these two moons as they were too small, and so the only way of guessing their size was to measure their brightness and assume their reflectivity was similar to that of some other body in the solar system. Using our Moon as a comparator yielded about 15 km diameter for Deimos and about 25 km for Phobos, which are very close to the true values. It was thought possible that these moons were captured asteroids (see Page 103), as the moons were as small as asteroids and one asteroid was known to have an orbit that crossed that of Mars.

An interesting feature of the orbit of Phobos is that it was the only moon then known to orbit its planet faster than the planet rotated about its own axis. Thus to an observer on Mars, he would see one of the moons moving across the sky in one direction, while the other moon would be seen to move in the opposite direction.

The Surface

Eighteen ninety-one dawned with the argument on the possible existence of canals on Mars, which had been started by Schiaparelli in 1877, still in full swing. This controversy lasted well into the twentieth century, fuelled mainly by Percival Lowell's observations. At times the disagreements became more personal than scientific, which was a pity because the observers were genuinely trying to establish the facts. The controversy had a more positive side to it, however, as far as planetary studies was concerned, as it significantly increased the effort devoted to planetary work, which had been reduced because of the enormous strides being taken in the field of stellar research.

Schiaparelli was prevented by failing eyesight from continuing his observations, but the millionaire Percival Lowell was determined to continue Schiaparelli's work. Lowell returned to the United States in late 1893, from a period of working in the Far East, and decided to set up an observatory to study Mars during the opposition of October 1894. He opened his small observatory at Flagstaff, Arizona in May 1894, where he installed a borrowed 18 inch (45 cm) refractor.

At the beginning, Lowell, like most other astronomers of the day, assumed that the large dark bluish-green areas on Mars were seas. William Pickering could find no evidence of polarisation in the light coming from them, however, suggesting that they could not be bodies of water. Lowell observed that these dark areas changed their appearance with the seasons, and he also saw clear linear markings on Mars with his telescope, just like Schiaparelli's canals. This led Lowell to conclude, in 1895, that the canals had been built by intelligent beings to transport precious water from the melting polar caps to the dark areas, which were areas of vegetation.

Antoniadi started a 20 year study of Mars during the opposition of 1909, using the Meudon refractor, and showed that, what had been seen by Schiaparelli and Lowell as linear features, could be resolved at very high magnifications into a series of small spots. Antoniadi confirmed that the large bluish-green areas changed with the seasons, and agreed with Lowell that they were probably areas of vegetation. In addition to these seasonal changes, there were non-seasonal changes that occurred in some of the dark areas over periods of a few years. Solis Lacus, for example, changed its size and shape dramatically between 1924 and 1926, but by 1928 it had resumed its previous appearance.

In 1909 the Russian astronomer Gavril Tikhov compared spectra of the various greenish-coloured regions of Mars with those of localities on Earth, and concluded that some Martian areas had vegetation similar to some subarctic regions on Earth. Vesto Slipher repeated the measurements of the dark areas of Mars at the Lowell Observatory in 1924, looking for evidence of chlorophyll. None could be found, although a number of astronomers then speculated that the dark areas of Mars could be covered in a very basic form of life, like moss or lichens, some of which were known to have no chlorophyll on Earth. These basic forms of vegetation were also consistent with organisms that had to survive with relatively small amounts of water, as the polar ice caps, for example, seemed to be very thin. In 1956, Gerard Kuiper confirmed the lack of chlorophyll lines from the dark areas of Mars, finding instead a resemblance to the spectrum of lichens, thus adding to speculation that these areas were covered in lichens and possibly moss.

So much for the dark bluish-green areas, but what about the lighter reddish ones? In 1924 Bernard Lyot measured the polarisation of light scattered by the planet, and concluded that the surface was probably covered with limonite sand (a hydrated iron oxide) which was disturbed by the atmosphere from time to time to produce dust storms. Donald Menzel found, in 1927, that the surface temperatures varied from 170 K (−103°C) at dawn to about 280 K (+7°C) at noon. Kuiper disputed Lyot's findings in 1956, concluding that the yellow regions were covered in felsitic rhyolite, which is an igneous rock. Help was almost at hand to establish the truth, however, as the spacecraft era was just around the corner.

The American Mariner 4, which was the first successful spacecraft to fly-by Mars, provided a big surprise to planetary scientists, when it discovered craters on Mars in July 1965. Instead of a planet with patches of vegetation, like moss or lichens, we were treated to a series of images of a cratered desolate world resembling the Moon, and the dark areas visible from Earth were found to be simply low albedo areas.

Whenever shock discoveries are made, there is always a search to see if they had been made before.

Craters on Mars had been seen before. Edward Barnard had seen them in 1892, using the Lick 36 inch (91 cm) refractor, but he did not publish his observations for fear of being ridiculed, and John Mellish had seen them in

1917 with the 40 inch (102 cm) Yerkes refractor, but he did not publish his observations either. In addition, D. L. Cyr had predicted their existence in 1944, as had Clyde Tombaugh, the discoverer of Pluto, and the Estonian astronomer Ernst Öpik. Notwithstanding these observations and predictions, when Mariner 4 visited Mars in 1965 virtually no-one expected to see craters on Mars, and their discovery was a major surprise.

Mariners 6 and 7 also sent back images of Mars when they flew past the planet in 1969, but, although they had higher resolution and flew over different areas, they showed a similar cratered planet as Mariner 4. Mars was seen to be a barren, lunar-like planet, with little but impact craters and a tenuous atmosphere. This image of Mars was due for a major revision two years later, however, when Mariner 9 was put into orbit around Mars.

Mariner 9 arrived at Mars in November 1971, in the middle of a planet-wide dust storm. When the dust had cleared, the spacecraft showed a very different world from the one seen during the brief encounters of its three predecessors. There were four enormous shield volcanoes, all of which were far larger than any on Earth. The largest, Olympus Mons, was found to rise 26,000 m (85,000 ft) above the surrounding plains (compared with only 9,000 m for Mount Everest on the Earth), have a base diameter of 540 km, and have a volcanic crater 60 km in diameter on its summit. Schiaparelli had, in fact, noticed a small whitish patch in this area in the late nineteenth century, and had christened it Nix Olympica, the Snows of Olympia. The other three large volcanoes were in the Tharsis region, where a number of observers had also seen small whitish patches. All these whitish patches were shown by Mariner to be due to clouds near the tops of the volcanoes.

Mariner 9 also found deep valleys, the largest of which is Valles Marineris (see Figure 4.4), being over 4,000 km long, 600 km wide and 6,000 m (20,000 ft) deep. These large valleys appeared to have been formed by the collapse of fault lines, rather than by running water, although Mariner 9 did find smaller features that appeared to have been produced by water (see Figure 4.5). In particular, there are large channels, generally on the boundary between the southern cratered terrain and the northern plains, that indicate that water may have flowed extensively in earlier Martian history.

How could these enormous volcanoes, deep valleys, and water channels, possibly have been missed by the three previous spacecraft? The earlier Mariners had only observed a total of 10% of the surface of Mars between them during their brief fly-bys and, as it happens, these new features were in the parts not seen by them.

Like the Moon and Mercury, Mars shows no evidence of plate tectonics, and the surface has not been significantly modified in the last 2 billion years. There are some small volcanoes on Mars, however, with no superimposed impact craters, indicating that a small amount of volcanic activity has occurred in relatively recent times, and may still be going on today.

The American Viking 1 and 2 spacecraft landed on Mars in 1976, taking photographs from the surface of the planet for the first time. Both spacecraft

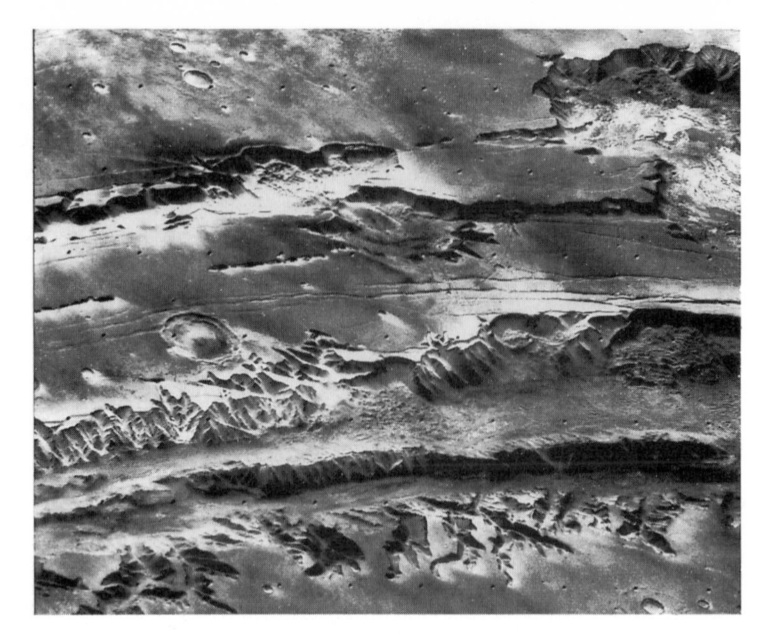

Figure 4.4 *Part of the Valles Marineris on Mars which is up to 6,000 m (20,000 ft) deep in places. (Courtesy* National Space Science Data Center, World Data Center-A for Rockets and Satellites, NASA; Experiment Team Leader, Dr. Michael H. Carr, US Geological Survey.*)*

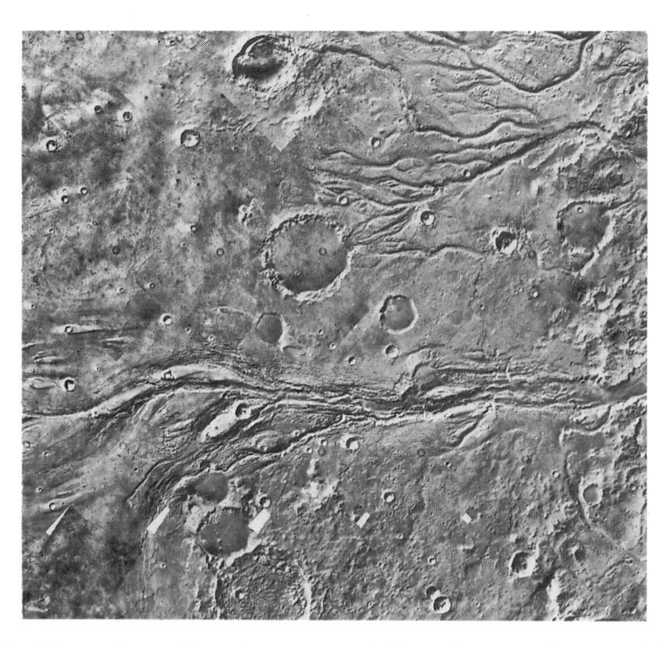

Figure 4.5 *Channels on Mars that appear to have been made some time ago by running water. (Courtesy* National Space Science Data Center, World Data Center-A for Rockets and Satellites, NASA; Experiment Team Leader, Dr. Michael H. Carr, US Geological Survey.*)*

showed a reddish-brown surface littered with boulders up to 2 m in diameter, and a light pinkish sky. Vikings 1 and 2 also searched for elementary forms of life, by analysing samples of Martian soil, but the results indicated that life was not present, at least in the areas of Mars tested. The Viking 1 lander also showed prominent sand drifts, evidence of sandstorms of the type that had affected the early observations of Mariner 9.

The Atmosphere

Occasional clouds had been seen in the atmosphere of Mars in the nineteenth century, and spectroscopic evidence had indicated the presence of oxygen and water vapour. Spectroscopic results in the first half of the twentieth century were confusing and contradictory, however (see Page 295), and the only clear evidence was of trace amounts of carbon dioxide found by Kuiper in 1952.

In 1924, W. H. Wright took photographs of Mars from the Lick Observatory through different colour filters using the 36 inch (90 cm) Lick telescope, and showed that the planet was about 150 km larger in diameter in violet light than it was in red light. As an atmosphere would scatter violet light much more than red light, this difference in diameter was considered to be proof of the existence of an atmosphere at least 75 km deep. The surface pressure was thought, in the early 1960s, to be about 100 millibars, that is about 10% of that on Earth.

In 1965, Mariner 4 found a much thinner atmosphere than anticipated, and almost completely made up of carbon dioxide. The surface atmospheric pressure was about 5 to 10 millibars. Evidence of water flows on the surface of Mars in its earlier history suggests, however, that the atmosphere was much denser in those days.

Ten years later, the Viking landers measured their local atmospheric conditions on the surface, finding temperatures that ranged from 180 K (−93°C) at dawn, to 240 K (−33°C) at about 2 pm local time, and atmospheric pressures that varied from 7 millibars in the northern summer to 10 millibars in the northern winter. The winds were found to be generally light at about 12 miles/h (20 km/h), although they reached 150 miles/h (250 km/h) during dust storms. Viking 1's latitude was 23°N, and Viking 2's was 44°N, and their own local conditions varied slightly about these average figures.

The Viking landers were the first spacecraft to measure the atmospheric constituents accurately, confirming a preponderance of carbon dioxide, at 95% (by volume). Both oxygen and water vapour are present but only in minute quantities, 0.13% and 0.03%, respectively.

The Polar Caps

The rapid melting of the polar ice caps puzzled nineteenth century astronomers, as Mars should be colder than the Earth as it is further from the Sun. A. C. Ranyard and Johnstone Stoney suggested, in 1898, that the polar caps consisted of frozen carbon dioxide, which could explain their rapid melting at low temperatures (as carbon dioxide has a lower freezing point than water). Unfortunately, there appeared to be a blue melt band at the edge of the melting ice caps in spring, and, at the low atmospheric pressure on Mars, carbon dioxide would not melt into a liquid, but would sublimate directly into the gaseous state, so water ice still seemed to most astronomers to be a more likely candidate. If the polar caps did consist of water ice, however, they must be relatively thin, or they would not melt sufficiently during the spring and summer.

Kuiper deduced, in the 1950s, that the caps must consist of frozen water, rather than frozen carbon dioxide, since both water ice and the polar caps appear nearly black in the infrared, while carbon dioxide ice appears white. So, prior to the arrival of spacecraft, the caps were thought to consist of very thin water ice, or a thick layer of hoar frost.

Mariner 9 showed that the polar caps extend down to about 55° latitude in winter, but recede to within about 5° of the poles in summer. At their greatest extent their surfaces are at the temperature of carbon dioxide ice, but the residual caps are intermediate in temperature between carbon dioxide ice and water ice. This indicated that the ice caps are mainly composed of carbon dioxide which freezes out of the atmosphere in winter, exactly as suggested by Ranyard and Stoney in 1898. Using Mariner data, it was concluded that the ice caps are only a few centimetres thick in the regions near the edges, whereas, at the poles, they may be up to several kilometres thick.

Ten years later, the more accurate measurements made by the Viking orbiters showed that the two polar caps were not the same. The thin layer of carbon dioxide ice disappears completely from the northern cap in its summer, exposing water ice beneath, whereas the carbon dioxide ice does not completely disappear from the southern cap during its summer. The Viking spacecraft also reduced the upper estimate of the thickness of the polar caps to tens of metres, from the previous estimates of possibly several kilometres.

In the southern summer the atmosphere of Mars was found to contain more dust than in the northern summer. It is thought that this dust restricts the solar heating of the southern ice sheet in the southern summer, thus explaining why the carbon dioxide ice does not sublimate completely from it in the summer.

So, our picture of Mars has changed completely over the last 100 years, from a planet with water, vegetation, a significant atmosphere and possible life, to a dead world, with no liquid water and a very tenuous atmosphere. There is evidence that water had existed on Mars earlier in its history, however.

Phobos and Deimos

Mariner 9 took photographs of both Phobos and Deimos, the two small moons of Mars, showing them to be irregular objects. Phobos was seen to be littered with craters, but the images of Deimos did not have enough resolution to show surface detail.

The images of both moons from the Viking orbiters were much sharper than from Mariner 9 and, at one stage, Viking Orbiter 2 was manoeuvred to within only 30 km of the surface of Deimos, producing images with a resolution of 3 m. Phobos was found to have a network of linear grooves, which appear to be fractures caused by the collision that created the 10 km crater called Stickney. This collision must have almost shattered Phobos, as the crater's diameter is about 1/3 of the moon's largest diameter. Deimos was seen to have a smoother surface than Phobos, and some of its craters have been partially filled by material.

5 The Gas Giants

Jupiter

Early Work

All the planets that have been considered so far are relatively small, solid bodies close to the Sun. Jupiter is completely different, being the first of the large non-rocky planets in the outer solar system with a density, in Jupiter's case, of not much more than that of water.

Markings on Jupiter are relatively easy to see in a telescope, with longitudinal bands and spots (see Figure 5.1) in a continuous state of change. These markings rotate around the globe at slightly different speeds, with the equator rotating about 5 minutes faster than the higher latitudes, so we are clearly looking at the atmosphere of Jupiter which is sufficiently dense for us never to see the true surface, if such a surface exists. Jupiter was found to rotate about its axis in about 10 hours, which is astonishing considering that its diameter is more than 10 times that of the Earth, and, because of this rapid rotation, and associated centrifugal force, its equatorial diameter is appreciably larger than its polar diameter.

Jupiter's Great Red Spot, which has a length of over three times the diameter of the Earth, was seen as an apparently new feature on the planet in 1872, although it was probably the reappearance of a somewhat smaller spot first recorded by the English scientist Robert Hooke in 1664 and the Italian astronomer Giovanni Cassini in the following year. In 1879 it was brick red in colour and very prominent, but between 1882 and 1884 it had almost disappeared, although by 1886 it was quite evident again.

The storms in the atmosphere of Jupiter, of which the Great Red Spot was the largest, were clearly much larger and much more violent than any on the Earth, even after allowing for its larger size, and yet Jupiter is much further away from the Sun than the Earth. This led to the conclusion that these

Figure 5.1 *Drawing of Jupiter made in 1878 by G. D. Hirst of Sydney, Australia, showing its longitudinal bands and spots. (From* The Sun: Its Planets and their Satellites, *by Edmund Ledger, 1882, Plate VIII.)*

storms could not be produced by solar heating but, if solar heating was not driving these storms, what was?

As Jupiter is much larger than the Earth, it would have lost heat from its original gaseous state much more slowly than the Earth, so it was concluded that the storms on Jupiter are driven from heat that is still within the planet. It was clear, however, that although Jupiter was thought to be hot, this internal heat was not enough to cause Jupiter to emit any appreciable light of its own, because the shadows of its moons cast by the Sun on the planet's surface were jet black.

Spectral analysis had shown the main absorption band of Jupiter's atmosphere was at a wavelength of 618 nm in the red, with several others in the red and yellow parts of the spectrum, but these bands did not appear to correlate with those of any known element, so the composition of Jupiter's atmosphere was a mystery.

There were suggestions that Jupiter, because of its size, its appearance, and the fact that its angular rotation is slightly faster at the equator than at higher latitudes, may have originally been like a miniature Sun in years gone by, and the Sun may, when it has cooled in years to come, look like Jupiter of today. It was also thought possible that Jupiter could become habitable when it had cooled somewhat. To quote Ball again (Reference 4) "The time will assuredly come when the internal heat must decline, when the clouds will gradually condense into oceans. On the surface dry land may then appear, and Jupiter be rendered habitable."

Galileo discovered the four large moons of Jupiter in 1610, but by 1890 no further moons were known. These four so-called Galilean moons were all observed to show variations in brightness, and William Herschel concluded in 1797 that they all kept their same sides permanently facing Jupiter (i.e. their axial rotation periods were synchronous with their orbital periods). In the nineteenth century there were conflicting views on whether their axial rotation periods were synchronous, although it was thought most likely for Callisto which, incidentally, seemed to be the most variable moon in brightness. Io was observed to be the brightest moon, and this was attributed to it having probably the most cloudy atmosphere of the four; it being taken for granted that all four moons had atmospheres.

All four moons are frequently eclipsed by Jupiter (i.e. Jupiter cuts off the Sun's light from that moon). It was noted by the Danish astronomer Ole Rømer in 1675 that the timing of these eclipses was different when Jupiter was close to the Earth, compared with when it was further away. He concluded that this was because the velocity of light, which until that date had been thought to be infinite, was finite, and this enabled him to make an estimate of its velocity, knowing approximately the diameter of the Earth's orbit around the Sun.* By 1890 the velocity of light had been measured more accurately using laboratory apparatus, enabling Rømer's calculation to be done in reverse to estimate the diameter of the Earth's orbit around the Sun.

The Atmosphere

In 1922 radiometric measurements were made of Jupiter by W. Coblentz and Carl Lampland at the Lowell Observatory at Flagstaff, Arizona which showed a cloud-top temperature of only 140 K. This was a big surprise as Jupiter had previously been thought to be quite hot because of the retained heat at formation still in its interior.

Infrared thermal emission measurements were made of Jupiter from high flying aircraft in 1969, showing that the planet radiates roughly twice as much energy as it receives from the Sun, and resurrecting the old idea that Jupiter had not completely cooled from its original formation.

In 1932, Rupert Wildt cleared up another mystery when he discovered that the atmospheric absorption lines found in the nineteenth century were due to methane (CH_4) and ammonia (NH_3). Two years later, Harold Jeffreys concluded that the atmosphere consisted of hydrogen, helium, nitrogen and methane with clouds of ammonia crystals. Hydrogen and helium are very difficult gases to detect at planetary temperatures but, at the end of the 1950s, molecular hydrogen (H_2) was detected in Jupiter's atmosphere, with similar proportions of hydrogen, carbon and nitrogen to those in the Sun. During the 1970s, water, ethane (C_2H_6), phosphine (PH_3), acetylene (C_2H_2) and a number of other molecules were also detected.

---

*The difference in distance between the Earth and Jupiter when they are closest together, compared with when they are furthest apart, is the diameter of the Earth's orbit, if the orbits of Jupiter and the Earth are circular and in the same plane as each other. The reality is a little more complicated than that, but the principle is basically the same.

Helium was identified for the first time in Jupiter's atmosphere by the American Pioneer 10 spacecraft in 1973, showing that Jupiter's atmosphere is about 99% hydrogen and helium, like the Sun, so it seems to have evolved relatively little since the planet was once part of the solar nebula. Pioneer also confirmed that Jupiter emits more heat than it receives from the Sun, either because of its retained original heat at formation, and/or because heat is being generated by the gradual contraction of the planet. The brightness temperature of 128 K, measured by the infrared radiometer on Pioneer, is very close to that measured in the 1920s.

Infrared measurements, from Earth and Pioneer, of the temperature of the various cloud belts and zones, showed that the dark belts are warmer than the white zones and are, therefore, at a lower altitude. It is thought that the white zones consist of ammonia cirrus clouds at the top of regions of ascending gas, and that the dark belts are layers of haze over the base of the descending columns, so the white zones are high pressure areas, and the dark belts are of low pressure.

The Great Red Spot had been observed on Jupiter for centuries, although there had been periods when it was invisible. In 1972, a much smaller spot was discovered from Earth, and 18 months later, it was shown by Pioneer 10 to be similar in shape and colour to the Great Red Spot. This small spot had disappeared a year later when Pioneer 11 arrived at Jupiter.

Pioneer showed that the cloud-tops of the Great Red Spot are colder, and probably higher, than the cloud-tops of the adjacent South Tropical Zone, so the Great Red Spot is an enormous, long-lived, high pressure hurricane.

In 1979, the American Voyager spacecraft (see Page 333) found that the cloud movements in the Jovian atmosphere were more complicated than had been anticipated. Instead of a relatively simple zonal rotation around the planet, with occasional vortices, there were vortices of various shapes and sizes all over the planet. The interaction of the zonal flow with the Great Red Spot, for example (see Figure 5.2), produced a complicated disturbed flow in which clouds were torn apart, and often incorporated into the spot itself. Many red, brown, white and blue spots were found interacting with the zones and belts, and the interfaces between the belts and the zones resulted in especially disturbed cloud motions. The atmosphere of Jupiter was clearly highly dynamic, as well as being colourful.

The Internal Structure

Harold Jeffreys suggested, in 1934, that Jupiter consisted of a core of rock, surrounded by a thick layer of water ice and solid carbon dioxide, which was, in turn, surrounded by an extensive atmosphere. Rupert Wildt calculated, 3 years later, that the inner core should have a diameter of about 70,000 km, or 50% of that of the planet, the surrounding ice being about 25,000 km thick, and the atmosphere 10,000 km deep. The atmosphere should have an average density of 0.78 g/cm^3, almost that of water on the Earth, and have a pressure at its base in excess of a million bars.

Figure 5.2 *The Great Red Spot and other features on Jupiter seen from Voyager 1 at a distance of 33 million km. (Courtesy* National Space Science Data Center, World Data Center-A for Rockets and Satellites, NASA; Team Leader, Dr. Bradford A. Smith.*)*

W. H. Ramsey of Manchester University in England proposed a completely different internal structure for Jupiter in 1951, with the majority of the planet being a core of hydrogen which is so highly compressed that it behaves as a metal. He suggested that this core was about 120,000 km in diameter, surrounded by an 8,000 km thick layer of ordinary solid hydrogen, and followed by the atmosphere.

The Pioneer data indicated that neither Wildt nor Ramsey were correct. Instead it was concluded that Jupiter probably has a 25,000 km diameter core of rock and ice, surrounded by a 45,000 km thick shell of liquid metallic hydrogen and liquid helium, followed by a 15,000 km thick shell of liquid molecular hydrogen and helium, and finally a 1,000 km deep atmosphere.

The Magnetosphere

The Americans Kenneth Franklin and Bernard Burke found, in 1955, that Jupiter emits bursts of radio waves, at a frequency of 22.2 MHz, which appeared to be generated in the planet's atmosphere. Franklin and Burke found a correlation with the position of the Great Red Spot, and the Australian C. A. Shain showed that there were similar radio emissions associated with one of the white spots in the South Temperate Belt. The signal's intensity was found to be modulated by Io, the innermost Galilean satellite. Two or three years later, another type of radio emission was discovered, at a frequency of about 1 GHz (i.e. 1,000 MHz). It had an almost constant intensity, and originated in a toroidal region around Jupiter, inclined at about 10° to the planet's equatorial plane. In 1959, Frank Drake and Hein Hvatum concluded that this 1 GHz signal was due to synchrotron radiation generated by relativistic electrons* trapped in Jupiter's magnetic field.

The Pioneer 10 spacecraft discovered, in 1973, that Jupiter's magnetic field appeared to consist of two different regions; one being a dipole field of strength 4 gauss, or about ten times that of the Earth, extending to 20 Jupiter radii (R_J), and the other, stronger field outside of 20 R_J, being more confined to Jupiter's equatorial plane. The axis of the dipole field was found to be inclined at about 10° to Jupiter's spin axis, and the centre of the magnetic axis was found to be displaced from the centre of the planet by about 0.1 R_J.

Pioneer 11 found that within 3 R_J the field is more complicated than a simple dipole field, and that intensity varies from about 3 to 14 gauss at the cloud-tops.

The source of the magnetic field was thought to be liquid metallic hydrogen in the interior of Jupiter, whose complicated circulation pattern produces the non-dipole structure.

Pioneer 10 first discovered energetic particles as far as 300 R_J from the planet, and outside of the bow shock,† which it crossed at 108 R_J. It was surprising to find particles from Jupiter in front of the bow shock, as this meant that they had to escape from the outer magnetosphere and traverse the shock wave. When these particles were discovered, the records of background cosmic ray electrons were examined from Earth-orbiting satellites, and an enhancement was discovered every 13 months, which is the same as Jupiter's synodic period. Thus some of Jupiter's high energy electrons even reach the Earth.

---

*Relativistic electrons are electrons travelling at near the speed of light.

†The bow shock is where the solar wind is deflected by a planet's magnetic field.

When Pioneer 11 arrived at Jupiter, one year after Pioneer 10, it found that the bow shock, at its closest point to Jupiter, was only half the distance from Jupiter found by Pioneer 10. Even so, at 50 R_J, it was still very much further from Jupiter than the Earth's bow shock is from the Earth, which is at 13 R_E, or 1.1 R_J. (This is because the pressure of the solar wind at Jupiter's orbit is only about 4% of its value at the Earth's, and Jupiter's magnetic field is also an order of magnitude greater than the Earth's.)

Pioneer found that Jupiter's radiation belts, where protons and high energy electrons are trapped in the planet's magnetic field, appear to be 10,000 times more intense than the Earth's Van Allen radiation belts, with the most intense part lying within 20 R_J. Inside 20 R_J, the magnetosphere behaves like the Earth's, but outside of that region, the centrifugal force caused by the rapid rotation of Jupiter causes the particles to concentrate in Jupiter's equatorial plane. Inside the orbit of Amalthea, at 2.5 R_J, Pioneer 11 found that the flux of energetic particles becomes more complex, just as the magnetic field does.

Prior to the Pioneer 10 encounter, there had been speculation about the effect of Jupiter's inner satellites on its radiation belts, as these inner satellites were thought, correctly as it turned out, to lie within the belts. The spacecraft found evidence of the sweeping effect of Io and Europa on the population of protons and electrons in the belts. Similar, but less pronounced effects were found for Amalthea, Ganymede and Callisto, so, when the number of particles was found to decrease close to the planet, it was suggested that Jupiter could have another satellite there, or possible a ring system. This ring system was discovered by Voyager 1 in 1979.

Pioneer 10 found in 1973 that, not only do Amalthea and the four Galilean satellites trap particles from Jupiter's magnetosphere, but Io is unique in that it produces and accelerates them. Io was found to have an ionosphere extending up to about 700 km above its surface on the day side, which gradually decays during Io's night. It also appeared to have a tenuous atmosphere, with a surface pressure of about 10^{-9} bar.

A torus of atomic hydrogen was found around Jupiter, at Io's distance from the planet, extending to $\pm$ 60° from Io. It was thought that it was produced by the sputtering of Io's surface by particles in Jupiter's magnetosphere. At about the same time, Robert Brown of Harvard University discovered, using a ground-based telescope, that Io was surrounded by a yellow sodium cloud, which was also thought to have been produced by surface sputtering.

In 1979, the Voyager spacecraft showed that Jupiter's plasma originates mainly from gas ejected by Io's volcanoes which, once ionised, forms the Io torus of low energy particles around Jupiter.

The Satellites

Edward Barnard, using the 36 inch (91 cm) refracting telescope of the Lick Observatory in 1892, discovered Amalthea orbiting Jupiter once every 12 hours just 110,000 km above the surface of the planet. This was the first satellite of Jupiter to be discovered since Galileo found the four large satellites almost 300 years earlier, and it also marked the end of an era as it proved to be the last satellite of any planet to be discovered visually. At about the turn of the century, Barnard discovered that the poles of Io, the innermost of the Galilean satellites, appear to have reddish caps, and in the first two decades of the twentieth century four more satellites of Jupiter had been discovered photographically.

By the time that Pioneer 10 arrived at Jupiter, in 1973, three additional satellites had been discovered, making a grand total of 12. Of these 12, the four Galilean satellites, plus Amalthea, orbit relatively close to Jupiter, then come three small satellites orbiting at about 11 million km from the planet, followed by four small satellites orbiting at about 22 million km from the planet. The latter group of four all orbit Jupiter in the opposite direction to the others, in what is called a retrograde sense. This, and the closeness of Jupiter to the asteroids, led Forest Moulton to suggest that the outside group were captured asteroids.

A. B. Binder and Dale Cruikshank, who were then at the University of Arizona, found, in the 1960s, that Io is a few percent brighter than usual for 15 minutes or so after it emerges from Jupiter's shadow. They suggested that Io may have an atmosphere and that, during the cold eclipses, clouds or frost could condense, which then disperse when Io is illuminated by the Sun once more. Later observations showed that this brightening only occurs for about half of the eclipses, and the low atmospheric pressure discovered by Pioneer 10 seemed to mitigate against Binder and Cruikshank's theory. An alternative theory, that the surface of Io is modified by low temperatures, was then proposed.

Pioneer did not image Jupiter's satellites at all well, and so they were still very much unknown quantities before the Voyager intercepts of 1979. Prior to Voyager, it was thought that Io would look like a reddish version of our Moon, but covered with sulphur-coated impact craters, and that the other three Galilean satellites, namely Callisto, Ganymede and Europa, would have ice-covered surfaces with probably few impact craters. It was thought that the ice would have flowed on the surfaces of these three Galilean satellites, largely eliminating their craters.

Imagine the excitement, therefore, when eight active volcanoes were discovered on Io by Linda Morabito and her colleagues at JPL using Voyager 1. The largest volcano and the first discovered was Pele, named after the Hawaiian volcano goddess. Four months later, Voyager 2 found that Pele had become dormant, the volcano Loki (see Figure 5.3) had become more active, and two new volcanic vents had appeared elsewhere on Io's surface. This surface appears to be made of sulphur and sulphur dioxide, and its volcanoes produce plumes hundreds of kilometres high, which deposit particles of solid sulphur and crystals of sulphur dioxide hundreds of kilometres from the volcanic craters. Thus the surface of Io is being regenerated at a very fast rate. Some of the ejected material is lost to Io, and was found to have become ionised to form the large torus of low energy particles around Jupiter.

Three days before the Voyager 1 encounter with Io, Stanton Peale, of the University of California, and Patrick Cassen and Ray Reynolds, of the NASA Ames Research Centre, actually predicted Io's active volcanism, but this was not generally known at the time of its discovery. They suggested that tidal interactions between Io and Europa and Ganymede would cause stresses in the surface of Io, which would lead to volcanic eruptions.

The Voyager spacecraft showed that Callisto, the Galilean satellite furthest from Jupiter, and the darkest of these satellites, has a surface of dirty water ice, and is the most heavily cratered body yet found in the solar system. The craters are quite shallow, however, because the ice melted on impact, and filled the craters more readily than if the surface had been rocky. Even after re-freezing, the ice would have had a plastic flow in the warmer temperatures of that earlier era. The outstanding feature of this very old surface is the large basin now called Valhalla, which is surrounded by numerous concentric rings of ridges up to a diameter of 4,000 km, formed by shock waves caused by the impact that formed the basin.

Ganymede, the largest of the Galilean satellites, also has an icy surface, but its structure is much more complex than that of Callisto. Not only does it have numerous impact craters, but some are at the centre of bright ray systems, and there are also bands of light grooved terrain. Some parts of Ganymede's surface appear to be as old as Callisto's, but some are noticeably younger.

Europa, the other Galilean satellite, was found to have the smoothest surface of any body in the solar system. Its albedo is higher than that of

Figure 5.3 *Voyager image showing the plume from the volcano Loki on Io. (Courtesy* National Space Science Data Center, World Data Center-A for Rockets and Satellites, NASA; Principal Investigator, Dr. Bradford A. Smith.*)*

Callisto or Ganymede, implying that it has a crust of relatively clean water ice. Europa has a complex system of linear features 20 to 40 km wide and up to thousands of kilometres long, which look as if they are surface cracks, that may have been caused by the outer part of the core expanding on freezing, if it consists of liquid water. There were few impact craters visible, indicating a relatively young surface.

Barnard's satellite, Amalthea, was found to be an extremely irregular object, of dimensions of 260 × 145 × 140 km. Its surface is heavily cratered, very dark and distinctly reddish, probably as a result of dusting with sulphur compounds ejected by Io. The craters are enormous, compared with the size of Amalthea, one being 90 km across.

Saturn

Early Work

The second largest planet in the solar system is a beautiful sight in any telescope because of its spectacular ring system (see Figure 5.4), which had first been recognised as a single ring by Christiaan Huygens in the seventeenth century. Shortly afterwards, Giovanni Cassini observed that Huygens' ring was two concentric rings (now called rings A and B) separated by a small dark gap (now called the Cassini division). A faint division was also found in the outer ring (ring A) by Johann Encke, the director of the Berlin Observatory, in 1837.

Figure 5.4 *Drawing of Saturn made in 1872 by L. Trouvelot of the Harvard College Observatory. The Enke division in the outer ring (A), and the Cassini division between rings A and B, are clearly shown. (From* The Story of the Heavens, *by Robert Ball, 1897, Plate I.)*

In the late eighteenth century there were indications that the rings were not uniform when seen almost edge-on, and it was by measuring the movement of a bright spot on the rings that William Herschel deduced a rotation period 10 h 32 min 15 s. This is now known to be the rotation period of about the middle of ring B and, as the closest moon to Saturn is outside of this ring, Herschel must have seen a bright mark on ring B itself.

In the mid nineteenth century, the Scottish physicist James Clerk Maxwell proved theoretically that the rings were composed of a great many small bodies in orbit around the planet, and Daniel Kirkwood showed that the Cassini division was caused by the perturbations of three of Saturn's moons, which have orbital periods of exactly two, three and four times that of a particle in the Cassini division.

Shading on the surface of the rings was sometimes observed in the latter half of the nineteenth century when the rings were open, and, although two divisions were clearly known by then (the Cassini and Encke divisions), a number of other divisions were also vaguely observed. Ledger, for example, reported in Reference 1 that "not only is a certain amount of shading often visible upon the ring surface, . . ., but indications are seen of numerous divisions, so that it is impossible to decide into how many rings . . . the whole may be divided." Some shading on rings A and B, and irregularities on the inner edge of ring A, are seen at the extreme left and right of the rings in Figure 5.4.

The rings are inclined at 27° to the plane of Saturn's orbit around the Sun, which Saturn traverses in 29 years. So for $14\frac{1}{2}$ years we see one side of the rings, for $14\frac{1}{2}$ years we see the other side, and every $14\frac{1}{2}$ years we see the rings edge-on. When, in the nineteenth century, the rings were seen to be exactly edge-on they completely disappeared even in the largest telescopes of the day, so they were known to be very thin. By comparing them with the smallest known moon, Hyperion, it was concluded that they could not be more than 300 km thick and were probably much thinner (present day values are 150 *metres* maximum).

In the middle of the nineteenth century, the French mathematician Édouard Roche had shown theoretically that if a fluid moon orbits within 2.44 mean radii of its primary, the moon will be destroyed by tidal forces. As the outer edge of ring A lies at 2.38 radii from the centre of Saturn, George Darwin concluded that Saturn's rings were the remnants of a fractured moon and, as such, they would gradually dissipate. Laplace, on the other hand, had concluded that the rings were the remnants of the nebula that had once surrounded Saturn and which had condensed to form its moons.

Mention has already been made of the moons of Saturn, the first and largest being Titan, which was discovered by Huygens in 1655. Over the next 200 years a further seven moons were found, the last of them, Hyperion, being discovered in 1848 by William Cranch Bond, the director of the Harvard College Observatory, and William Lassell, an English amateur astronomer and ex-brewer.

Discoveries are quite often made almost simultaneously by people in different places without any collaboration. The case of Le Verrier and

Adams in their parallel discoveries of Neptune is described below on Pages 87–89, and in the case of Saturn simultaneous discoveries have happened twice. The eighth moon to be discovered was first seen independently by Bond on 16th September 1848, and Lassell on 18th September of the same year, and both men realised on exactly the same day, namely 19th September 1848, that the faint body they had seen was a moon orbiting Saturn.

The second parallel discovery was that of the dark crape* ring (ring C) in 1850 by Bond in the USA and, independently, a fortnight later, by the English clergyman and amateur astronomer W. R. Dawes. Although the crape ring is quite dark, it is of considerable width, and it was surprising that it had not been discovered until 1850, leading late nineteenth century astronomers to wonder whether it was still in the process of formation. Some time after its discovery, however, it was found to have been seen in the seventeenth century by both Campani, an Italian telescope maker, and Robert Hooke, the famous English physicist. This C (crape) ring, which lies inside the two bright rings A and B, was known to be partially transparent, as Saturn could be seen through most of its width and Edward Barnard was able to observe the moon Iapetus through it in 1889.

Saturn's moon Iapetus was known by both Cassini and William Herschel to have an irregular surface reflectivity, because they observed it to vary in intensity from time to time. Herschel, in fact, found that these variations were perfectly regular with a period equal to the moon's orbital rotation period around Saturn, thus indicating that the axial and orbital rotation periods of Iapetus were the same (i.e. its axial rotation was synchronous). The brighter hemisphere was observed to be 4.5 times brighter than the darker hemisphere.

Very little was known about the planet Saturn itself. The surface did not have markings that were as clearly defined as those on Jupiter, and generally all that could be seen were faint longitudinal bands. Spots were seen from time to time, however, and on 7th December 1876, Asaph Hall saw a brilliant white spot near the equator which gradually, over the next few weeks, became more and more elongated until it finally dispersed into a faint long streak. It rotated about Saturn once every 10 h, 14 min, 24 s, which was the best estimate of Saturn's rotation period for many years.

Detailed observation of the various surface markings showed that they rotated around Saturn with slightly different speeds. This indicated that astronomers were, as in the case of Jupiter, observing the cloud layer, rather than the surface of the planet itself. Spectroscopic analysis showed similar absorption bands to those of Jupiter, but their source was unknown also. Because the density of Saturn was so low (less than that of water), it was thought that the planet was largely, or maybe completely, gaseous.

---

*So-called as, in Victorian times, a crape was a black veil, or a black band worn on the sleeve or hat in times of mourning.

Although Saturn is appreciably smaller than Jupiter, it is still very much larger than the Earth, and so it was assumed that it had retained a fair amount of its original heat to maintain a heated interior. As it is smaller than Jupiter, however, it should not be quite as hot, and this was consistent with the less obvious atmospheric circulation patterns on Saturn. A cooler gaseous planet should be more compact than a hotter one, and so it was considered strange that Saturn was the less dense of the two.

The Atmosphere

Over the last hundred years or so, prominent white spots have been found to recur about every 30 years on Saturn, alternating between equatorial and middle latitudes. Those spots that appeared near to the equator were found to rotate around Saturn in about 10 h 14 min, whereas those at middle latitudes took about 10 h 40 min. This clearly showed Saturn's differential rotation; the rotation taking longer, the higher the latitude, just like the Sun and Jupiter.

The sidereal period of Saturn is 29 years, which is approximately the interval between successive appearances of the white spot. In fact, every time the white spot appeared, the north pole of Saturn was tilted at virtually its maximum angle to the Sun. Whether there is any significance in the fact that the spot appeared in equatorial regions on every other appearance, and at higher latitudes on intermediate occasions, is unknown. In all cases, it became gradually elongated before disappearing after a few weeks.

Radiometric measurements of Saturn by W. Coblentz and Carl Lampland in 1922 showed a cloud-top temperature of only 120 K. This was a surprise (like a similar result was for Jupiter) as it had been thought that Saturn was probably quite warm, due to it still retaining in its interior some of the heat present at its formation. However, infrared measurements made in 1969 showed that Saturn, like Jupiter, was emitting about twice as much energy as it was receiving from the Sun. So there seemed to be heat still in Saturn; it just was not as much as had been thought originally.

In the nineteenth century it was known that the atmospheres of Saturn and Jupiter produced similar absorption lines in their spectra, but their source was not understood until 1932 when Wildt showed that they were due to methane and ammonia. The ammonia clouds were thought to be less dense on Saturn, however, than on Jupiter, as the ammonium lines were weaker.

The images of Saturn received during the Pioneer 11 satellite fly-by in 1979 were disappointing, showing only the subtle banded structure seen from Earth, with no spots to assist in understanding the atmospheric dynamics. The images of Titan, Saturn's largest satellite, were even more disappointing, as they showed no features at all.

Just over a year later, the much higher resolution Voyager 1 images finally showed the detailed atmospheric structure on Saturn. Various spots and markings were seen that enabled the velocity of the various zonal flows to be determined. The equatorial wind current was found to be 80,000 km wide, and have a maximum speed of an amazing 1,700 km/h*, showing that it is much larger and much more energetic than the 30,000 km wide, 400 km/h equatorial current of Jupiter. This was a surprise, since Saturn receives much less heat from the Sun than Jupiter and, as it is a smaller planet, it should have cooled more rapidly after its formation. The cause of this more rapid flow on Saturn is still unclear.

The Magnetosphere

Jupiter was known to have a magnetic field before the Pioneer spacecraft arrived there in 1973, because of Jupiter's radio emissions, but it was not known whether Saturn had a magnetic field, or not, until Pioneer 11 visited the planet in 1979. Prior to the encounter, it was thought that Saturn probably did have a magnetic field, as the internal structures of Saturn and Jupiter were thought to be broadly similar, although it was expected that Saturn's rings would stop the associated radiation belt from coming within the outer edge of ring A.

The first evidence for a magnetic field on Saturn came when Pioneer detected the bow shock at 24 R_S as it approached Saturn (compared with about 100 and 50 R_J, measured by Pioneers 10 and 11, respectively, for Jupiter, and 13 R_E for the Earth). As expected, the charged particle flux showed an abrupt cut-off as Pioneer passed beneath the outer edge of ring A.

Saturn's magnetic moment was found by Pioneer to be 550 times that of the Earth, but 35 times less than that of Jupiter. The magnetic axis of Saturn was found to be almost exactly the same as the spin axis, and its magnetic centre is only 0.04 R_S from its geometric centre. The charged particles in Saturn's outer magnetosphere (beyond 7 R_S) tend to congregate in an equatorial sheet, although the effect is not as pronounced as in the case of Jupiter.

Voyager 1 found, in 1980, that Saturn appeared to emit radio waves in the kilometre waveband, with a period of 10 h 39 min 25 s. This was assumed to be the rotation period of the core of Saturn, and of its magnetic field.

Voyager also showed that Titan, whose orbit lies just within Saturn's outer magnetosphere, loses nitrogen gas from its atmosphere to Saturn's magnetosphere, where it is ionised and retained. Titan and the other large satellites inside the magnetosphere, namely Rhea, Dione, Tethys, Enceladus and Mimas, absorb electrons and protons from the magnetosphere, in a similar way to the large inner satellites of Jupiter, but electrons orbiting Saturn at the same velocity as a satellite are not swept up by that satellite when they migrate across its orbit. A similar effect is not observed in the magnetospheres of the Earth or Jupiter, nor is it observed for protons in Saturn's magnetosphere.

---

*Relative to the internal structure of the planet, determined by observation of periodic radio emissions.

The Rings

Gustav Müller and Paul Kempf showed in 1893 that Saturn's rings were composed of particles, as predicted by Maxwell almost 40 years earlier, by measuring their phase coefficient using a photometer (see Page 299). Two years later, James E. Keeler, at the Lick Observatory, put the slit of his spectrograph along Saturn's equator, and found that the absorption lines were inclined, because of the different Doppler shifts for points along the equator. The inclination indicated an equatorial velocity of 10 km/s for the surface of the planet, which was consistent with its observed rotation rate, giving him confidence in his technique. He also measured a similar effect for the rings, giving velocities of 16 and 20 km/s at the outer and inner edges, respectively, which was the most convincing evidence available at the time of their particulate nature.

In the nineteenth century, Saturn was known to have three rings (designated A, B and C), but it was not until 1966 that Walter Feibelman photographed a very faint ring (called ring E) extending from outside of ring A outwards to beyond the orbit of Dione, about 400,000 km from the centre of Saturn. In 1980, when the Earth passed through the ring plane, ring E was found to extend almost to the orbit of Rhea, some half a million kilometres from Saturn. Pierre Guerin claimed, in 1969, to have discovered a faint ring inside ring C. This ring (now called ring D) was confirmed by Voyager, but whether Guerin really detected it is uncertain, because it appears to be too faint to be seen from Earth.

The C ring was known in the nineteenth century to be transparent, and the A ring was independently found to be transparent in 1917 by the English amateur astronomers M. A. Ainslie and J. Knight, who observed a seventh magnitude star as it passed behind the ring. They also noted that the star showed two increases in brightness during its passage behind the A ring, indicating that the ring was not uniform. Similar variations in transparency were observed by the South African amateur W. Reid in 1920 when the B ring passed in front of a star.

Unexpectedly strong radar echoes were received from Saturn's rings in 1972 and 1973, when a 400 kW beam was aimed at Saturn. These strong echoes indicated that particles in the rings may be up to a metre in diameter, with the majority falling into the 4 to 30 cm range. Some of the particles were thought to be metallic because of their high reflectivity. Radio and infrared measurements indicated that the rings also contained ice and silicate particles, with a mean radius of about 1 centimetre.

Pioneer 11 found a narrow ring, called the F ring, just outside the A ring (between the A and E rings), during the spacecraft's 1979 encounter with Saturn. Pioneer not only imaged the new ring using its photopolarimeter, but it also detected its effect by measuring a reduced number of charged particles in the magnetosphere at that distance from Saturn.

The Voyager spacecraft surprised everyone with its images of the rings (see Figure 5.5), with Richard Terrile making the surprise discovery of spoke-like markings on ring B, 6 weeks before the spacecraft's closest

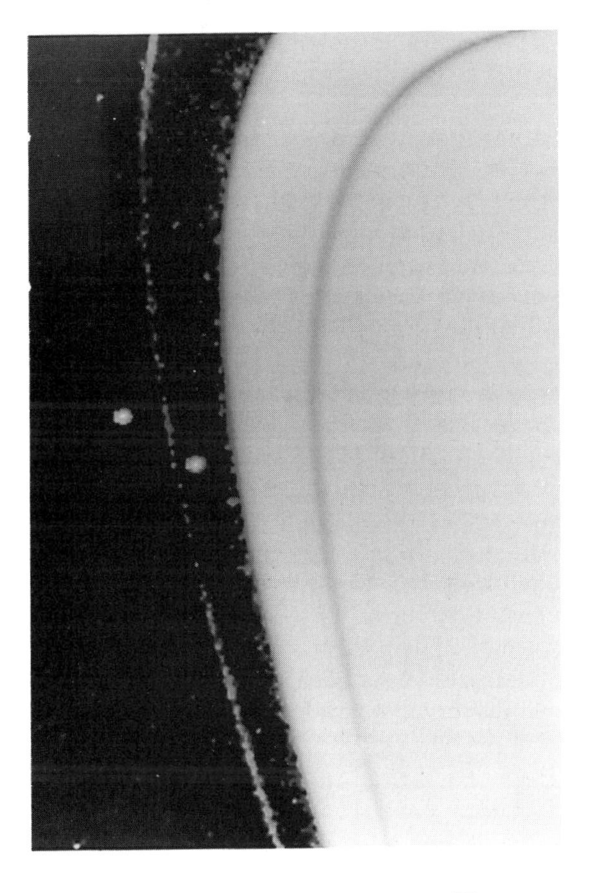

▲
Figure 5.5 *A Voyager image showing the complex nature of Saturn's rings. (Courtesy* National Space Science Data Center, World Data Center-A for Rockets and Satellites, NASA; Principal Investigator, Dr. Bradford A. Smith.*)*

◀ ***Figure 5.6*** *Voyager image showing the satellites Prometheus and Pandora shepherding the F ring. (Courtesy* National Space Science Data Center, World Data Center-A for Rockets and Satellites, NASA; Principal Investigator, Dr. Bradford A. Smith.*)*

approach. Shadings on ring A, and occasionally on ring B, had been observed in the nineteenth century, but it was never clear if they were real or not.

Andy Collins and Richard Terrile also found two new satellites, Prometheus and Pandora, using Voyager 1, one just inside and one just outside the narrow F ring, apparently shepherding, or stabilising, the ring (see Figure 5.6). Terrile also discovered another small satellite, Atlas, just outside of ring A, apparently restricting its outward expansion. Shepherding satellites had first been proposed by Peter Goldreich of Caltech and Scott Tremaine of the University of Toronto in 1979, to explain the narrow rings of Uranus, but these shepherding satellites of Saturn were the first such satellites to be discovered.

The detailed structure of the rings shown by Voyager was astonishing. In all, the ring system of Saturn was found to consist of over 1,000 narrow rings, some of which were eccentric. Some of the rings were located where major resonances with satellites should have produced gaps, and the Cassini division itself was found to contain over 100 narrow rings. The narrow, newly-discovered, F ring was found to consist of three intertwined or braided rings, which indicated that they may be controlled by magnetic forces, as well as gravitational ones. The narrow rings in both ring C and the Cassini division were found to have some regularity to their spacing, whereas those in the B ring appeared to be completely random. Voyager also found a new tenuous ring, ring G, between ring F and the inner edge of ring E.*

Before Voyager arrived at Saturn, it was generally thought that the ring structure was understood, with the gaps caused by resonances with Saturn's satellites, Mimas in particular. These resonances do have an effect, and shepherding satellites have also been found to play a key role, but the rings are far more complex than anyone had imagined, with over 1,000 individual rings being discernible. The spokes on the rings also baffled astronomers for some time, until they realised that they rotated at the same rate as Saturn itself and must, therefore, be associated with its magnetic field.

The smallest known satellite is about 20 km across, and most of the ring particles are less than a metre in size. There may well be a number of undiscovered, very small satellites, intermediate in size between the satellites and ring particles, producing some of the gaps between the rings.

Solid satellites can exist within the Roche limit for fluid satellites, because solid satellites have a significant strength, but even they are broken up if they orbit too close to the planet. The rings could have been formed by a satellite that was ruptured by such tidal forces, or by one that broke up on impact. There are relatively few asteroids in the vicinity of Saturn, and so the most likely source of an impact would have been a comet. Mimas shows evidence of almost having been destroyed by an impact (see below), and

---

*So the order of the rings is (from Saturn outwards) D, C, B, A, F, G and E. Of these, D, F and G are narrow, the others are broad.

both Iapetus and Dione have one hemisphere much darker than the other, indicating that they may once have passed through a cloud of dark material. The impact theory thus appears to have much to recommend it.

Laplace had suggested an alternative theory in the eighteenth century, that the rings are the remnants of the nebula from which Saturn was formed. Unfortunately, the rings appear to be very young (less than 100 million years old) because they are made of relatively bright particles and, if this is correct, Laplace's theory is untenable. No doubt there will be many more twists in the story before their origin is fully understood.

The Satellites

Phoebe, the ninth satellite of Saturn to be found, and the first satellite in the solar system to be found photographically, was discovered by William Pickering in April 1899, on plates taken 7 months previously. It was at the huge distance of 13 million km from its primary, and was the first satellite in the solar system to be found to have a retrograde orbit, although the outer satellites of Jupiter were later found to also have retrograde orbits.

J. Cornas Solá noticed in 1908 that Titan, Saturn's largest satellite, showed a pronounced limb darkening effect, and he suggested that this meant that it had an atmosphere. The existence of this atmosphere was confirmed in 1944, when Gerard Kuiper detected methane lines in Titan's spectrum. Voyager 1, while confirming that methane is a constituent of Titan's atmosphere, surprised astronomers by finding that the majority of the atmosphere, like that on Earth, is made up of molecular nitrogen. Titan's surface atmospheric pressure is also very similar to that on Earth, but its surface temperature of 94 K is very low, and only a few degrees above what it would have been if there were no atmosphere at all. This temperature is close to the triple point* of methane, and so methane probably exists on Titan in both a liquid and solid state, in addition to being present as a gas in its atmosphere. In this sense, methane on Titan may be like water on Earth, which also exists in all three states. The cloud layer on Titan is opaque to visible light and is entirely featureless, although Voyager did find that the northern hemisphere is noticeably darker than the southern one.

Andouin Dollfus, at the Pic du Midi Observatory in France reported, in 1966, the discovery of the first new satellite of Saturn this century, which was given the name of Janus. There was some doubt expressed about the orbit attributed to this satellite, however, and, in 1978, John Fountain and Stephen Larson reanalysed the 1966 results and found two so-called co-orbiting satellites in the same orbit, instead of just one. One of these satellites was then named Janus, and the other Epimetheus, although which was the original Janus was unclear.

In March 1980, Jean Lecacheaux and P. Lacques of the Pic du Midi

---

*The triple point is where a substance can exist in its solid, liquid and gaseous phases all at the same time.

Observatory discovered a small satellite, Helene, co-orbiting with Dione, but near the 60° Lagrangian point ahead of it. So Saturn was known to have 12 satellites before Voyager 1 arrived in November 1980.

The thirteenth and fourteenth satellites, Telesto and Calypso, were discovered by Bradford A. Smith and his colleagues on the Voyager 1 imaging team at the Lagrangian points preceding and following Tethys. Three other satellites (Prometheus, Pandora, and Atlas) were also found that had key interactions with the rings. Finally, in July 1990, Mark Showalter of NASA discovered the eighteenth satellite, Pan, from Voyager images, as the cause of the wavy edges to the Encke gap.

Cassini and Herschel had deduced, many years ago, that Iapetus has one side which is very much darker than the other. Voyager proved this, and also showed that Dione has one hemisphere which is smooth and dark, and the other which is bright with many impact craters. Dione and Tethys were both found to have valleys, hundreds of kilometres long and tens of kilometres wide, together with a large number of craters, some of which are apparently young in geological timescales. The cratering density on both satellites is variable across the surface, indicating that some surface restructuring has occurred in their early history.

Enceladus showed a somewhat similar surface structure to Dione and Tethys, but there are more valleys and fewer craters. In fact, some of the surface of Enceladus has no craters at all, indicating that that area has been resurfaced in the past 50 million years or so. The whole surface of Enceladus is so uncontaminated that it has the highest albedo of any body known in the solar system, with a reflectivity close to 100%. The main cause of its resurfacing appears to be tidal flexing, produced by the much heavier Dione, that has an orbital period double that of Enceladus.

The most noticeable feature of Mimas is a large 135 km diameter crater, which is the remains of an impact that came close to breaking up the entire 400 km diameter satellite.

Finally, one of Voyager's strangest discoveries involves the co-orbiting satellites Janus and Epimetheus. Their orbital periods were found to differ by only 30 seconds and, once every 4 years, they would approach so close to each other that they would swop orbits!

Uranus

Early Work

There could hardly be a greater contrast in the telescopic appearance of two objects; comparing the spectacular Saturn with its beautiful rings and the dull indistinct Uranus, the first planet to be discovered since ancient times.

William Herschel had discovered Uranus in 1781 when he was undertaking a general sky survey and, by the late the nineteenth century, Uranus was known to possess four moons. Observing the movement of these moons enabled the density of the planet to be determined, which appeared to be similar to that of Jupiter. Spectroscopic observations of Uranus showed absorption lines of unknown origin, namely the 618 nm line seen on Jupiter and Saturn, and four other unidentified lines.

Uranus was much smaller and further away from the Sun than Jupiter and Saturn, and so it was thought probable that it had little of its original heat remaining. So, although the atmospheric constituents were thought to be similar to those of Jupiter and Saturn, the proportions were probably different. Because Uranus is so far away from the Earth, it was very difficult to see any surface markings, and consequently its axial rotation period was unknown, although Buffham claimed to have measured a value of 12 hours in 1872 from the movement of faint bands that he believed he had seen on the planet.

Titania and Oberon, the two largest moons of Uranus, had been discovered by William Herschel in 1787. Ten years later he also announced the discovery of four more moons, but although three of these have since been proved to be spurious the fourth may have been an early sighting of Umbriel, although there is considerable doubt of this. The other two known moons, Ariel and Umbriel, were discovered by William Lassell in 1851.

The orbits of the four moons of Uranus are very unusual as they are almost perpendicular to the orbit of Uranus around the Sun (see Figure 5.7). Moreover, all four appeared to be in exactly the same plane as each other, and their orbits appeared to be exactly circular. The orbits were tilted at more than 90° (98° in fact), and so all four appeared to rotate around Uranus in a retrograde motion (i.e. the opposite sense to the rotation of Uranus around the Sun). It was thought that, as the main moons of Jupiter and

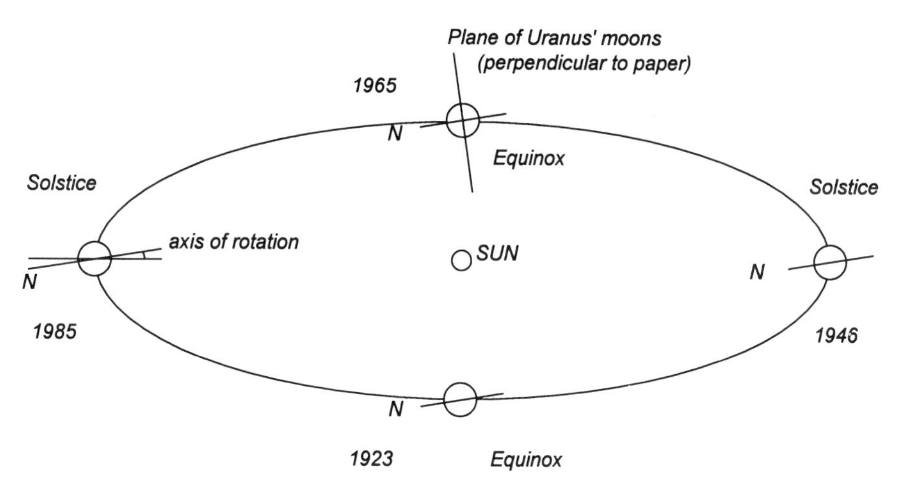

Figure 5.7 *The relative positions of the spin axis of Uranus are shown as it orbits the Sun in 84 years. The Sun was overhead at Uranus's equator in 1923 and 1965, almost overhead at its North pole in 1946, and almost overhead at its South pole in 1985. The Uranian moons orbit the planet in its equatorial plane which is perpendicular to the plane of the paper.*

Saturn revolve around their primary in approximately the same plane and direction as their planet's equator, the same would probably be the case for Uranus. In that case, Uranus would be a unique planet with both a retrograde spin rotation and an axis almost in the plane of its orbit around the Sun.

The Atmosphere

In 1932 Rupert Wildt showed that the unknown atmospheric absorption lines discovered on Uranus in the nineteenth century were due to methane. Unlike in the cases of Jupiter and Saturn, however, no ammonia lines were present, presumably as ammonia had been frozen out of the atmosphere by the intense cold. Then, almost 20 years later, molecular hydrogen was discovered in Uranus' atmosphere by Gerard Kuiper and Gerhard Herzberg at the Yerkes Observatory near Chicago.

Vesto Slipher measured the Doppler shift of the Uranian spectral lines in 1911, and deduced an axial rotation rate of 10 h 40 min. A large series of visual observations were then made of Uranus in 1916, showing a regular 15% variation in brightness with a period of 10 h 49 min. This was very close to Slipher's result, but observations of the brightness of the planet made in the following year could find no such period. Maybe the clouds responsible for the brightness variations in 1916 had dispersed. Over the next two or three decades some visual observers found variations close to the 10 h 49 min value, but others could detect no variation at all. Photoelectric observations between 1927 and 1950 showed no brightness variations either, but these devices had a much bluer response than the eye, which may explain their negative results.

In the 1930s J. H. Moore and Donald Menzel repeated Slipher's measurement of the Doppler shifts, and deduced a period of 10 h 49 min, the same as the visual observers. This seemed to be conclusive and this period was accepted for many years, until longer periods of 15 to 17 hours were deduced in the 1970s from intensity fluctuations.

The first image of Uranus to show any atmospheric structure was made by Bradford Smith, James Janesik and Larry Hoveland in 1976 using a charge coupled device, or CCD, on the 1.5 m telescope of the University of Arizona. They found evidence for thin clouds of ice, or photochemical particles, high in the planet's atmosphere, above the methane layer. Ten years later, as the Voyager 2 spacecraft approached Uranus, a few spots were discovered in its atmosphere which, after image enhancement, indicated a rotation period of the cloud layer of about 16 to 17 hours. Then, just before encounter, Voyager discovered radio emissions from Uranus that showed that the interior of the planet rotates with a period of 17 h 14 min.

Uranus' sidereal period (i.e. its year) is 84 years, and its equator makes an angle of 98° with the plane of its orbit around the Sun, so it orbits the Sun virtually spinning on its side (see Figure 5.7). The Sun can thus illuminate one pole within 8° of the polar zenith, and then gradually, over the next

42 years, move until it is almost overhead at the other pole. Half way between these two positions, the Sun is directly overhead at the equator. Because one pole can be illuminated by the Sun for a number of years, while the other is in darkness, it was assumed that there would be a difference of maybe 5 to 10 K in the temperatures of the cloud-tops over the two poles, when one was in sunlight and the other in darkness.

It so happens that Voyager arrived at Uranus when the Sun was almost overhead at the south pole, but, rather than measure a temperature difference of the cloud-tops over the two poles of 5 K or so, a difference of only 2 K was measured, indicating very good atmospheric mixing. Another surprise was that the prevailing winds blew east-to-west at the equator, rather than north-to-south, thus showing that the rotation of the planet was more important to the atmospheric circulation than the solar heating.

Other measurements indicated that Uranus, unlike Jupiter and Saturn, has no significant internal heat source, and so solar heating, weak as it is for a planet so far from the Sun, is by far the most important thermal energy source for the atmosphere.

The maximum wind speeds measured on Jupiter and Saturn are in the equatorial region, where they blow in the same direction as the planet's axial rotation (i.e. their axial rotation periods about the planet's axis are faster than that for the planet as a whole). Put another way, the equatorial winds on Jupiter and Saturn were westerlies. On Uranus, however, the equatorial winds, at least during the Voyager fly-by, had a slower axial rotation period than the planet as a whole, and so were easterlies blowing at about 350 km/h. At high latitudes on Uranus, where the maximum wind speeds of 700 km/h were measured, the wind flow had reversed direction to become westerlies.

The Magnetosphere

In the early 1980s, Pioneer 10 showed that the solar wind extends beyond Uranus' orbit and, at about the same time, the IUE satellite, in orbit around the Earth, observed aurora-like emissions from Uranus' atmosphere. Voyager showed that Jupiter and Saturn had extensive magnetospheres, and so it was not surprising that Voyager discovered, in 1986, that Uranus also had an extensive magnetosphere, and confirmed the aurora-like emissions in the upper atmosphere.

Voyager found Uranus' magnetic moment to be about 50 times that of the earth, or 10% that of Saturn, but the big surprise was the orientation of the Uranus' magnetic axis. It was tilted at a massive 60° to the rotation axis, and the magnetic centre was displaced by 0.3 R_U from the geometric centre (see Figure 5.8).

All of Uranus' satellites lie within the magnetosphere and, as at Jupiter and Saturn, the larger, inner satellites, namely Miranda, Ariel and Umbriel, and the particles in Uranus' rings, were found to absorb energetic particles from Uranus' radiation belts. The 98° orientation of the equator of Uranus to the Sun–Uranus line, the 60° inclination of the magnetic axis to the spin axis,

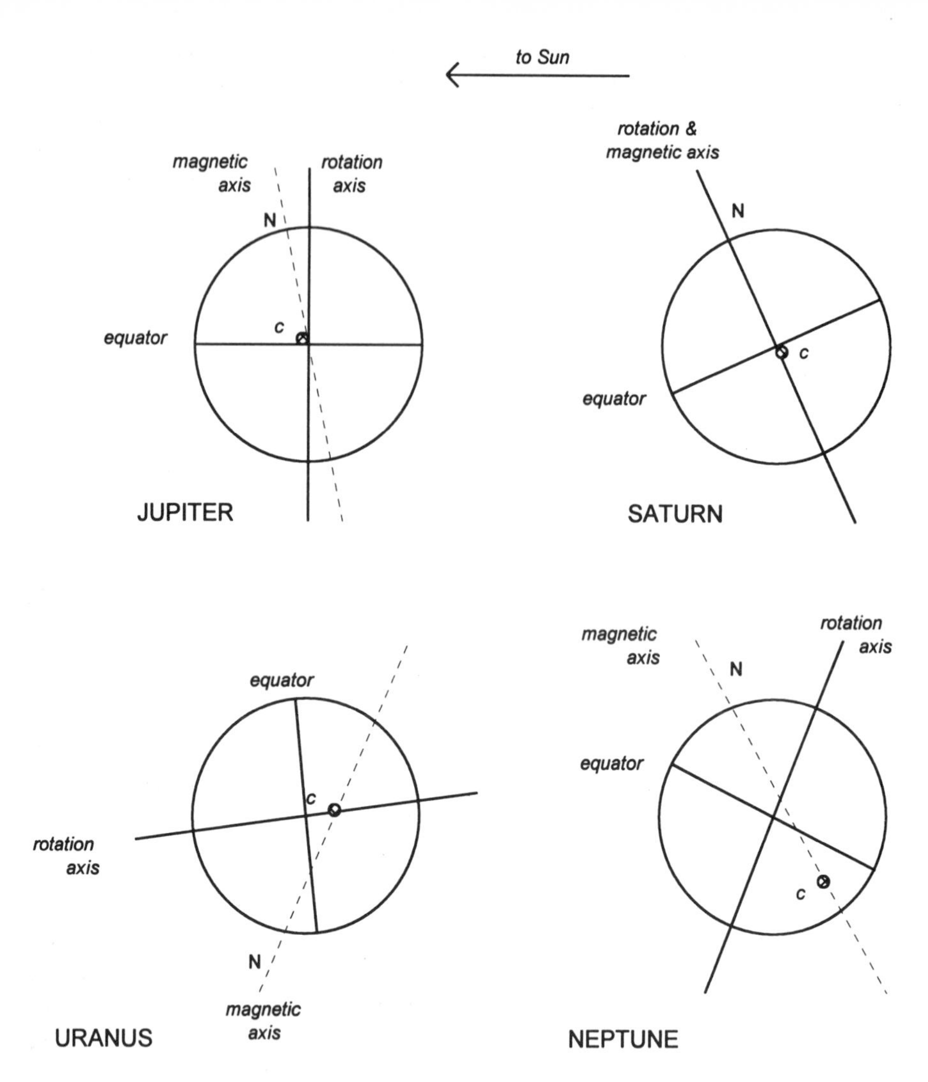

Figure 5.8 *The magnetic axes and magnetic centres of the gas giants, relative to their rotational axes and the direction of the Sun. (The magnetic axes are indicated by the thin lines, and the magnetic centres by the letter 'c'. The position of the north magnetic pole is indicated by the letter 'N'.)*

and the relatively fast axial rotation rate of just over 17 hours, were found to cause a complex diurnal variation of the belts.

The Interior

The strange orientation and position of the magnetic axis of Uranus was explained, immediately after its discovery by Voyager, as the result of a glancing impact early in Uranus' lifetime. This had caused the planet to tip on its side, and had also changed the position of its spin axis with respect to the structure of the planet. The theory could explain the observed axial orientations, at least in general terms.

An alternative theory suggested that Uranus was in the process of a reversal in magnetic polarities, of the sort known to happen on Earth about every 500,000 years.

Unfortunately, both theories only lasted 3 years, until a similar magnetic–spin axis arrangement was discovered for Neptune. These orientations are now thought to be produced, for both Uranus and Neptune, by convection in ionised water or ammonia in their mantles.

The Satellites

Uranus was known to have four satellites in the late nineteenth century, and the fifth satellite, Miranda, was discovered by Gerard Kuiper, in 1948, using the 82 inch (2.1 m) reflector at the McDonald Observatory in Texas.

Voyager discovered Uranus' sixth satellite, Puck, on the last day of 1985, and obtained one image of it, from a distance of 500,000 km, on 24th January 1986, showing it to be approximately spherical, with an albedo of only 0.07. Three craters could be seen, although the Voyager resolution of 10 km, at such a large distance, was too poor to show much detail on a satellite whose diameter was only 150 km.

In all, Voyager discovered ten satellites, of which Puck was the largest, all orbiting Uranus between Miranda and the planet. Two of these satellites, Cordelia and Ophelia, were found to be shepherding the epsilon ring.

The densities of Uranus' four largest satellites were, much to most astronomers' surprise, found to be higher than for similar-sized satellites of Saturn, indicating that the Uranian satellites were composed of less water ice and very little, if any, of the lighter ices of methane and ammonia. Torrance Johnson calculated that water ice constitutes about 40% to 50% of the mass of Uranus' large satellites, compared with 60% to 65% for those of Saturn. This could partially explain why Saturn's larger satellites, with albedos of from 0.6 to 0.8, are brighter than those of Uranus, with albedos of from 0.2 to 0.4. Because the surfaces of Uranus' large satellites are dark, they absorb more solar radiation, and are therefore warmer than the tops of Uranus' cloud layer.

The four largest satellites of Uranus are all cratered. Umbriel and Oberon appear to have the oldest surfaces, as they are the most heavily cratered, whereas the surfaces of Ariel and Titania have fewer craters and show evidence of resurfacing. Titania has extensive fault structures up to 1,600 km long, 50 km wide and 5,000 m (17,000 ft) deep, and some of the surface has been smoothed by volcanic flows. Ariel, on the other hand, has a global network of faults, with valleys which in some places are over 10,000 m (30,000 ft) deep. Its surface has fewer craters than Titania, because it has been more recently resurfaced by volcanic activity, although this was still probably 2 or 3 billion years ago.

The most important consideration in deciding the trajectory of Voyager through the Uranian system, was to ensure that it flew close enough to Uranus to get the required sling-shot effect to take it on to Neptune. Fortunately, this took it very close to Miranda, which Voyager showed to have the most bizarre surface of any body in the solar system (see Figure 5.9), with its chevron-shaped feature, two large ovoids and enormous cliffs 15,000 m (50,000 ft) high, or ten times the height of those of the Grand Canyon. The white chevron shape, which was detected when Voyager was over 1 million km from Miranda, was surrounded by a large trapezoid-shaped feature that had grooves intersecting at right angles. The two large ovoids, which were on opposite sides of Miranda, had concentric sets of ridges and grooves that made them look like large dirt racetracks.

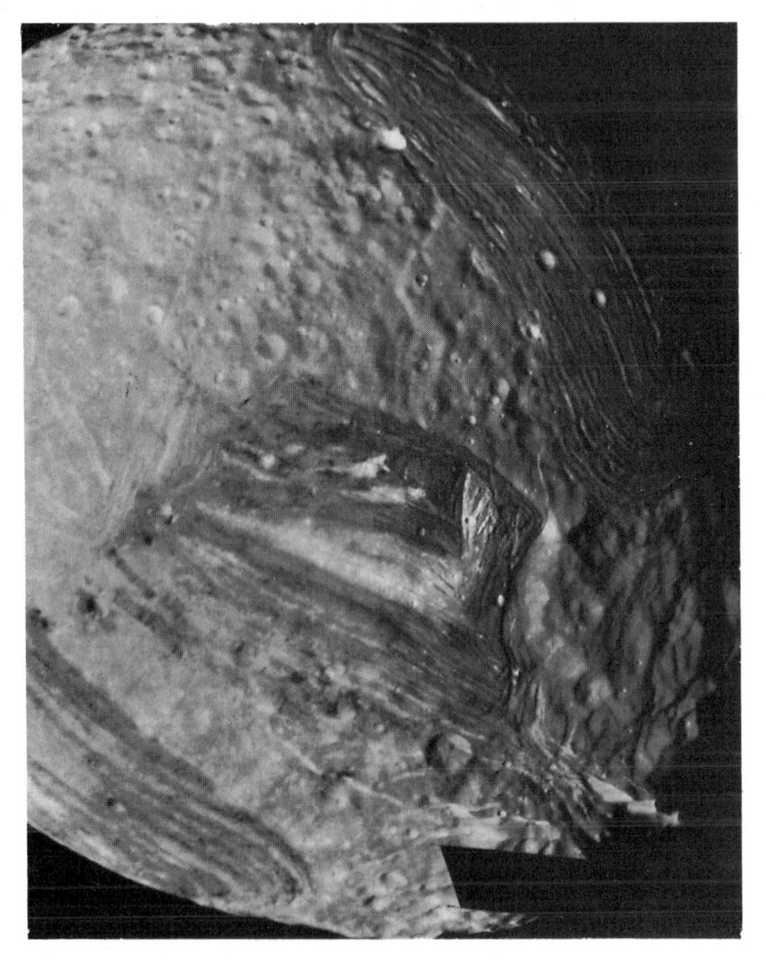

Figure 5.9 *Voyager image of Miranda showing the bright chevron feature (just below the centre), and the two ovoids (near the bottom left and top right). (Courtesy* National Space Science Data Center, World Data Center-A for Rockets and Satellites, NASA; Principal Investigator, Dr. Bradford A. Smith.*)*

The Rings

The discovery of a ring system around Uranus was a big surprise. It was known that on 10th March 1977 Uranus would occult star SAO158687, which is of magnitude 8.8 in the constellation of Libra, and James Elliot, Edward Dunham and Douglas Mink were observing the occultation to obtain a better estimate of the planet's diameter. Elliot's team, who were observing from the Kuiper Airborne Observatory high over the Indian Ocean, were surprised to observe that, starting 40 minutes before the occultation was due, there was a series of five reductions in the intensity of the star. After the occultation was over, they again found five reductions in intensity, which could only mean that Uranus was surrounded by five rings, later designated α, β, γ, δ and ε. Analysis of the fluctuations in the star's intensity increased the number of rings to nine, with the widest ring, the ε ring, being eccentric and variable in width.

Numerous stellar occultations by the rings were observed over the 9 years, separating the date of their discovery from that of the Voyager encounter. These occultations showed that the rings are very narrow, with all but the ε ring being 12 km, or less, in width.

Voyager imaged the rings in both forward and backscattering illumination in 1986. The forward illumination showed the nine known rings and little else, but the backscattering illumination showed a wealth of extra detail with many new rings, both wide and narrow. The nine original rings are thought to be composed of rocks a metre or so in size, while the rings that are only seen in backscattering appear to be mainly composed of dust. The main rings are dark and thought to be much older than Saturn's rings, whereas the dust rings are thought to be transitory, changing their appearance relatively quickly.

Neptune

The Discovery

Uranus had been discovered by William Herschel in 1781 but, shortly after its discovery, it had been found that Uranus had been observed as long ago as 1690, although it had not been recognised at that time as a planet. Alexis Bouvard, who undertook numerical analysis for Laplace, found that it was impossible to correlate these earlier orbital positions with later ones, and as the nineteenth century progressed, it was noticed that Uranus was deviating once again from the path predicted by Newton's laws of planetary motion.* By 1844, the deviations amounted to 2 arcmin and were clearly very much greater than observational error. Were Newton's laws flawed in some way, or was there another body close to Uranus disturbing its path?

---

*These laws describe how a planet moves under the gravitational influence of the Sun and other bodies of the Solar System.

In 1843, John Couch Adams who had just graduated at Cambridge University began to calculate the orbit of the hypothetical planet, and in September 1845 he produced a calculated position. He immediately went to see Professor James Challis, who was responsible for the Northumberland refractor at Cambridge, and asked if he would start a search for the planet. Challis declined and suggested, instead, that Adams contact the Astronomer Royal, Sir George Airy, but the Astronomer Royal asked for more information before he was prepared to ask for a search to be made. The young Adams, somewhat disillusioned by what he saw as the over-cautious reaction of the astronomical Establishment, took nine months to produce the data. In the meantime, Urbain Le Verrier of the École Polytechnique had started to work out the position of the hypothetical planet, in ignorance of Adams' work. Le Verrier published his results on 1st June 1846, and the Astronomer Royal then noticed that Le Verrier's predictions and those of Adams agreed to within 10. On 9th July the Astronomer Royal asked Professor Challis to begin a search, which he did on 29th July.

On 18th September Le Verrier, having failed to interest the Paris Observatory, wrote to the Berlin Observatory with his detailed orbital predictions for the new planet, and asked that they commence a search. His letter arrived on 23rd September and Johann Galle, an assistant at the observatory, started his search that same night. Fortunately for Le Verrier, the Berlin Observatory had just received a new map of the sky that had been commissioned by the Berlin Academy of Sciences some years earlier to assist in the search for asteroids (see Page 104). This map was much better than that available to Challis at Cambridge, as it showed all the stars down to the ninth magnitude, and so when Galle began looking for the eighth magnitude Neptune, he found it without undue difficulty on the first night, less than 1° from its predicted position. When Challis heard of the discovery a few days later, he looked back through his own data which he had not had time to analyse thoroughly, and found that he had actually seen Neptune on both 30th July and 12th August.

The discovery of Neptune was a key event in nineteenth century astronomy, as it showed that Kepler's and Newton's laws of planetary motion were basically correct. It also had a large psychological effect as theoreticians had, for the first time, been able to predict both the existence and position of an unseen member of the solar system.

There are two interesting postscripts to the story of the discovery of Neptune.

First, it transpires that Neptune had been seen and recorded by Galileo, when it was very close to Jupiter in 1612, and by Joseph Lalande, the director of the Paris Observatory, in 1795. Neither astronomer realised that they had seen a planet, however, and it took many more years before it was discovered for what it was by Le Verrier and Adams in 1846. Galileo's observations were only found in 1980, so at the end of the nineteenth century the first observations of Neptune were thought to had been made by Lalande.

Second, both Le Verrier and Adams had had to assume an orbital radius for the planet that was perturbing Uranus, in order to work out its possible position in the sky. They both assumed that it was further from the Sun than it actually was but, fortunately, Neptune happened to be relatively close to Uranus in 1846, the year of its discovery, so the effect of this wrong assumption was relatively unimportant. The situation would have been a good deal worse a decade or so later.

Early Work

In 1890, Neptune was known to be of virtually the same size as Uranus, and of similar density but, like Uranus, no markings could be seen on its surface, and so its axial rotation period was unknown.

In the late nineteenth century Neptune was only known to possess one moon, namely Triton, which was discovered by William Lassell less than a month after the discovery of Neptune itself. Triton had an orbit inclined at 35° to the ecliptic, and was rotating around Neptune in a retrograde motion, so it was concluded that Neptune probably had an equator inclined at this 35° to the ecliptic, and was spinning on its axis in a retrograde fashion. In this case, something had probably happened to Neptune to change the direction of its axis of rotation after the planet had been formed. The alternative proposition, that Neptune had been formed rotating on its axis in the opposite way to all the other planets (with the possible exception of Uranus), could not be explained if it had been originally part of the solar system nebula.

Triton was generally thought to be the largest moon in the solar system, as it appeared to be very bright, considering how far away from the Sun and Earth it was.

Some observers thought that they had seen rings around both Neptune and Uranus, but by the late nineteenth century the observations had been put down to optical illusions. (The instruments available then could definitely not have detected the rings that we now know exist.)

The Atmosphere

In the early twentieth century Vesto Slipher discovered that the atmosphere of Neptune produced similar absorption lines to those of Uranus and, in 1933, it was shown that these lines were due to methane on both planets.

Maxwell Hall in Jamaica observed in 1883 and 1884 that Neptune varied in brightness with a period of just under 8 hours, but the variations died out, so they could not be accepted as proof of the planet's rotation period. In 1922 and 1923, Ernst Öpik and Livländer at the Tartu Observatory in Estonia found similar fluctuations, with a period of 7 h 50 min. A few years later, however, Joseph Moore and Donald Menzel of the Lick Observatory measured the Doppler shift of Neptune's spectral lines, and deduced a rotation period of 15 h 50 min $\pm$ 1 hour, or twice that observed from the light fluctuations. This led to the conclusion that two areas of the surface, on opposite sides of the planet, had been brighter than the remainder of the planet in 1883/4 and 1922/3, giving light fluctuations in those years at twice the spin rate. The 15 h 50 min rotation period was the generally accepted until, in 1980, Belton, Wallace and Howard deduced a period of 18 h 10 min, using spectroscopic and photometric measurements.

As early as 1948, Bernard Lyot observed faint markings on Neptune and, in the early 1970s, the planet's brightness was found to vary from night to night. This variation was most noticeable when using filters that transmitted the red and infrared light that methane absorbs, so the brightness variations were attributed to clouds of methane ice crystals above the methane atmospheric layer. Bradford Smith took CCD images of Neptune, in 1979, in the light of the methane absorption band, and showed the existence of these bright clouds. The best pre-Voyager images taken of them were by Heidi Hammel, using the University of Hawaii's 88 inch (2.2 m) telescope on Mauna Kea. Then in July 1988 she detected a large bright cloud, centred on −30° latitude, which rotated around Neptune with a period of 17 h 40 min.

During the Voyager 2 spacecraft encounter with Neptune in 1989, the planet showed transitory white clouds, and a consistent large dark feature, now called the Great Dark Spot (GDS), centred at −22° latitude, and shaped like the Great Red Spot (GRS) on Jupiter. The GDS is very similar to the GRS, as they both rotate counter-clockwise about their centres south of the equator, making them high pressure features, and they are both of a similar size relative to their planets. A Small Dark Spot (called D2) was also found on Neptune, at about −55° latitude, and white cirrus clouds of methane ice crystals were discovered over both Dark Spots. High altitude methane cirrus clouds were seen near the terminator of the planet, casting shadows on the blue atmosphere 50 to 75 km below. Simultaneous observations from Voyager and the Earth showed that the white cloud detected from Earth is that associated with the GDS.

The GRS rotates around Jupiter only 8 seconds slower than the interior of the planet, whereas the GDS rotates around Neptune over 2 hours slower than the interior (determined from periodic variations in radio signals). The GDS is thus moving westward at about 1,200 km/h, and smaller features just south of the equator were found to be moving westwards at about 2,000 km/h. These are the fastest atmospheric winds in the solar system, which is surprising considering that Neptune receives such a small amount of solar radiation, so the winds are thought to be driven by Neptune's internal heat source, detected in the 1970s, which provides more energy than Neptune receives from the Sun. The D2 spot rotates at virtually exactly the same rate (16.1 hours) as the interior, and, in this respect, it is more like the GRS on Jupiter, than is the GDS.

When the D2 spot was discovered at −55° latitude, it had a rotation period around Neptune of 16.0 hours. Its period then slowed to 16.3 hours, as it moved north to −51°, before returning to −55° with a period of 15.8 hours. It seems to be less constrained in latitude than similar features on Jupiter and Saturn.

Size

Edward Barnard measured the diameter of Neptune as 52,900 km in 1899, and this value was accepted as the best value for many years until, in 1949, Gerard Kuiper produced a much lower value of 44,600 km. Both of these figures were based on measuring the diameter of the disc but, in the 1950s, an alternative approach was considered of timing the occultation of a bright star by Neptune. The first suitable occultation occurred on 7th April 1968, when observers in Australia, New Zealand and Japan observed the occultation of the star BD-17° 4388, which lasted for about 45 minutes, resulting in a new value for the equatorial diameter of Neptune of 50,950 ± 170 km.

The Magnetosphere

Voyager found that the position and orientation of Neptune's magnetic axis was even more bizarre than that of Uranus, and this caused a rejection of some of the theories previously put forward to explain the magnetic characteristics of Uranus. The magnetic axis of Neptune was found to be inclined at 47° to the spin axis (see Figure 5.8), but it was off-centre by about 14,000 km, or about 0.6 R_N. The magnetism was attributed to convection in a highly compressed (200,000 bar) ionized water layer at a temperature of 2,600 K, 5,000 km beneath the cloud-tops. The periodicity of radio emissions enabled a rotation period of 16 h 7 min to be deduced for the interior of the planet.

The field strength at the surface of Neptune in the southern hemisphere, was found to be much larger than in the north, because of the strange offset and orientation of the magnetic axis. The magnetic moment was about half that of Uranus, and the magnetosphere was comparatively empty of high energy particles.

The Satellites

In the nineteenth century, the only known satellite of Neptune was Triton, which orbited Neptune in a retrograde sense. This led astronomers to wonder if Neptune itself had a retrograde axial rotation, but this was disproved in 1928 when Moore and Menzel measured the Doppler shift of the spectral lines. So Triton was seen to rotate in the opposite sense to the spin of its primary, Neptune, the first major satellite in the solar system to be observed to do so.

Dale Cruikshank and Peter Silvaggio detected methane on Triton in 1978, most of which was expected to be deposited on the surface as ice, but a small amount was thought to form a tenuous atmosphere. If so, Triton would be only the third satellite in the solar system to be shown to possess an atmosphere. (The others are Titan and Io). Then in 1983 Cruikshank and Roger Clark of the University of Hawaii and Robert Brown of JPL detected nitrogen on the surface of Triton, indicating that it is also present in Triton's tenuous atmosphere.

Gerard Kuiper discovered Nereid, Neptune's second satellite, in 1949, and found that it orbited Neptune in the opposite sense to Triton. Triton's orbit is almost circular, with a radius of 350,000 km, but Nereid's is highly elliptical, with an apogee of 9.7 and a perigee of 1.3 million km. If Triton was orbiting Neptune in a prograde sense, and Nereid was orbiting in a retrograde sense, then Triton would be seen to be a normal member of Neptune's family, and Nereid would be thought of as a captured asteroid. But given the fact that they both orbit Neptune in the opposite senses to these, maybe Triton is the captured asteroid, and Nereid is a normal satellite. Maybe they have both been captured, as their orbits are both inclined at more than 20° to Neptune's equator, and their orbits are highly inclined one to the other. It was hoped that the anticipated discovery of further satellites by the Voyager spacecraft should help to solve these questions.

Stephen Synnott of JPL discovered Proteus, Neptune's third satellite, using Voyager 2 images, on 7th July 1989. Its orbit was found to be prograde, of low inclination and inside that of Triton and, at about 400 km diameter, the satellite was found to be larger than Nereid.

Larissa had been discovered in 1981, earlier than Proteus, from stellar occultation observations (see Pages 93–94), but its existence had been disputed until Voyager proved its existence in a prograde orbit inside that of Proteus. Voyager also discovered four more satellites, Galatea, Despina, Thalassa and Naiad, all of which were within the orbit of Larissa. They also rotated about Neptune in a prograde sense, with low orbital inclinations, indicating that both Triton and Nereid may be captured satellites, for the reasons outlined above.

Until relatively recently, Triton was thought to be the largest satellite in the solar system, but more recent estimates had gradually reduced both its size and its mass. Voyager found Triton to have a diameter of 2,705 km, and a density of just over 2.0 g/cm^3, the largest density seen since Voyager had left Jupiter. In fact, Triton appears to be virtually the same size and density as Pluto, adding to the idea that Triton is a captured asteroid.

Triton's albedo estimates had to be increased, as its diameter estimates were reduced, resulting in the current value of about 0.7. With such a high reflectivity, Triton absorbs little solar energy, resulting in the temperature measured by Voyager of only 38 K, by far the lowest of any body yet measured by a spacecraft, and far too cold to allow nitrogen to exist in a liquid form on its surface.

Triton's surface showed a network of fractures, which appear to have been caused when the water ice in Triton's mantle froze and expanded. Superimposed on this is a dimpled terrain, unlike anything seen anywhere else in the solar system. There are very few impact craters, indicating that the surface is relatively young, maybe because Triton had melted during its capture by Neptune.

Triton's orbit is inclined at 23° to Neptune's equator, and Neptune's equator is inclined at 29° to the plane of its orbit around the Sun. Thus the Sun is at the zenith on Triton at 52° latitude on midsummer's day. As Neptune's year is 165 years long, there is plenty of time for one hemisphere of Triton to heat up and the other to cool down, the poles each being without sunlight for 82 years.

Voyager arrived at Triton towards the end of the spring in the southern hemisphere, and found that the edge of the highly reflective pinkish-white southern ice cap was retreating towards the pole, revealing a red surface beneath. The retreating and advancing polar ice caps, have ensured that the surface of Triton is still subject to change, which was a big surprise considering how cold the satellite is. Dark streaks up to 160 km long were seen on the highly reflective icy surface and, as the surface melts every 165 years, these streaks were clearly very recent. Their appearance suggested that material had been ejected at one end of the streak and carried by the wind, in Triton's tenuous nitrogen and methane atmosphere, as it fell to the ground. The streaks appeared to provide evidence of some form of volcanism.

A month after the encounter, Lawrence Soderblom and Tammy Becker were examining the Voyager images stereoscopically, when they were amazed to discover an eruption in progress, throwing material to a height of 8 km. The plume extended horizontally for about 150 km, being blown downwind. They had found the cause of the black streaks, and then a second active volcano was discovered. These two volcanoes, and three other suspected active volcanoes, were all near to the sub-solar point, indicating that they are caused by solar heating.

Did Voyager really see volcanoes? Surely Triton cannot remain molten just beneath its surface, when the satellite is subjected to such very low temperatures? Current views are that the eruptions are not volcanoes, emitting molten material from the interior of Triton, but are more like geysers. They are caused by nitrogen ice being sublimated just below the surface, and generating gases that break through the surface with explosive force.

The Rings

Rings had been discovered around Uranus in 1977, and Jupiter in 1979, so did Neptune have rings? The rings of Uranus were virtually perpendicular to our line of sight when they were discovered, because of the orientation of Uranus' spin axis, but the maximum inclination of Neptune's rings would be about 30°, and their maximum angular width would be only half that of Uranus' rings, as Neptune is almost twice as far away from Earth. Neptune also moves more slowly against the stellar background. For all of these reasons, the chances of finding a suitable stellar occultation are clearly much worse for Neptune than for Uranus.

Neptune almost occulted a star in 1981, and that encounter was observed by Harold Reitsema, William Hubbard, Larry Lebofsky and David Tholen. They noted that, as Neptune approached, the star was eclipsed for

8 seconds, but there was no eclipse observed as Neptune receded. These astronomers concluded that the occultation could not have been caused by a ring, as that should have caused eclipses on both sides of the planet, and so they announced that they had discovered a probable new satellite of Neptune. The discovery of a satellite was not completely certain, however, because it seemed too much of a coincidence for a small satellite to be just in the right place, at the right time, to occult the star.

A similar occultation opportunity to look for Neptune's possible rings occurred in 1983, but no eclipse of the star was observed. On 22nd July 1984, however, André Brahic of the University of Paris observed a 35% reduction in the intensity of a star, that lasted for about one second, as Neptune almost occulted it. One hundred kilometres away, William B. Hubbard of the University of Arizona found exactly the same result. This virtually proved that the effect could not have been due to a satellite, as the glancing occultation of a satellite should show up differently at the two different observing sites. The 1984 result indicated that Neptune was surrounded by a partial ring, or ring arc, about 25 km wide and 67,000 km from the centre of Neptune. The eclipse observation at the earlier 1981 occultation could also be explained by another ring arc situated at 70,000 km from Neptune, although Reitsema still maintained that the eclipse was due to a satellite. A further occultation in 1985 also showed that there was probably a third ring arc, 56,000 km from Neptune.

The question of how ring arcs could exist for, presumably, millions of years, without forming a continuous ring, or completely dissipating, was difficult to answer. Peter Goldreich, Scott Tremaine and Nicole Borderies suggested that the ring arcs could each be shepherded by a satellite, and Jack Lissauer of the University of California suggested that they could each be at the Lagrangian points 60° in front or behind a satellite.

Voyager 2 solved the ring arc problem during its fly by of Neptune in 1989, by showing that Neptune's rings were all continuous. Four rings were found, the outermost of which, called Adams, was the brightest. The Adams ring had a remarkable degree of clumpiness along it, and, at three places, its density was about ten times that of the remainder of the ring. It was this clumpiness that had caused ground-based astronomers to think that the ring was not complete.

Further analysis showed that the ring arcs discovered in 1984 and 1985, at 67,000 and 56,000 km from Neptune, were part of the Adams ring at 62,900 km. The reason why the initial distances were wrong, was because the angle of Neptune's spin axis to its orbital plane had been estimated slightly incorrectly. More surprising, however, was the discovery that the first ring arc to be found, in 1981, was no ring arc at all. The occultation had been caused by a satellite that was discovered by Voyager 73,600 km from Neptune. So Reitsema had been right all along and, in recognition of the prior discovery by him and his team, the International Astronomical Union invited them to name the satellite from names connected with Neptune mythology. They chose Larissa as both Reitsema and Hubbard have daughters named Laurie, which is a near fit.

Each of the two thin rings were found to have a shepherd satellite (Galatea and Despina) on their inside, but no shepherd moons have yet been found on the outside of these rings to stop them from expanding outwards.

6 Small Bodies of the Solar System

Pluto

The Discovery

Neptune had been discovered in 1846 because of irregularities in the orbit of Uranus, but, in many ways, its discovery was fortunate, as it was much closer to Uranus than either Le Verrier or Adams had assumed. Le Verrier was not convinced that Neptune was the limit of the solar system, and said so to Galle on 1st October 1846, but he recognised that finding a yet more distant planet would be considerably more difficult than finding Neptune.

David Todd of the US Naval Observatory analysed deviations in the motion of Uranus and concluded, in 1877, that there was a planet, located in the constellation of Virgo, orbiting the Sun at a distance of 52 AU.* His search for it was unsuccessful. Two years later, the French astronomer Camille Flammarion suggested a similar orbit, based on an analysis of cometary orbits, but there was no significant progress in the search for the new planet until the twentieth century.

Percival Lowell had been interested in the idea of finding a new planet ever since Benjamin Peirce, his tutor at Harvard, claimed that the discovery of Neptune was a fortunate accident. In 1905, Lowell recruited an assistant, William Carrigan, to analyse the residual motions of the outer planets, to try to predict the position of the unknown planet. Without waiting for the results of Carrigan's analysis, however, Lowell and his assistants at his observatory at Flagstaff, Arizona, started a photographic search of the region around the ecliptic.

By 1908, neither the analytical efforts of Carrigan, nor those of the photographic team, had yielded any concrete results. To make matters

*1 Astronomical Unit (AU) is the average distance of the Earth from the Sun.

worse, later that year Lowell attended a presentation by William Pickering of the Harvard College Observatory, in which he described his search for the unknown planet. Lowell was horrified to discover that Pickering appeared to be further down the road than he was, so he dismissed Carrigan, on the grounds that he was being too meticulous and slow, and decided to lead the calculations himself.

Lowell started his analysis in 1910, assisted by Elizabeth Williams and, in the following year, he requested Carl Lampland, his assistant observatory director, to begin a new photographic survey. On Lampland's advice, Lowell bought a blink comparator to facilitate comparing plates, looking for the movement that would give the planet away. Lowell published his analytical conclusions in 1915, in which he predicted that Planet X, as he called it, would have a mass of 6.6 times that of the Earth, and would be found in Gemini, near its border with Taurus. Lowell never found Planet X, and he died of a heart attack the following year.

William Pickering also tried to find Planet O, as he called it, publishing various predicted positions for it between 1908 and 1928, but he could find no such planet on his photographs. In 1911 he predicted that there were three more planets beyond Planet O, one of which, Planet Q was in a highly elliptical polar orbit, with a mass of 63 times that of Jupiter.

After Lowell's death in 1916, the operation of his observatory was put on a minimum cost footing, while the ownership of his estate could be sorted out. Eventually, in 1927, the litigation was completed and Lowell's brother, Abbott Lowell, who was president of Harvard, agreed to provide funds for a 13 inch (33 cm) photographic refractor, to enable the search for Planet X to be recommenced with a better telescope than before. Vesto Slipher, the director of the observatory, then recruited Clyde Tombaugh to take the photographs.

The selection of Clyde Tombaugh, the farm boy from Kansas, was an inspired, if somewhat unconventional, choice. Tombaugh, a keen amateur astronomer who made his own telescopes, had contacted the observatory to ask for an opinion about some drawings that he had made of Mars and Jupiter. Slipher was so impressed with his obvious determination and precision, that he offered him a job, as an assistant observer on a trial basis.

Tombaugh started the search on 6th April 1929 by photographing the areas of Gemini and Cancer, but he had a number of problems at the beginning with both the drive of the telescope, and the plates which cracked in the intense cold. He solved the problems, but by September the planet had not been found, so he decided to photograph a band along the complete ecliptic. By January 1930 he had photographed the whole of the zodiac and was back at Gemini again. He took three plates of the Delta Geminorum region, on 21st, 23rd and 29th January and began blinking the plates on 15th February. Three days later, at about 4.00 pm, he saw a 15th magnitude object move when he blinked the latter two plates (see Figure 6.1). He checked the plate of the 21st and found that the object moved there as well. He had found Planet X!

Slipher, the director, insisted on more evidence before making a public

Figure 6.1 *Small section of the discovery plates of Pluto taken on 23rd January 1930 (left) and 29th January 1930 (right) by Clyde Tombaugh. The position of Pluto, which has clearly moved, is marked. (*Lowell Observatory Photograph.*)*

announcement; the delay giving the observatory a head-start in calculating the orbit of the new planet. Just over 3 weeks later, he was ready to announce the discovery of the planet, sending a telegram at midnight Boston time on 12th March, to the Harvard College Observatory. The discovery was made public on the following day, being the 75th anniversary of Lowell's birth, and the 149th anniversary of William Herschel's discovery of Uranus.

Slipher's next priority was to establish the orbit of Planet X, and here the Lowell Observatory was very unpopular, as it released only one position of the planet, and not the series of positions it had covering almost 2 months. In spite of this, the first orbital estimate was published on 7th April by Armin Leuschner, Ernest Bower and Fred Whipple of the University of California, who had deduced a distance of 41 AU and an orbital inclination of 17°, both very close to the truth. Lowell's Observatory produced their first estimate on 12th April, but it was wildly out, with an orbital period of 3,000 years. It was not until Andrew Crommelin of the Royal Observatory of Belgium found that the observatory had photographed Planet X on 27th January 1927, that a reasonably accurate orbit could be calculated.

Various names were proposed for the new planet. Mrs Lowell suggested Zeus, Percival, Constance (her own name) or Planet X, but each of these suggestions were rejected. Slipher preferred Minerva, but that was already the name of an asteroid. The observatory considered many other names, some of their own choosing, and some sent in by the enthusiastic public. Among them was Pluto, which the observatory was considering, when it was also suggested by Venetia Burney, an 11 year-old schoolgirl from England, who thought that the name of the god of the underworld was suitable for a planet so far from the Sun. The observatory agreed, and Slipher officially proposed on 1st May 1930 that the planet should be called Pluto.

Both Uranus and Neptune had been seen before their discovery, but they had not been recognised as planets. The British astronomer Andrew Crommelin found that Pluto had been photographed before its discovery, in Belgium in 1927, but had it been detected by anyone else?

Subsequent investigations showed that Pluto had been recorded on two plates taken on 19th March and 7th April 1915 by Lowell's assistant Thomas Gill, but it had been missed because the plates contained too many stars for the fainter-than-expected planet to be noticed. It had also been recorded on two plates taken by Milton Humason on Mount Wilson in 1919, in support of William Pickering's search, but on one plate it had fallen on a plate defect, and on the other it was too close to a bright star. Tombaugh had also unknowingly photographed Pluto before. It was found on plates taken on 11th April and 30th April 1929, just after he started his search, but the first plate had cracked, and Pluto was too faint to be easily seen on the second.

Was Pluto the Planet X predicted by Percival Lowell? Yes and no. Yes, because his orbital prediction was quite accurate, and no, because Pluto was much smaller than he had calculated, and was far too small to have perturbed Uranus by anything like the extent that he had assumed.

Neptune had been discovered within 1° of its predicted position, whereas Lowell was 6° out with his prediction for Pluto. On the other hand, Neptune's mean distance from the Sun is 30 AU, instead of the 37 AU assumed, whereas Pluto has a mean distance of 39.5 AU compared with Lowell's prediction of 43 AU. So Lowell's orbital prediction appeared to be quite good.

Pluto did not show a disc in even the largest telescopes in 1930, and it did not appear to have a satellite, making it impossible to produce an accurate estimate of its mass. Even if Pluto was made completely of iron, however, it could not be much more than the mass of the Earth, and it was considered more likely that it was somewhat lighter. Over the years since its discovery, the estimate of Pluto's mass has been gradually reduced, until today it is only 0.002 times that of the Earth, or 20% of that of the Moon.

As Pluto is so small, its discovery was a fortunate accident, and if something really is disturbing Uranus to the extent deduced by Lowell, it must be a larger planet than Pluto. Clyde Tombaugh was asked by Slipher to try to find it, and spent the next 13 years in an unsuccessful search.

Pluto's orbit is highly eccentric, with an aphelion of 30 AU and a perihelion of 49 AU, so that, at its closest to the Sun, its orbit lies just within that of Neptune. Pluto's orbit is inclined to the ecliptic at an inclination of 17°, which is by far the largest inclination of any planet. In fact, Pluto's large orbital eccentricity and inclination are very reminiscent of the orbits of some of the outer satellites of Jupiter and Saturn. A similarity that Lowell had predicted in 1915.

In 1955, Merle Walker and Robert Hardie of the Lowell Observatory observed regular fluctuations of 20% in the brightness of Pluto, with a period of 6 d 9 h 17 min, and deduced that this is the axial rotation period of the planet.

It was suggested on Page 92 that Triton, Neptune's largest satellite, may be a captured asteroid. Alternatively, as Pluto's orbit crosses that of Neptune, albeit not in the same plane, there has also been speculation that maybe Pluto was once a satellite of Neptune, like Triton, and that the event that dislodged Pluto from Neptune, could also have caused Triton's orbit to become retrograde. Either way, it seems more than a coincidence that the size and density of Pluto and Triton are very similar.

Surface and Atmosphere

Dale Cruikshank, David Morrison and Carl Pilcher discovered in 1976 that Pluto had methane ice on its surface, using the 4 m telescope on Kitt Peak. This meant that Pluto had a much higher albedo than previously thought, which, in turn, meant that a smaller surface area was required to produce the observed intensity of the planet as seen from Earth. This was one reason for a significant reduction in the estimated size of Pluto since its discovery, and a diameter less than that of Mercury was proposed in 1976.

As Pluto approached perihelion in the 1970s and 80s, it was observed to become darker, consistent with the sublimation of the methane frost, uncovering a darker surface. In 1983, the IRAS satellite found that Pluto's equator is appreciably darker and slightly warmer than the poles, indicating that the methane ice does not presently cover the equatorial regions.

The first occultation of a star observed by Pluto took place on 9th June 1988, when the star was observed to fade gradually in intensity, and increase gradually again, indicating the presence of a tenuous atmosphere, some tens of kilometres deep, with a density of 3×10^{-3} millibar, which is similar to that of Triton. It is thought that the atmosphere, which is probably produced by the surface ice sublimating, exists for only a few decades around perihelion (which was in 1989 for the current orbit). In 1992, Tobias Owen and Dale Cruikshank discovered that the main gas in Pluto's atmosphere, as in Titan's and Triton's, is nitrogen, leading to the view that the ice on Pluto is mainly nitrogen, as on Triton, rather than methane as previously thought.

Charon

Gerard Kuiper and Milton Humason looked for moons of Pluto in 1950 by taking long exposure photographs of the planet, reasoning that the moons, if they existed, would be faint and some distance from Pluto. They found nothing; then almost 30 years later, James Christy of the US Naval Observatory stumbled across a moon very close to the planet.

Christy was examining plates of Pluto on 22nd June 1978, when he noticed that the images of the planet were not circular, although those of the stars were, indicating that the cause was not a telescope tracking problem. Pluto appeared to have a bulge, which was in the same position on plates taken on 13th and 20th April, but was in a different position on the plates taken on 12th May. Christy then examined a series of five plates taken during 1 week in 1970, which showed the bulge moving clockwise around Pluto with a period of about 6 days. He immediately realised that Pluto had a satellite. The photographs that Christy used had been taken, by an interesting coincidence, at the Naval Observatory at Flagstaff, Arizona, just 6 km from the Lowell Observatory where Pluto had been discovered.

Christy's colleague, Robert Harrington, computed the orbit of the new satellite, and found that, not only did one side of it appear to be permanently facing its primary, as do most satellites in the solar system, but the Pluto system was unique in that Pluto also appeared to keep its same face pointing at the satellite. He found that the satellite orbited Pluto every 6 d 9 h, the same as Pluto's axial rotation period. The centre-to-centre distance of Pluto to its satellite turned out to be only 20,000 km (or 1/20th of the distance of the Earth to the Moon), and the diameters of Pluto and its satellite were estimated at about 3,000 and 1,000 km, respectively, making the satellite the largest, relative to its primary, of any in the solar system. The orbit of Charon showed that the angle of Pluto's equator to the plane of Pluto's orbit around the Sun is similar to that of Uranus, with both planets almost spinning on their side, with a retrograde spin direction.

Christy chose the name Charon for Pluto's satellite, which is the name of the ferryman who takes the souls of the dead across the River Styx to Pluto's underworld, in Greek mythology. He also chose Charon, as his wife is called Char.

The discovery of Charon enabled the total mass of the Pluto–Charon system to be determined as less than that of the Earth's Moon alone. Between 1985 and 1990, Pluto and Charon eclipsed each other while Charon orbited Pluto, because of an orbital alignment that occurs only every 124 years (half Pluto's year). These mutual eclipses enabled Pluto's diameter to be accurately estimated, for the first time, as 2,300 km, and Charon's as 1,200 km. So Pluto is a little smaller than Triton, although its density of 2.0 g/cm^3 is almost exactly the same, implying that it has a rocky core, like Triton. Charon, on the other hand, with a density of 1.2 g/cm^3, appears to have much more ice in its constitution, like the large satellites of Saturn.

Spectroscopic observations of Pluto and Charon, taken during their period of mutual eclipses in 1987, showed that Charon has water ice on its surface, but no methane or nitrogen ice like Triton.

The discovery of Charon made it less likely that Pluto was once a satellite of Neptune, as any disturbance powerful enough to pull Pluto from Neptune would almost certainly have separated Pluto and Charon. Thus most astronomers today think that Pluto, Charon and Triton are planetesimals, left over from the formation of the solar system, and that Triton was captured by Neptune. There are some, however, who think that the event that ejected Pluto from its supposed orbit around Neptune, could have fractured Pluto into a number of pieces, the largest of which were Pluto and Charon.

A New Planet X?

Pluto was too small to have perturbed Uranus by the amount deduced by Lowell, so was there another planet, the real Planet X, that was causing the disturbances?

Clyde Tombaugh was the first to try to find the tenth planet of the solar system, the real Planet X. He had continued with his photographic survey of the zodiac, once he had confirmed the existence of Pluto, and was then asked by Slipher, at the end of May 1930, to continue searching for more planets. Over the next 13 years he covered the whole of the sky visible from the Lowell Observatory, and, as a result, blinked over 40 million objects down to magnitude 16.5, without finding another planet. He found many other objects en route, however, including over 700 asteroids.

Many other attempts at finding Planet X have been undertaken, for example:

(i) Charles Kowal of the Lowell Observatory undertook a photographic survey, starting in 1977, using the Palomar 49 inch (1.2 m) Schmidt Telescope. He searched an area up to 15° on either side of the ecliptic, but he did not find Planet X. In the process, he did find the strange asteroid Chiron, however.

(ii) Robert Harrington and Thomas Van Flandern of the US Naval Observatory looked for a planet that they thought had ejected Pluto from its supposed orbit around Neptune. Various searches were made starting in 1979, all to no avail. In 1988, Harrington produced a suggested orbit for Planet X with a period of 600 years, based on Uranus' residuals, but the search was abandoned, because of funding cuts, after only 30% of the expected location area had been surveyed.

(iii) The IRAS satellite completed an infrared sky survey in 1983, and Thomas Chester and Michael Melnyk of JPL have been using this to try to find brown dwarf stars, which are intermediate in mass between Jupiter and the Sun. Chester and Melnyk are comparing the objects in the IRAS catalogue, with those in the Palomar Sky Survey, to find those that have moved. Although their main objective is to find brown dwarfs close to the solar system, they may also find planets in orbit around the Sun. So far none have been found.

(iv) Conley Powell of the Lowell Observatory re-examined the discrepancies between the predicted and observed positions of Uranus, and predicted the existence of a planet three times the mass of the Earth, at a distance from the Sun of 61 AU, and having a period of 494 years. The Tombaugh photographs were re-examined in 1987, for the area of sky in which Powell predicted the planet would be, but nothing was found. Powell's more recent calculations have resulted in the prediction of an Earth-sized planet in an orbit similar to Pluto's, but no search has yet been made.

(v) John Anderson of JPL used residuals in the orbital data for Uranus and Neptune that had been observed since 1910, and the unaffected trajectories of the two Pioneer spacecraft, to predict, in 1987, that

Planet X was in a highly inclined, highly elliptical orbit, with a period of about 700 to 1,000 years. He suggested that this planet is now approaching aphelion, which is why the planetary deviations are much smaller this century than last.

One key question is whether the orbits of Uranus and Neptune can be completely explained by the gravitational pull of all the known planets in the solar system. If all the measured positions of Uranus and Neptune are 100% accurate they cannot, but clearly all measurements have an error of some magnitude, and the older the measurements, the larger the likely error. Sometimes errors can creep in, in transposing the raw observations to celestial co-ordinates, and sometimes the reference catalogues used by observers have errors in them as well.

Myles Standish of JPL has analysed the older raw observations made at the Greenwich Royal Observatory, the US Naval Observatory, and the Paris Observatory, in an attempt to correct them for as many errors as possible, and then see if the orbits have any characteristics that remain unexplained. He concluded, in 1993, that the apparent perturbations in the motion of Uranus are mostly due to an incorrect value being used for the mass of Neptune, which was 0.5% too high. He attributed the remaining discrepancies in Uranus' position to observational errors, and systematic errors in the catalogue used by the US Naval Observatory.

On the other hand, A. Brunini of the Universidad Nacional de La Plata in Argentina reported, in 1992, that he could obtain good agreement between the theoretical and actual positions for Uranus, between 1782 and 1897, and between 1897 and 1985, but not if he put both groups of data together. He suggested that the cause of the discrepancy could be a very small planet, smaller than Pluto, orbiting the Sun just outside the orbit of Uranus, and perturbing Uranus when they were closest together in 1897.

It is too early to say definitively which, if any, of the above analyses is correct.

The Asteroids

Early Work

The first minor planet or asteroid was discovered in 1801 by Giuseppe Piazzi at Palermo in Sicily orbiting the Sun between the orbits of Mars and Jupiter. The diameter of this asteroid, called Ceres, was still uncertain in 1890, but a value of about 500 km was deduced from photometric measurements. This is only about 10% of the diameter of Mercury, the smallest planet known at the time.

It was a surprise that Ceres was so small, as astronomers had been looking for a planetary-sized body in the large gap between the orbits of Mars and Jupiter. When Pallas the second asteroid was discovered in 1802, however, Wilhelm Olbers, a doctor and amateur astronomer from Bremen, suggested that both Ceres and Pallas may have been part of a larger body that had exploded. If that were the case there may be other fragments whose orbits should all intersect at the place where the explosion had taken place, and so he suggested looking for more asteroids in orbits that went through the point of intersection of the orbits of Ceres and Pallas. Shortly afterwards, Juno was discovered, and was found to have an orbit consistent with Olbers' theory, but Vesta, discovered by Olbers himself in 1807, had an orbit that did not fit with his theory, and so he abandoned it.

As Simon Newcomb observed in 1898 (Reference 6), Olbers appeared to have forgotten that the orbits of the asteroids would have been drastically changed over time by the gravitational attraction of the planets, Jupiter in particular, and so it was hardly surprising that their orbits did not all intersect in one place. The fact that they did not do so now did not disprove Olbers' theory, and so it had been dropped prematurely, although his theory became less tenable when many more asteroids were discovered later in widely different orbits.

After the discovery of Vesta no more finds were made for almost 40 years, although astronomers were convinced that there must be many more. In order to assist in the search, the Berlin Academy, at Bessel's suggestion, organised the production of star maps showing all stars down to the ninth magnitude in the band near the ecliptic. Astraea was discovered in 1845, and by 1890 there were 287 asteroids known, although most of them were very small with diameters of only about 10 to 20 km. All of these asteroids orbited the Sun between the orbits of Mars and Jupiter, although the orbit of Aethra near perihelion just crossed the orbit of Mars. The orbital inclinations and eccentricities of the asteroids, although generally higher than those of the planets, were still appreciably less than those of the comets as shown by the following:

		Planets	Asteroids	Comets
Maximum orbital inclination		7°	35°	162°
Eccentricity*	Min.	0.007	0.005	0.463
	Max.	0.21	0.38	>1

In 1866 the American astronomer Daniel Kirkwood noted that, although there were by then 88 asteroids known, there were no asteroids with orbital periods of one third, two fifths and one half that of Jupiter. These so-called Kirkwood gaps were caused by the disturbing action of Jupiter, based on a theory first proposed by the French mathematician Joseph Lagrange in 1776.

The total mass of all the asteroids was estimated at less than 0.025% of the mass of the Earth and, although all but one of the orbits lay between those of

*Orbits are either circular, elliptical, parabolic or hyperbolic, with eccentricities of 0, <1, 1 or >1, respectively. Of these, the hyperbolic orbit is open, i.e. the object traversing it can do so only once.

Mars and Jupiter, these orbits were very different in size, shape and inclination one from another. So Olbers' planetary explosion theory seemed unlikely to be correct. Instead, the asteroids were thought to be left over from the original nebula that had condensed to form the solar system, having not condensed to form a planet because of their small total mass and the gravitational perturbations of the massive Jupiter close by.

Laplace, in his theory of the origin of the solar system, proposed that the original nebula had condensed to form flat rings (like the rings of Saturn) and that these rings had further condensed to form the planets. If the asteroids were left over from this process, however, they should have orbits with only small inclinations to the ecliptic, like the planets. The fact that their orbital inclinations were as high as 35° was a problem, and various modifications were proposed to Laplace's theory, including, for example the idea that the original nebula had condensed into toroidal instead of flat rings. Such modifications improved the theory, but did not make it entirely convincing.

If the asteroids were the remains of the original solar system nebula, it was thought that there may well be other asteroids orbiting the Sun between the orbits of Jupiter, Saturn, Uranus and Neptune, but nineteenth century telescopes would not be able to see them, as they would be too small and too far away.

A number of astronomers thought that some of these asteroids, or minor planets, may have complex surface features, together with an atmosphere of sorts. For example, in 1897 Professor Ball wrote (Reference 4) "It may be, for anything we can tell, that these (minor) planets are globes like our Earth in miniature, diversified by continents and oceans." Later on he explained that the very low surface gravity on such bodies would probably mean that their atmospheres would be "in an extremely rarefied condition, though possibly of enormous volume." Ledger wrote (Reference 1) of the possible sightings of "nebulous atmospheres" around these bodies. Clearly the concept of escape velocity was not properly understood.

Surface Characteristics

The German astronomer Max Wolf introduced a new method of searching for asteroids in 1891, when he started photographing star fields with exposures of several hours. He set his telescope to track the stars, so the asteroids showed themselves by moving relative to the star field during the exposure to leave small streaks. This resulted in the number of asteroids found increasing to 450 by 1900, and by 1932 Wolf and his assistant Karl Reinmuth had discovered over 500 from the Königstuhl Observatory in Heidelberg. In 1938 the total number of asteroids had reached 1,500, and by 1991 over 4,500 had had their orbits determined and many more asteroids had been detected.

The sizes of the asteroids had originally been deduced from photometric measurements, as the discs of even the largest asteroids were too small to be measured in early telescopes. Edward Barnard was finally able to measure the disc sizes of the four largest objects in 1895, and his results were generally accepted as the best available for over 70 years. Ceres is still the largest known asteroid, but its diameter, using the best estimate available today, is only $\frac{1}{4}$ that of the Moon.

Barnard's measurements led him to attribute an incredibly high albedo of 0.74 to Vesta which could not be explained by any known rock, and it was thought that it could not be due to snow as Vesta was too small to retain water vapour in a tenuous atmosphere. This dilemma was only solved in the 1970s when it was realised that Vesta's diameter is 40% larger than Barnard's estimate and this, together with a slight revision in its intensity, resulted in a much lower albedo of about 0.30. It is now thought to have a surface composed of igneous rocks.

Many of Barnard's diameter estimates have been increased in the last 20 years or so, and this has resulted in current albedo estimates for many asteroids as low as 0.05. This low albedo is usually associated with carbonaceous, or C-type, asteroids, which have flat featureless spectra similar to those of carbonaceous meteorites. C-types account for about 75% of all asteroids. Next most numerous are the reddish silicaceous, or S-type, asteroids, with albedos ranging from 0.15 to 0.25, generally thought to resemble metal-bearing chrondite meteorites. Various other types of asteroids have been defined, but the chemical characteristics of asteroids are still not altogether clear.

In 1977, Larry Lebofsky of the University of Arizona detected an absorption band due to water molecules on the surface of Ceres, and the spectrum of organic compounds was detected on Electra by Dale Cruikshank of the University of Hawaii and R. H. Brown in 1987. Only a few other asteroids give clear spectra.

So far, the only asteroids to have been viewed from close range are Gaspra and Ida, which were photographed by the Galileo spacecraft in October 1991 and August 1993, respectively, en route to Jupiter. Gaspra was found to be an irregularly-shaped 19 × 12 km asteroid, pock-marked with craters. There are a number of grooves, about 300 m across and 10 to 20 m deep, which traverse the surface for up to a few kilometres, resembling the fractures seen on the Martian satellite Phobos. Ida was found to be 52 km long, irregular in shape, and littered with craters of all sizes. In fact, virtually all of the small bodies of the Solar System seen so far, whether they be planetary satellites, asteroids, or the nuclei of comets, have two features in common, they are irregular in shape and have numerous craters.

Orbits

In 1906, two asteroids were discovered which travelled in similar orbits to Jupiter, and over the next few years similar asteroids were found travelling in two groups, one about 60° in front of Jupiter and the other about 60° behind. The Trojans, as they came to be called, were oscillating about the Lagrangian libration points which are stable points along Jupiter's orbit.

• *Near Earth Objects*

Although most of the asteroids have orbits lying between Mars and Jupiter, a number do not. In particular, on 14th August 1898 Gustav Witt at Berlin discovered an eleventh magnitude asteroid with an orbit that would bring it to within about 22 million km of Earth every 37 years. This asteroid, now called Eros, was used to estimate the Astronomical Unit during its oppositions of 1901* and 1931, and gave the most reliable measurements prior to the use of radar in 1961.

As the number of asteroids increased, so objects were found that passed ever closer to Earth. In 1932, Eugene Delporte at Brussels discovered Amor which approaches to within 16 million km of the Earth every 8 years, having a perihelion just outside the Earth's orbit. In the following month, Karl Reinmuth discovered Apollo, whose orbit crosses not only the orbit of the Earth but even that of Venus, and which passed within 11 million km of Earth on 25th May 1932. Five years later, a new asteroid called Hermes was found to pass within 780,000 km of Earth. Although Hermes was very small, probably only 1 km in diameter, the impact of such a body with the Earth would have caused a major catastrophe. In 1991 an asteroid, temporarily called 1991 BA, passed even closer than Hermes, at 170,000 km, and, although its diameter was only about 9 m, if it had hit the Earth, it would have released the energy equivalent of an atomic bomb. Then, on 20th May 1993, a 6 m diameter asteroid, 1993 KA_2, missed the Earth's surface by only 140,000 km.

The asteroids that cross the orbit of Mars, or have orbits lying completely within the orbit of Mars, are now called Near Earth Objects (NEOs). They are classified into three families:

Amor Orbits cross that of Mars, but not the Earth's.
Apollo Orbits cross the Earth's orbit.
Aten Average distance from the Sun is less than 1 AU. A number of these of these asteroids have orbits that lie completely within the Earth's orbit, as does Aten itself, which was discovered by Eleanor Helin in 1976 as the first such object.

By the end of 1992, 91 Amor asteroids had been found, plus 104 Apollos and 12 Atens.

One of the most well-known Apollo asteroids is Icarus, which was discovered by Walter Baade in 1949, as an object of the sixteenth magnitude. Herrick calculated its orbit and found it to have a perihelion which lies well within the orbit of Mercury, and an aphelion that lies outside of the orbit of Mars. Gilvarry pointed out a few years later that it has a relativistic advance of its perihelion of 11 arcsec per century (compared with 43 arcsec per

---

*Unfortunately, Eros was discovered just after one of its closest approaches (in 1894). At its opposition of 1901 it was just over twice as far away as its minimum distance.

century for Mercury), giving further proof of Einstein's general theory of relativity.

Phaethon, discovered in 1983 by J. Davis and S. Green, using the IRAS satellite, is about 5 km in diameter, and has a perihelion even closer to the Sun than that of Icarus. It was known in the nineteenth century that some meteor showers are associated with comets (see Page 112), but no comet had been discovered that has the same orbit as the Geminid meteors. It was, therefore, considered very significant when it was found that Phaethon's orbit is virtually the same as that of the Geminids. Whether Phaethon is the remains of an old comet is thought to be doubtful, but it does appear to be the source of the Geminid meteors.

• *Distant Asteroids*

So much for those asteroids that come close to the Earth and Sun, but what about very distant asteroids?

In 1920, Walter Baade discovered the first asteroid (Hidalgo) whose orbit intersected that of Jupiter, and almost reached out to Saturn. No other asteroid was known with such a distant aphelion until, in 1977, Charles Kowal of the Lowell Observatory discovered Chiron, using the Schmidt telescope on Palomar Mountain. Chiron was found to have a perihelion just inside the orbit of Saturn, and an aphelion of 18.5 AU which is nearly as far as the orbit of Uranus (19.2 AU).

Chiron is a strange object, and it is not even certain that it is an asteroid. It suddenly increased in brightness in 1988 and, in the following year, Karen Meech of the University of Hawaii and Michael Belton of the American National Optical Astronomy Observatories found it to have a fuzzy coma, using the 4 m telescope on Kitt Peak. Then, in January 1991, Bobby Bus and Ted Bowell of the Lowell Observatory and Mike A'Hearn of the University of Maryland detected cyanogen (CN) gas surrounding Chiron to a distance of 50,000 km. Cyanogen was known to be a constituent of the ionised gas tails of comets, but this was the first time that it had been detected at such a large distance of 11.3 AU from the Sun. In the following month, Mark Sykes and Russell Walker, using both ground and IRAS satellite data, estimated that Chiron could have a diameter as large as 370 km.

Is Chiron a comet or an asteroid? No previous comets have been found with such a very large nucleus, and so it is thought that Chiron may be more like Pluto, with icy deposits on its surface which are evaporating to form a temporary atmosphere as it nears perihelion (due in 1996).

Two other distant asteroids have recently been discovered, 1991 DA and Pholus. The aphelion of 1991 DA at 22 AU is further from the Sun than Uranus, but that of Pholus at 32 AU is even further than Neptune.

• *The Kuiper Belt*

Gerard Kuiper suggested, in 1951, that the small outermost satellites of Jupiter and Saturn may be captured asteroids from a belt, too faint to detect, that orbits the Sun, from just outside the orbit of Neptune (at 30 AU distance) to well beyond the aphelion of Pluto (at 49 AU). He proposed that these planetary satellites had originated from near the inner edge of this belt, where they had had their orbits modified by the gravitational pull of the four giant planets, Jupiter, Saturn, Uranus and Neptune. Other astronomers suggested that this Kuiper belt, as it came to be known, may also be the source of short-period comets. The Kuiper belt was assumed to be the remains of the nebula that had condensed to form the planets.

The discoveries of Chiron, 1991 DA, and Pholus made Kuiper's theory more tenable, but the most significant development was the discovery, in 1992 and 1993, of a number of asteroids currently even further away from the Sun than Pluto.

The British astronomer David Jewitt and Jane Luu of the University of California had been looking for asteroids in the outer solar system for 5 years when they discovered 1992 QB1, or Smiley, on 30th August 1992, using the 2.2 m telescope on Mauna Kea. Brian Marsden, of the Central Bureau for Astronomical Telegrams at Cambridge, Massachusetts, calculated a preliminary orbit of this 23rd magnitude object, before its discovery was announced on 14th September. Smiley was at a distance of 41 AU from the Sun, and appeared to be about 200 km in diameter, assuming that its albedo is the same as that of cometary nuclei, i.e. about 0.04. In March 1993, Jewitt and Luu found a second asteroid, 1993 FW, or Karla, at about the same distance from the Sun. Like Smiley, it appeared to be reddish in colour, and both Smiley and Karla were assumed to have surface temperatures of about 50 K (warmer than Triton, because they are both thought to be much darker, and therefore more effective at absorbing heat from the Sun). Since then four more such objects have been found. Jewitt and Luu discovered 1993 RO and 1993 RP in September 1993, and Iwan Williams of Queen Mary College, London and Alan Fitzsimmons discovered 1993 SB and SC in the same month. These four asteroids appear to be about 32 to 35 AU from the Sun, or just outside the orbit of Neptune.

These new findings have confirmed the suspicion of those nineteenth century astronomers who thought that distant asteroids ought to exist, and the Kuiper belt theory first proposed over 40 years ago.

Comets

Early Work

The solar system was known to consist of eight planets, 20 moons and 287 asteroids in 1890. In addition, there were those mysterious and nebulous bodies known as comets, swarms of meteors which the Earth passed through on about the same dates each year to produce "shooting stars", and occasional meteorites that came crashing to Earth. We will consider first the comets.

The visible structure of comets was very different from one comet to another, and even from one day to another with the same comet. Generally speaking, however, they were observed to have a bright core, and a tenuous head and tail, and to have a highly eccentric orbit around the Sun. They were assumed to have a solid nucleus, at the centre of the bright core, but such a nucleus had never been seen. The orbital planes were at all angles to the ecliptic, with comets orbiting the Sun in prograde and retrograde directions.

In the early eighteenth century, the English astronomer Edmond Halley published his *Synopsis of Cometary Astronomy* in which he analysed the orbits of 24 comets. In particular, he showed that the orbit of the comet of August 1682 and that recorded by Johannes Kepler in 1607 were the same. This led him to suggest that the comets of 1378, 1456, 1531, 1607 and 1682 were all probably different appearances of the same comet, orbiting the Sun with a period of about 76 years, and he correctly forecast its return in 1758. It has since been called Halley's comet in recognition of his work.

Some comets, like Halley's comet, followed elliptical paths around the Sun and made periodic returns, while others were in hyperbolic orbits making only one appearance. As comets are such diffuse and presumably light bodies (nobody had been able to measure the mass of a comet), they are easily disturbed by planets, particularly the massive Jupiter. This sometimes caused periodic comets to have their orbits perturbed so much that they were never seen again.

The French astronomer Jean Louis Pons discovered a comet from Marseilles Observatory in 1818 that was shown by Johann Encke, in the following year, to have the very short orbital period of 3 years and 4 months. The orbit of this comet, now called Encke's comet, has an aphelion within the orbit of Jupiter, and a perihelion well within the orbit of Mercury. Because it has such a short period, its orbit could be accurately determined, and deviations on its close approaches to Mercury could be used to estimate the mass of the planet; Mercury being a difficult planet to "weigh" because it is small and has no moon.

Encke's comet gave late nineteenth century astronomers a problem, however, as its orbital period was observed to be getting shorter by about 2½ hours every orbit. Although this reduction was very small, when compared with its period of 1,210 days, it was known to be a real reduction. Suggestions were made by Encke that the comet's period may be reducing because of tenuous material near the Sun, which could be either the resisting medium in interplanetary space proposed by Olbers, or the material that caused the solar corona or the zodiacal light.

Faye's comet was the next shortest period comet (period about 7 years) whose orbit was reasonably well known, but no change in its orbital period had been detected, and no effect had been observed on the orbit of comet 1882 II that had passed within only 500,000 km of the Sun. So the idea of a resistive medium causing the effect with Encke's comet was rejected, and the real cause of that effect was still a mystery.

Attempts were made to estimate the density of the heads of comets by observing the intensity and position of faint stars as comets passed in front.

Not only could quite dim stars still be seen through the heads of comets, but there was no clear evidence of refraction either, indicating that comets were generally very tenuous objects.

When comets were far from the Sun their spectra were found to be only the reflected spectra of the Sun. In 1864, however, Giovanni Donati the director of the Florence Observatory observed the spectrum of Temple's comet when it was near to the Sun, and found three faint luminous bands. Four years later, William Huggins identified these bands with those emitted by various hydrocarbon compounds when they are made luminous in the laboratory by electrical discharges. Meteorites, when heated in experiments, also gave off hydrocarbons, and so it appeared that comets and meteorites were made of the same basic elements. The tails of short period comets were observed to be shorter at each perihelion passage, so the solid nucleus of a comet was thought to partially evaporate when it was close to the Sun, apparently giving off these hydrocarbons to form the head and tail of the comet.

A little later, cyanogen CN was discovered in the head of a comet, and then, in 1882, it was found that the hydrocarbon band spectrum of Wells' comet (1882 I) disappeared when it was close to the Sun, being replaced by the yellow sodium line. A similar effect was seen later that year with comet 1882 II, but when it was very close to the Sun the sodium line was also accompanied by several lines of iron. As comet 1882 II receded from the Sun, however, the sodium and iron lines faded and the usual hydrocarbon bands reappeared.

Not all comets have tails. The large naked-eye comets usually do, but the much more frequent smaller telescopic comets often do not. Where tails do exist they gradually become longer as the comet gets nearer to the Sun, but the tail points away from the Sun no matter in which direction the comet is travelling. In the seventeenth century, Kepler had suggested that this may be due to the pressure of radiation emitted by the Sun, but this theory had been dismissed in the nineteenth century because light was then thought to be a wave motion, exerting no pressure. The direction of the tails was, instead, thought to be the result of electrical repulsion between the Sun and the comet.

Very often comets had more than one tail, each tail having a different length and making a different angle to the solar–nucleus line. In 1877, the Russian physicist Fëdor Bredikhin proposed that, of the three basic types of tail, one was due to hydrogen, one to hydrocarbons, and one to heavy elements like iron or chlorine. He calculated that, in the case of hydrogen, the electrical repulsion of the Sun was about 15 times as strong as its gravitational attraction, and so hydrogen formed the long straight tails. For hydrocarbons, the electrical repulsion was about one to two times as strong as the gravitational attraction, and so they formed the curved scimitar-type tails, whereas for iron, the gravitational attraction far outweighed the electrical repulsion, thus producing short strongly-curved tails. At the time, hydrocarbons and iron had been detected in the spectra of comets, but hydrogen had not.

Laplace had suggested, in the late eighteenth century, that comets originally came from interstellar space and, in the nineteenth century, the Sun was known to be moving in space towards the star λ Herculis (see Page 208). Carrington and Mohn suggested, therefore, that, if non-periodic comets were in interstellar space, the Sun should sweep up more of them in its forward direction, and those comets should, on average, have a higher velocity relative to the Sun, than those behind it. No such effect could be found, however, and so it appeared as though non-periodic comets must be an integral part of the solar system.

• *Meteors*

Groups of meteors were known to follow tracks with different orbital periods. When the Earth crossed the track of one of these groups there was a display of shooting stars, and when the Earth went through the group itself there was an exceptionally fine display. Sometimes the groups were so long that the Earth could go through the head one year and the tail in the following year, as was the case with the Leonid meteor swarm that was over 1,000 million km long.

In 1866 Schiaparelli showed that meteors move in nearly parabolic orbits. He also proved that comet Swift-Tuttle crossed the Earth's orbit very close to where the Earth intercepted the Perseid meteors on about 9th or 10th of August every year. He then calculated the orbital elements for the Perseids and Swift-Tuttle, and showed that they were in the same orbit with a period of 119 years. Theodor von Oppolzer of Vienna independently calculated a period of 124 ± 10 years, so Swift-Tuttle was expected to return in about the year 1986 ± 10.

Following Schiaparelli's discovery, there were many other associations found between comets and meteor showers, of which one of the most interesting was that connected with Biela's (or Gambart's) comet.

In 1826, Wilhelm von Biela, who was an Austrian army officer and amateur astronomer, had discovered a comet that had a period of about 6 years 8 months. On 28th November 1845, during its 1845–46 return, Biela's comet looked perfectly normal, but on 19th December it appeared to be pear-shaped, and by 29th December it had become two objects. These gradually separated so that they were 250,000 km apart when they disappeared from view in April 1846. On its next appearance in 1852 the two comets were about 2 million km apart, and it was never seen again. In 1872, however, in place of the comet, the Earth was treated to a display of meteors, and on 27th November 1885, two orbits later, there was another spectacular display. So some meteor showers have clearly been produced by comets which no longer exist.

Composition and Structure

Halley's comet, the most famous periodic comet, has made two appearances this century. In 1910, it was very bright (see Figure 6.2) but, in 1986, it

Figure 6.2 *Halley's comet in 1910. (Courtesy* California Institute of Technology.*)*

was a disappointing sight, because it reached perihelion when it was on the other side of the Sun from the Earth. Incidentally, the 1910 appearance is often confused with the Great Daylight Comet of that year, which was the brightest comet seen so far this century.

Since 1910 there have been few great comets, and a number of comets that were predicted to be impressive have turned out to be disappointing sights, showing how difficult it is to predict their development and appearance. A classic example was Comet Kohoutek, which was seen to be exceptionally bright when it was some distance from the Sun in 1973, but which failed to materialise into the great comet predicted.

Faint stars had been seen through the heads of comets in the nineteenth century, so their density was clearly very low, but how low? Karl Schwarzschild and E. Kron examined Halley's comet during its 1910 apparition, and concluded that only 150 g/s of material was being emitted from the solid nucleus into the tail, giving a density in the tail of only one molecule per cubic centimetre, which is clearly extremely low. So when molecules are evaporated from the nucleus and stream away in the tail, they would generally not be subjected to collisions with other molecules.

The discovery that radiation produced a pressure on objects in its path, caused a re-evaluation to take place, in the early twentieth century, of what causes comets' tails to stream away from the Sun. This had been attributed, in the nineteenth century, to the electrical repulsion of the Sun, but now radiation pressure was thought to be the primary cause, with electrostatic repulsion between particles in the tail simply affecting its shape.

In the 1860s, Huggins had shown that the spectra of gases in the heads of comets were the same as those produced by hydrocarbon compounds made luminous by electrical discharges and by meteorites heated in the laboratory. In the following decade, the cyanogen molecule, CN, had also been discovered in the head of a comet. When molecular spectra were better understood in the 1920s, C_2, CH, CH_2, CO^+, NH, OH and N_2^+ were found in comets. When meteorites were heated in the laboratory they gave off hydrogen, nitrogen, carbon dioxide and hydrocarbons, and when these gases were illuminated by sunlight they produced C_2, CH, CO^+, CN and CO, confirming the idea that comets and meteorites are made of the same basic elements, namely carbon, hydrogen, oxygen and nitrogen, in similar molecular arrangements. In 1951, Fred Whipple suggested that the nucleus of a comet was like a dirty snowball, consisting of ices of water, ammonia (NH_3), and methane (CH_4), with grains of embedded meteoric material.

Did the nucleus of a comet look like a dirty snowball? Dale Cruikshank suggested in 1985 that the nucleus of Halley's comet had an albedo of 0.04, using ground-based infrared measurements, while Neil Divine proposed an albedo of 0.06. Whipple's dirty snowball would have an albedo higher than either of these estimates, so was it correct?

Our knowledge of comets has been greatly enhanced by the fleet of spacecraft that examined Halley's comet at close range in 1986 (see Pages 339–341), although it is clearly unwise to assume that all comets have the same properties as Halley's.

The European Giotto and the Soviet Vega spacecraft found that the dirty snowball theory was broadly correct for Halley's comet. The inner coma had a very high water content, and the nucleus had a density of about 0.2 g/cm^3, very much the same as a snowball. The inactive part of the nucleus had an albedo of only 0.03, however, so it was a very dirty porous snowball. The dark surface was a good absorber of solar radiation, producing a temperature of 330 K, which is appreciably higher than the temperature required for the sublimation of ices. This is thought to take place, therefore, below the surface.

Observations made by the American satellite OAO-2 in 1969, at the Lyman α wavelength of 121.6 nm, showed that comet Tago-Sato-Kosaka was surrounded by a spherical cloud of neutral hydrogen over a million kilometres in diameter. At a distance of 1 AU from the Sun this comet was producing hydrogen atoms at the rate of 10^{29} per second. Similar clouds were also found around Bennett's comet in 1970, and Kohoutek's comet in 1973, which, in the latter case, was found to be over 10 million km in diameter. Halley's comet was also found by the Japanese Suisei spacecraft to be surrounded by a spherical corona of neutral hydrogen atoms, extending some 10 million km from the nucleus at the time of the spacecraft intercepts. These neutral hydrogen atoms become ionised when they had travelled about 10 million km, so at that distance they are controlled by the electric and magnetic fields associated with the solar wind.

The fleet of spacecraft found that the bow shock of Halley's magnetosphere was 400,000 km from the nucleus at its closest point, almost exactly as predicted. The solar wind was found to decrease in velocity from 400 km/s in interplanetary space, to 60 km/s about 150,000 km from the nucleus. The interplanetary magnetic field dropped abruptly to zero inside the ionopause which was some 4,700 km from the nucleus. Inside the ionopause the solar wind was non-existent, and a stream of neutral molecules and cold ions were found flowing away from the nucleus at 1 km/s. It is these neutral molecules, which drag dust particles away with them, to form the coma or head of the comet.

Prior to the Halley spacecraft encounters, no-one had imaged the nucleus of a comet nor knew its size. David Jewitt and Edward Danielson estimated a diameter of 8 km for the nucleus of Halley's comet from the comet's intensity at recovery in October 1982, Dale Cruikshank suggested 20 km, and Neil Divine 6 km. In fact, the Giotto spacecraft found the nucleus to be about 16 × 8 × 8 km in size, with bumps and hollows, so that it resembled a potato or peanut in shape (see Figure 6.3). Bright jets could be clearly seen streaming towards the Sun. Measurements using the Giotto spacecraft indicated that dust is being emitted by Halley's nucleus at about 3 tons/s. Although this is many of orders of magnitude greater than Schwarzschild and Kron's estimate in 1910, it is low enough to allow about 1,000 orbits of Halley's comet before it would run out of material.

A few months prior to the spacecraft encounters with Halley's comet, Zdenek Sekanina of JPL and Stephen Larson of the University of Arizona deduced a rotation period for the nucleus of 2.2 days, by carefully analysing photographs taken during the 1910 appearance of the comet. This period was confirmed by Kaneda, using data from the Japanese Suisei spacecraft by measuring the brightness variation of the hydrogen corona, and various other observers confirmed this period using data from the other spacecraft to intercept the comet. Then Robert Millis of the Lowell Observatory and David Shleicher estimated a period of 7.4 days, using a different technique for analysing the variations in the corona's brightness. This period was also confirmed by other astronomers using ground based data and information

Figure 6.3 *The nucleus of Halley's comet as imaged by the Halley Multicolour Camera on board ESA's Giotto spacecraft. (Courtesy* Dr. H. U. Keller, Max-Planck-Institut für Aeronomie, Lindau/Harz, Germany.*)*

provided by the IUE spacecraft. Unfortunately, the spacecraft data are not clear-cut, and it is possible that both the 2.2 and 7.4 day rotation rates are correct, describing rotations about two different axes.

The inner coma of Halley's comet was found to consist mainly of water (H_2O), with lesser amounts of carbon monoxide (CO), carbon dioxide (CO_2), methane (CH_4), ammonia (NH_3), and polymerised formaldehyde ($(H_2CO)_n$). The dust particles consisted of the elements carbon, hydrogen, oxygen and nitrogen, the so-called CHON particles, and stony particles, consisting mainly of silicon, magnesium, iron and oxygen, similar to those found in meteorites.

The relative abundances of key elements in the material emitted by Halley's comet are close to those in the Sun, rather than in the Earth or in meteorites, indicating that comets consist of very primitive material, which is depleted only in the volatile elements of hydrogen and nitrogen.

On 12th February 1991, Halley's comet suddenly increased in brightness by a factor of 300. It was too far away from the Sun for this to be due to a spontaneous explosion caused by solar heating, so the sudden increase was generally attributed to the impact of a large meteorite or small asteroid, although such a collision would be very rare, considering how small Halley's nucleus is. Calculations showed that the impacting body need only be about 30 m in diameter.

Origin

It appeared, in the nineteenth century, as though non-periodic comets were part of the solar system, although the evidence was not altogether convincing at that time. If comets had hyperbolic orbits before they entered the gravitational fields of the planets, they must have come from interstellar space, however, so were those orbits originally hyperbolic?

Strömgren showed, in 1914, that those comets that approached the Sun with hyperbolic orbits, had been in elliptical orbits before they had been perturbed by the planets. Thirteen years later, G. Van Biesbroeck confirmed this for Comet Delavan, in particular, which followed a hyperbolic orbit near the Sun, showing that it had originally moved in an elliptical orbit with a semimajor axis of 2.7 light years.

In the last decade of the nineteenth century, H. A. Newton of Yale and the Frenchmen Tisserand and Callandreau analysed the orbits of short-period comets. They showed that those comets with a period of less than 9 years had probably had their orbits substantially modified by Jupiter during one or more close encounters. Newton showed, in 1891, that a comet would be more likely to have its trajectory affected by Jupiter if it orbited the Sun in the same direction as the planet, and it was thought to be no coincidence that all of the short-period comets known at that time orbited the Sun in that direction.

In the 1940s, the Dutch astronomer Jan Oort carefully analysed all the available data on the orbits of non-periodic comets, and found that the most usual semimajor axis size, prior to their being perturbed by the planets, was about 2.4 light years, or about half the distance to the nearest star. This led him to conclude, in 1950, that non-periodic comets originate from a large cloud at a distance of from about ½ to 2 light years from the Sun, at the edge of the region dominated by the Sun's gravitational field. From time to time, some of the objects in this cloud are disturbed in their distant orbits around the Sun by a passing star, or by an interstellar gas cloud, and some of them then enter the planetary system where they become comets as they swing by the Sun. Generally these comets leave the planetary system after perihelion, but their orbits can be disturbed by a planet, either before or after perihelion, to change them into the elliptical orbits of the periodic comets.

Oort suggested that the Oort cloud, as it is now called, may have originated, together with the asteroids and meteorites, from an explosion* of a planet that was between the orbits of Mars and Jupiter. Those fragments that had almost circular orbits, became members of the planetary solar system, losing their gaseous constituents, because of their continuous exposure to solar radiation, and becoming asteroids and meteorites. Those fragments with elliptical orbits, on the other hand, had their orbits perturbed by Jupiter and the other major planets. A number of these fragments were given hyperbolic orbits, and were thus lost to the solar system, but a significant percentage were given orbits with aphelia of about ½ to 2 light years. Stellar perturbations then distorted these orbits near aphelia, making them more circular, thus producing the Oort cloud.

Kuiper hypothesised in 1951 that comets are the result of the condensation of the original solar nebula outside of the orbit of Neptune. These condensations had lost their highly volatile material, but were otherwise hardly changed since the origin of the solar system. Kuiper further proposed that the major planets had caused the orbits of these condensations to become highly elliptical, causing them to be injected into the Oort cloud, where their orbits had been made more circular by neighbouring stars.

We have no direct evidence for the existence of the Oort cloud even today, but it is widely believed to be present in some form. Kuiper's concept of comets being the remnants of the original solar nebula, rather than part of the debris of an exploded planet, is the preferred theory of the origin of comets, but it is by no means proven.

If either Oort or Kuiper are correct, a large number of comets must have been lost to the solar system completely, and now be in interstellar space. If the same process has occurred with other stars, there ought to be interstellar comets visiting the solar system, from time to time, originating from other stars. Possible encounter rates are estimated to be of the order of one interstellar comet every hundred years visiting the solar system, so it will be very difficult to identify them, if they exist.

---

*This is a development of an idea originally proposed by Olbers in the early nineteenth century, to explain the origin of the asteroids.

Meteorites

Early Work

In addition to hundreds of comets (and asteroids) there were also known to be numerous smaller bodies in the solar system, which varied from meteorites weighing a few hundred kilograms,* to meteors weighing a few grams or milligrams. The meteorites orbited the Sun on their own, but the meteors tended to orbit the Sun in groups.

When the Earth intercepted a meteorite in its orbit, the frictional heating in the Earth's atmosphere caused the meteorite to glow and partially evaporate before hitting the ground. Such bodies were seen in the atmosphere as fireballs. The less massive meteors were completely evaporated by such heating, however, and were seen in the atmosphere as shooting stars. (No meteor that orbited in a group had ever been known to reach the ground.)

Few meteorite specimens had been found on the Earth's surface before 1800, but in the nineteenth century interest escalated and a number of collections were built up. The British geologist Henry Sorby undertook a detailed analysis of samples, concluding that they had been formed out of the same gaseous nebula that had produced the Sun and planets. After the nebula had condensed into small particles he suggested that "These . . . collected together into larger masses, . . ., and (were then) broken up by repeated mutual impact, and often again collected together and solidified. . . . The study of the microscopical structure of meteorites reveals to us the physical history of the solar system at the most remote period of which we have any evidence."

An alternative theory had been proposed by Tschermak that meteorites had a volcanic origin. It was thought that they had originally been thrown out of volcanoes on the Earth many millions of years ago, when these volcanoes were much more powerful than they are now.

Later Work

The largest meteorite found in the nineteenth century was discovered at Melville Bay in Greenland. Knives made of meteoritic iron had been given to Captain John Ross by Eskimos as long ago as 1818, but it was not until 1894 that the American Arctic explorer Robert Peary was shown the area where the original meteorite fell. The largest mass, weighing 31 tons (31,000 kg), is now preserved in the American Museum of Natural History. An even larger meteorite, weighing about 60 tons, was found at Grootfontein, South West Africa, in 1920.

---

*The largest find made prior to 1890 was a mass of about 700 kg found in 1749 at Eniseisk, Siberia.

On 30th June 1908 there was a gigantic explosion in Siberia which was recorded on seismographs all over the world, and which caused pink glows in the sky after sunset in many parts of Europe, including the UK. The first scientific investigation to find out what had happened in Siberia was led by Leonid Kulik, of the Russian Academy of Sciences, in 1921. He spoke to people in the Krasnoyarsk region, close to the source of the disturbance, who described seeing a bluish-white light streak across the sky, and a sudden blast of wind that knocked people to the ground. The site of the devastation was thought to be over a hundred kilometres away, but he found it impossible to get there by road because it was completely blocked by upturned trees.

Kulik returned to the region in 1927 and, after a journey of 3 months, mostly by river, finally reached the devastated area which was in a marshy forest. He found that virtually all trees within a radius of 30 km of the centre of the devastation had been flattened, and there were numerous holes up to 50 m in diameter, although there was no central crater. In all, 80 million trees had been uprooted, and the noise of the explosion had been heard up to 1,000 km away. Kulik returned to the site in each of the next 2 years, but was unable to find any meteorite fragments. The explosion has now been attributed to the nucleus of a small comet that exploded a few kilometres above the ground.

As the twentieth century progressed, a number of large meteorites were found, and more and more evidence of large meteorite craters was found on Earth. The most well-known and best preserved crater being Meteor Crater in Arizona, where samples of meteoric origin were discovered in 1891. This 1.2 km-diameter, 200 m deep crater is thought to have been produced about 40,000 years ago by a 25 m diameter meteorite, weighing about 100,000 tons, and travelling at 15 km/s. The 62 km wide Manicouagan* structure in Quebec, where a great thickness of solidified rock was found, would probably have required a dense object a few kilometres in diameter travelling at the same speed, whereas the 180 km diameter crater discovered beneath the Yucatan Peninsula in 1992 (see Page 51) was probably caused by a dense object about 10 km in diameter.

The first accurate pre-encounter orbit for a meteorite was calculated from photographs taken from a number of different sites in Czechoslovakia of a fireball observed in 1959. The orbit had a perihelion of 0.8 AU, and an aphelion of 4 AU, which is typical of the orbit of an Earth-crossing asteroid.

---

*The 62 km diameter ring structure has been shown by Skylab and Landsat images to be surrounded by a depression 145 km in diameter. Geologists have found that the bedrock of the ring structure shows evidence of shock metamorphism about 210 million years ago.

7 Stellar Evolution and Stellar Structures

Early Work

It is often thought that very little was known about the stars in the nineteenth century because most astronomy books of the time tended to concentrate on the planets. This was far from the truth, however, although nineteenth century knowledge of the stars was purely observational, with no understanding of their internal processes. In fact, at the time astronomers found themselves somewhat frustrated by what they often perceived as scant reward for their considerable efforts.

Stellar Distances

Because the Earth moves in an orbit around the Sun once a year, those stars closest to the Earth should show an apparent movement against the background stars that are much further away. Measurement of this parallax motion would enable the distance of the nearest stars to be estimated, given the diameter of the Earth's orbit around the Sun.

The Englishmen, James Bradley and Samuel Molyneux tried to use this effect in the early eighteenth century to measure the distance of the stars, but their telescope was not accurate enough, and so parallax motion could not be detected. William Herschel also failed to detect such motion in the early years of the nineteenth century for the same reason (see Page 156) but, in the 1830s, Bessel, Henderson and Struve finally succeeded.

In 1792 Giuseppe Piazzi, the founding director of the Palermo Observatory, had found that the fifth magnitude star called 61 Cygni was moving through space at the rate of 5.2 arcsec/year. Friedrich Bessel confirmed this so-called "proper" motion in the early nineteenth century, by comparing his own stellar position measurements with those of Bradley made in the eighteenth century. When he decided to try to measure stellar parallax a few years later, therefore, he chose 61 Cygni, as such a large proper motion indicated that it was close to the Earth.

Joseph Fraunhofer had designed a heliometer for the Königsberg observatory, which used a split objective lens to enable stellar separations to be made with exceptional accuracy. It was completed in 1829, 3 years after Fraunhofer's death, so Bessel, who was the director of the Königsberg Observatory, was able to use this excellent instrument to try to measure the parallax of 61 Cygni. In December 1838 he was successful, measuring a parallax of 0.31 arcsec, giving a distance from Earth of about 10.4 light years (in today's units). After further observations in 1839 and 1840 he modified the distance estimate to 9.4 light years.

Thomas Henderson, the director of the Cape Observatory in South Africa, had measured the position of the first magnitude, southern star α Centauri in 1832/3, analysing his measurements to see if it exhibited any parallax, on his return to his native Scotland in 1834. Much to his surprise, he found a small effect, but it wasn't until Bessel announced his measurements for 61 Cygni 4 years later that Henderson decided to publish his results. In January 1839 he announced a parallax of 0.93 arcsec for α Centauri, which corresponds to a distance of 3.5 light years.

At the same time at the Russian University of Dorpat, F. G. W. Struve chose to measure the parallax of Vega, as it was a first magnitude star with a measurable proper motion. His observations, using a filar micrometer attached to his telescope, lasted on and off from 1835 to 1838, and although he completed his analysis shortly before Bessel, he did not announce his results until 1840. He found Vega to have a parallax of 0.262 arcsec, equivalent to a distance of 12.5 light years.

In 1853 Struve also measured the parallax of 61 Cygni, the same star as Bessel, deducing a distance of about 6.5 light years, and in 1886 the Oxford astronomer Charles Pritchard measured the distance of 61 Cygni as about 7.5 light years, using measurements from 200 photographic plates. Although these measurements were not very accurate in today's terms (present estimates of the distance of 61 Cygni are 11.2 light years), their order of magnitude was correct.

By 1890 the approximate distances of about 20 to 30 stars were known.

Stellar Classification

Fraunhofer had begun investigating the spectra of stars in the early nineteenth century from Munich, and had noticed that the absorption lines in the spectra of the stars Castor and Sirius did not have the same pattern as in the Sun. In the 1850s Gustav Kirchhoff and Robert Bunsen, working together at Heidelberg, were able to explain the origin of absorption lines in general (see Page 291), and in 1863 William Huggins identified hydrogen, sodium, iron, magnesium and calcium in stars at his own private observatory near London. This showed that stellar atmospheres are composed of the same elements found on the Sun and Earth.

The spectroscope added a vital new dimension to the study of stars. They

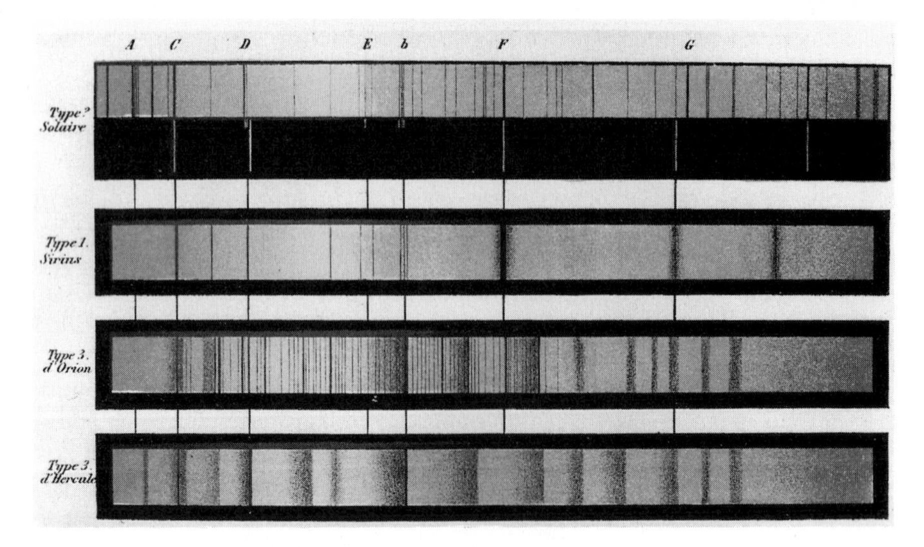

Figure 7.1 *The spectra of typical stars in three of the four Secchi categories. From top to bottom they are: type 2, the Sun; type 1, Sirius; type 3, Betelgeux; and type 3, α Herculis. (From* Les Étoiles, *by Angelo Secchi, Vol. 1, 1879, Plate VII. Image supplied by the Royal Astronomical Society.)*

had previously been catalogued by position, intensity and colour but now, with this powerful new tool, they could also be categorised by spectral type. This would eventually, in the twentieth century, prove of fundamental importance in the understanding of processes going on in the interior of stars.

The Italian Jesuit astronomer, Angelo Secchi produced a system of stellar spectral classification in 1863, after observing the spectra of over 4,000 stars, by dividing them into two categories. Three years later he introduced a third category, and in 1868 added a fourth to produce the first recognised system of stellar classification. His four classes were:

- I White or blue stars like Sirius or Vega, with very strong hydrogen lines, but barely perceptible metallic* lines.
- II Yellow or orange-yellow stars like Capella, Aldebaran or the Sun, having less strong hydrogen lines, but more pronounced metallic lines.
- III Orange-red stars like Betelgeux or Antares, with many metallic lines, but no hydrogen lines.
- IV Red stars like R Cygni with prominent carbon lines.

The spectra of typical stars in some of these categories are shown in Figure 7.1. In addition to the above, two stars, β Lyrae and γ Cassiopeiae, were known with bright hydrogen and helium emission lines in their spectra (see Pages 154 & 158), but these were left uncategorised.

---

*In astronomical terminology, "metallic" refers to any element other than hydrogen and helium.

Figure 7.2 *Edward Pickering. (Courtesy* Harvard College Observatory.*)*

Hermann Vogel produced a different stellar classification system in 1874 at Potsdam, based on his theory of stellar evolution in which stars cool with age,* and in 1890 Edward Pickering (Figure 7.2), director of the Harvard College Observatory, introduced a more comprehensive scheme, the precursor of what is today called the Harvard system. In it Pickering subdivided the first two Secchi categories, and added categories O for Wolf–Rayet stars (see below), P for planetary nebulae and Q for miscellaneous stars, giving the following:

A	White stars like Sirius or Vega with strong hydrogen absorption lines.
B	Bluish-white stars like Rigel with hydrogen and helium lines.
C	Stars with double lines. (The double lines were later shown to be erroneous, and the C category eliminated).
D	White or blue stars with emission lines.
E	Yellow stars with certain hydrogen absorption lines.
F & G	As E but with weaker hydrogen lines and stronger metallic lines, e.g. Canopus, Capella and the Sun.
H	As F but weak lines at blue wavelengths.
I	As H but with additional lines.
K	Orange stars like Arcturus and Aldebaran with emission lines.
L	Stars with peculiar lines.

*Vogel recognised that stars must have an initial heating phase, as they evolve from a cold nebula, but thought that it was so short that no stars could currently be seen in it.

M	Orange-red stars like Betelgeux or Antares with virtually no hydrogen lines and a complex spectrum (i.e. Secchi type III).
N	Red stars like R Cygni with prominent carbon lines (i.e. Secchi type IV).
O	Wolf–Rayet stars. (Faint stars with broad emission lines. See Page 128)
P	Planetary nebulae.
Q	Miscellaneous.

Henry Draper of New York, who had been the first person to successfully photograph a stellar spectrum, died in 1882 at the early age of 45. Mrs Anna Palmer Draper, who wanted to establish a memorial to her husband, decided to approach Edward Pickering, with the idea of setting up a stellar spectroscopy programme which she would fund at Harvard. Her money made a substantial contribution to the Harvard programme and, in recognition of this, Pickering's catalogue of 1890 containing the spectra of 10,351 stars down to −25° declination, was called the Draper Memorial Catalogue.

Early Evolutionary Ideas

A number of astronomers suggested various schemes of stellar evolution, in parallel with these first attempts at cataloguing stellar spectra. J. K. F. Zöllner of Leipzig proposed in 1865, for example, that stars start their life as gaseous planetary nebulae which contract and cool to become liquid spheres with a solid crust. These then fracture periodically to release hot liquid as variable stars, but eventually they stabilise and cool, changing colour from blue to yellow to red. Zöllner's theory that blue stars were the youngest* and red the oldest was accepted by many astronomers for a number of years.

Twenty years later, Arthur Ritter of Aix-la-Chapelle developed a stellar evolution theory in which he added an initial heating phase, while the star contracted as a perfect gas sphere, before cooling takes over. Stars thus start off as large cool red stars, that become hotter as they contract to become small blue stars, and then, when cooling becomes dominant, the small blue stars cool, via yellow stars, to become small red stars. So there are two types of red stars, large red stars that are young, and old small red stars. Ritter also proposed that the maximum temperature that a star could reach depended on its mass, with the heaviest stars reaching the hottest temperatures, as these heavier stars would contract more under their own gravity, producing a higher internal pressure and, consequently, a higher temperature. Ritter's theory was more or less forgotten for a number of years, until it was dusted off in the 1890s by William Huggins and the American George Ellery Hale.

---

*Even today, long after this theory has been discounted, some astronomers still use the term "early type" to refer to hot blue (and white) stars.

The first of a series of papers on stellar evolution were presented by Norman Lockyer in England to the Royal Society on 17th November 1887, in which he developed his "meteoritic hypothesis" for the birth of stars. In this theory, the universe is filled with streams of meteorites which collided with each other to produce gaseous nebulae or comets, which then condense to form stars. Initially, as the stars condense, they get hotter, but eventually the heat generated by contraction is not enough to balance the heat lost by radiation, and the stars then start cooling.

In many ways Lockyer's theory is like Ritter's, following the same colour sequence as the stars cool, namely red, yellow, blue, yellow, red, followed by star death, but Lockyer believed that all stars follow exactly the same evolutionary path, irrespective of mass, while Ritter thought that the maximum temperature reached by a star was dependent on its mass.

In the process of trying to understand the evolution of stars, Lockyer had suggested in 1878 that, as a star becomes hotter, the compounds of which it is made break down to form atoms, and when the star becomes hotter still, these atoms break down to form, what he called, proto-elements. He had first noticed in laboratory experiments that lines, which he attributed to these proto-elements, were of greater intensity in the high-temperature high-tension spark spectrum of an element compared with the lower-temperature arc spectrum. This concept of proto-elements, as being the constituent parts of atoms, was remarkable, as Lockyer proposed it some 20 years before the discovery of the electron. Eventually it was shown, in the twentieth century, that Lockyer's proto-elements were, in fact, ionised atoms (i.e. atoms stripped of one or more electrons).

The meteoritic part of Lockyer's theory was the subject of much scepticism and some dispute, although the Englishman P. G. Tait had suggested that the Sun had gained much of its earlier mass and energy from meteoric bombardment. Similar meteoritic theories to those of Lockyer and Tait were put forward by the English physicists Lord Kelvin (W. Thomson) and James P. Joule, and a German doctor Julius R. Mayer, but Lockyer was the only person to develop the idea into a full-blown stellar evolution theory. Ritter and Lockyer were relatively unusual in believing in the red–blue–red sequence, as most astronomers favoured the linear cooling sequence from blue to yellow to red.

The Luminosity* of Stars

Astronomers had believed, until the second half of the nineteenth century, that the luminosities of stars (excluding novae) were quite similar to one

---

*The luminosity of a star is defined as the total energy emitted per second. So, in order to calculate it, we need to know the star's brightness as seen from Earth, its distance, and the magnitude of any intervening absorption. In the early twentieth century, the latter was generally assumed to be zero (see Pages 224–225).

another but, in 1862, this was shown to be wrong, when the companion of Sirius was found to have a luminosity of only 1/5,000th that of the main star. It was not until the turn of the century, however, that distance estimates of the closest stars were sufficiently accurate to enable the first general estimates to be made of the range of stellar luminosities. It turned out to be about 11.9 in absolute magnitudes,* which is equivalent to a luminosity range of about 60,000.

It was much more difficult to measure the masses of stars than their luminosities, as it was only possible to estimate the masses of stars in binary systems, but it was clear, at the turn of the century, that the range of stellar masses was very much less than the range of luminosities. For the Sirius binary, for example, the mass ratio of the pair was only about 2:1, compared with the luminosity ratio of 5,000:1. This difference in ranges of stellar masses and luminosities was clearly highly significant, and had to be explained in any theory of stellar evolution.

As the twentieth century progressed, the measured luminosity range of stars gradually increased, until today, when, excluding novae and other highly irregular stars like η Carinae (see Page 126), the most luminous star known is S Doradus with an absolute magnitude of −8.9, and the least luminous is MH18 with an absolute magnitude of about 16. Thus the ratio of luminosities known today is about 10,000 million to 1, with the Sun being an average star in luminosity terms.

The Harvard Classification

Edward Pickering decided, after publication of the Draper Memorial Catalogue in 1890, to have a more detailed analysis carried out of the spectra of the brightest† stars. Miss Antonia Maury, Henry Draper's niece, was given the task for 681 bright stars north of −30° and was allowed to devise her own classification system. She used a somewhat complex scheme which was not adopted by other astronomers, but she did introduce into it a classification of the sharpness of spectral lines for the first time. Type a was for stars with the least sharp lines, b for medium sharpness and c for stars with the sharpest lines. The sharp-line stars also had some metallic lines of unusual strength but, of the 681 stars that Miss Maury evaluated, she put only 18 into the c category. These c-type stars were the subject of considerable interest over the next few years. Norman Lockyer, for example, thought that the sharp lines were caused by very high tempera-

*The absolute magnitude is the magnitude that a star would appear to have if it was at a distance of 10 parsecs, or 32.6 light years, from us. Magnitude −5 stars are 100 times *brighter* than magnitude 0 stars which are, in turn, 100 times brighter than magnitude 5 stars.
†According to their apparent, not absolute, brightnesses.

tures, and Ejnar Hertzsprung concluded that c-type stars were highly luminous.

The Harvard stellar classification had been designed both to categorise stars, and to indicate their place in the evolutionary process, by running from the hottest to the coolest stars. Originally, it was thought that A stars were the youngest, followed by B, C etc. and so on alphabetically down to O. In Antonia Maury's work on northern stars, which was published in 1897, she reassessed this evolutionary sequence, concluding that it should start with O stars, followed by B stars and then A stars.

Miss Annie Jump Cannon was given the job in 1896 of classifying 1,122 southern stars, from plates taken using the 13 inch (33 cm) Boyden telescope at the Harvard Southern Station at Arequipa, Peru. In completing her task in 1901, she not only confirmed the start-up evolutionary sequence O, B, A of Antonia Maury, but completed the sequence with F, G, K and M. She also listed, P for planetary nebulae (only one listed), and Q for peculiar stars (three listed). In addition, she found that she was able to divide the Harvard classes O to M into subclasses.

No N-type (carbon) stars had been included in Annie Cannon's catalogue of 1901, but a number were included in Williamina Fleming's survey of peculiar spectra that was published posthumously in 1912. In the meantime, in 1908, Pickering had discovered a new category of stars, that had similar absorption bands to those of the red N-type carbon stars, but with far more blue light. Pickering designated these as R-type stars and listed 51 examples, which was extended to 61 in Mrs Fleming's paper of 1912.

The final extension to the original work of the Draper Memorial Catalogue was published by Miss Cannon in 1912, when she extended Antonia Maury's work on northern stars to include a number of stars omitted from Miss Maury's list. This detailed analysis by Maury and Cannon, supervised by Pickering, had yielded very detailed analysis of the spectra of about 5,000 stars in both the northern and southern sky in the 22 years since the original catalogue had been published. Furthermore, in 1913, the International Solar Union accepted the Harvard classification system as a provisional standard, and in 1922 the newly-formed International Astronomical Union (IAU) accepted the Harvard classification as the official system.

In the meantime, on 11th October 1911 Miss Cannon started the monumental task of classifying the spectra of 225,300 stars for the new Henry Draper (HD) Catalogue, which she basically completed four years later. It was published between 1918 and 1924 in nine volumes, introducing in Volume 8 a new category of stars, S-types, with a very complicated absorption and emission line spectrum between 450 and 470 nm.

Annie Cannon continued analysing stellar spectra until her death in 1941, classifying in all an astonishing 395,000 stars, of which 46,850 were published in the Henry Draper Extension (HDE) Catalogue between 1925 and 1936, and 86,932 in HDE charts, the final volume of which was dedicated to her and published posthumously in 1949.

Initial Evolutionary Ideas

In 1867 the French astronomers C. J. F. Wolf and Georges Rayet had discovered three faint stars in Cygnus that had broad emission lines on a continuous background. One of the stars, which are now called Wolf–Rayet stars, had a broad line centred on 468.6 nm, whereas the other two had a line centred on 465.0 nm. In 1872, L. Respighi, who was professor of astronomy in Rome, and E. H. Pringle, an English amateur astronomer based in India, independently found a bright Wolf–Rayet star in the southern sky, and by the mid 1880s a total of 13 Wolf–Rayet stars were known. Pickering included only one of these in his Draper Memorial Catalogue of 1890. It was the sole representative of the so-called O category.

Pickering made a very important discovery in 1896, when he found that the Wolf-Rayet star called ζ Puppis has a most unusual spectrum. It exhibited a series of lines that alternated with the Balmer hydrogen lines, having half integer values (i.e. $n = 2\frac{1}{2}, 3\frac{1}{2}, 4\frac{1}{2}$ etc.) instead of the integer values ($n = 3, 4, 5$ and 6) associated with the Balmer lines (see Page 293). Pickering concluded that these new lines in ζ Puppis must correspond with an unknown abnormal condition of hydrogen, while Antonia Maury concluded that Wolf-Rayet (O category) stars in general were stars in a very early stage of development. In Annie Cannon's 1901 classification she subdivided the O-type stars into pure Wolf–Rayet stars with no Pickering lines as Oa, through Ob, Oc, Od (ζ Puppis), Oe, to Oe5B (no emission but some Pickering lines).

Norman Lockyer also developed a completely different stellar evolutionary sequence between 1887 and 1914, to that of the gradual cooling of stars proposed by Vogel and extended by Pickering and his colleagues. Lockyer's theory was based on his "meteoritic hypothesis" in which he suggested that colliding meteorites produce gaseous nebulae or comets. These then condensed to form red stars like Antares (now classified as type M), which gradually got hotter to become white stars like ε Orionis (type B), followed by cooling to end their life as red carbon stars (type N), prior to their demise into obscurity. His theory was largely discounted by other astronomers of the time because of his radical meteoritic hypothesis, and because of the difficulty that he had in deciding which stars to place on the rising temperature part of the sequence and which on the cooling part.

Lockyer had earlier suggested that elements were dissociated into what he called proto-elements when they were subjected to very high temperatures. These transitions would cause very sharp strong lines, which Lockyer suggested, in 1900, were the same as those seen by Antonia Maury in her Division c stars (see above). After Pickering's discovery of ζ Puppis, Lockyer modified his stellar evolution theory to include this star as one of the highest temperature stars, attributing the Pickering lines to proto-hydrogen.

There the matter of ζ Puppis stood until, in 1912, Alfred Fowler in England, one of Norman Lockyer's pupils, produced the so-called Pickering lines seen in ζ Puppis by experimenting in the laboratory with a mixture of hydrogen and helium. Fowler still thought that they were caused by hydrogen, but the Danish physicist Niels Bohr explained that they were due to ionised helium.

In 1892, the Irish amateur astronomer W. H. S. Monck examined the proper motions of stars in each of the Harvard categories, and found that, on average, the proper motions increased from the blue to yellow stars, but then reduced, going from the yellow to red stars. As, on average, the stars with the largest proper motions are the nearest to us, this implied that there were more yellow stars close to us than those of any other colour. When the observed brightness of the stars was also included in the analysis, it was found that the yellow (i.e. G) stars had, on average, the least luminosity.

The fact that the Sun was located in a group of predominantly yellow stars was of no great surprise, as the Sun itself is yellow, but the fact that yellow stars were, on average, less luminous than red, was a surprise, causing a major problem with the generally-accepted theory of stellar evolution of the time. In this Vogel–Harvard-type theory, hot blue stars were thought to change to medium-temperature yellow stars which, in turn, change to cool red stars, it being assumed that stars became less luminous (and smaller) as they became older and redder. Monck suggested, in 1893, that his observations could be explained if stars evolve from blue to red to yellow, but this seemed unlikely, as to get from blue to red spectroscopically involves passing through yellow. Then, in the following year, he suggested that there may be two different classes or "populations" of yellow stars, one being of low luminosity and near to us like α Centauri, and the other of high luminosity and far away like α Aurigae (Capella).

The problem of the apparently low luminosity yellow stars was cleared up in 1905 by Ejnar Hertzsprung, who found that Antonia Maury's c stars of various colours had very small proper motions and parallaxes compared with other stars of the *same* colour, indicating that the c stars were, on average, much further away, with a much higher luminosity. So, Hertzsprung concluded that not only were there two different populations of high and low luminosity yellow stars, as Monck had proposed, but there were two different populations of high and low luminosity orange and red stars also.

Unfortunately Hertzsprung, who was still an amateur astronomer in Copenhagen, published his findings in an obscure magazine *The Journal of Scientific Photography*, and his results became known only gradually. Hertzsprung realised that if stars of the same colour and, therefore, of the same actual temperature, have different luminosities, they must have different surface areas. So the two different populations of high and low luminosity stars corresponded to large and small stars, or "giants" and "dwarfs", to use Hertzsprung's terminology.

In the late nineteenth century, the Irish astronomer J. E. Gore had estimated that Arcturus had a diameter about the size of the orbit of Venus, assuming that Arcturus had the same surface brightness as the Sun, although he suggested that this was only an order-of-magnitude estimate, because Arcturus' parallax was uncertain. Later, in 1906, Hertzsprung used

Wien's law of radiation* to estimate the diameter of Arcturus, and found it to be similar in diameter to that of the orbit of Mars! So Hertzsprung suggested that stars can evolve along two alternative paths, both of which start with hot, bright, bluish-white stars which gradually become dimmer and redder, one path ending with red dwarf stars, the other ending with red giants.

Henry Norris Russell, from Princeton University Observatory, moved to Cambridge University in 1902, and worked with A. R. Hinks on the trigonometric parallaxes of stars. He returned to the United States in 1905 and, 5 years later, published a paper discussing the relationship between luminosity and spectral type, in ignorance of Hertzsprung's work. In August 1910, Russell presented his ideas to the Astronomical and Astrophysical Society of America meeting at Harvard, also suggesting, as had Hertzsprung, that there were two different populations of G, K and M stars, one being of large bright stars, the other of small dim ones. Karl Schwarzschild, who attended the meeting, told Russell of the similar earlier work undertaken by Hertzsprung, who had, in the meantime, joined Schwarzschild in Göttingen.

Although Russell, like Hertzsprung, had recognised these two different stellar populations, his proposed evolutionary path for stars was different from that proposed by Hertzsprung. Russell suggested that the process starts with red giants, which contract and become hotter, according to the perfect gas equations, to develop into bluish-white type A and B stars, at densities that are then too high to behave like a perfect gas. The stars then cool and contract, progressively changing colour from bluish-white, to yellow, to orange, to red, ending their stellar lives as red dwarfs.

Russell recognised that his theory was similar in some respects to the evolutionary theories put forward by Ritter and Lockyer in the nineteenth century (see above), but his proposal had the advantage of clearly differentiating between those stars on the heating part of the curve, and those on the cooling part.

In 1911, Hertzsprung published colour-magnitude diagrams for stars in the Pleiades and Hyades star clusters, using both "effective wavelengths" (deduced from spectral work) and "colour indices" (deduced by comparing photographic and visual magnitudes of stars, see Page 302) as measurements of colour. Russell, in ignorance of Hertzsprung's new work, presented a paper to the Royal Astronomical Society, in June 1913, where he showed his first "Russell" colour-magnitude diagram, as it became known. In this diagram (see Figure 7.3) Russell used the Harvard spectral type, in the order defined by Miss Cannon, as the index of colour.

---

*Wien's law enables the temperature of a star to be determined by measuring the wavelength at which it emits the most energy. From this temperature, Stefan's law enables the total energy emitted per unit area to be calculated. If the distance of a star is known, the energy emitted by the star can be measured, and the surface area can then be calculated.

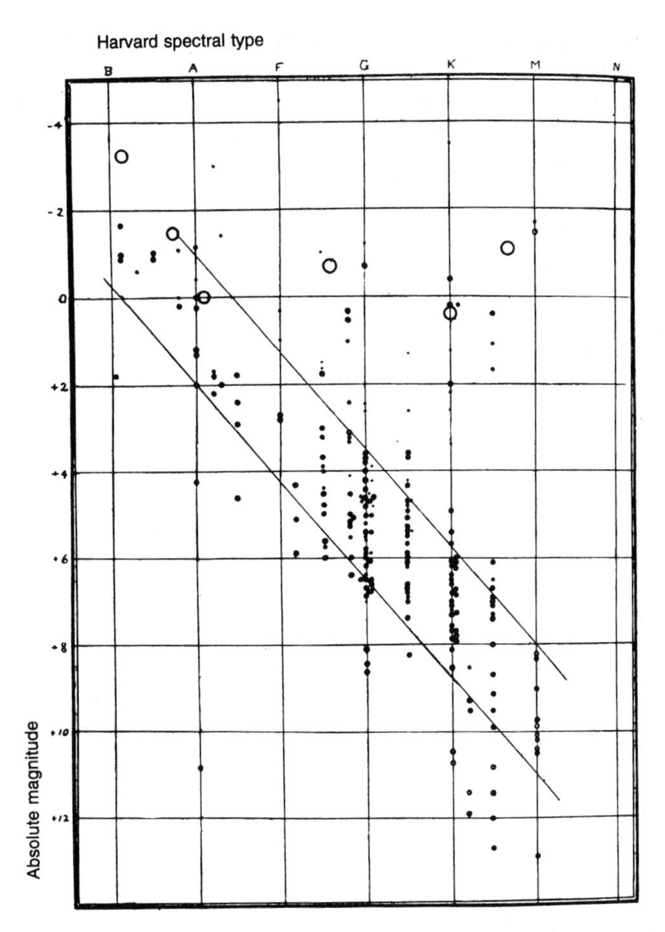

Figure 7.3 *The first H-R diagram produced by Henry Norris Russell. The Main Sequence is indicated by the two inclined parallel lines. (From* Nature, *Vol. 93, (1914), p. 252; reprinted with permission from* Nature. *Image supplied by the Royal Astronomical Society.)*

Russell's diagram was much easier to interpret than that produced by Hertzsprung two years earlier, but Russell was quite happy to acknowledge Hertzsprung's prior analysis. In 1933, the Scandinavian astronomer Bengt Strömgren referred to the "Hertzsprung–Russell diagram" when discussing Russell's results, and that (or the H–R diagram) is what it has been called ever since. Hertzsprung, when asked of his view of the late recognition of his work, is reputed to have said "Why not call it the colour-magnitude diagram? Then we would all know what it is about."

The H–R diagram shows a "main sequence" of stars running from giant bluish-white B stars with an absolute magnitude of about −1, to dwarf red M stars with an absolute magnitude of about 10. In addition, there are giant stars (now called supergiants) of all colours with absolute magnitudes of about 0, these being the c stars listed by Antonia Maury, containing what Lockyer thought to be proto-elements.

Russell found, in 1912, by analysing the light curves of eclipsing binary stars, that the average masses of stars on the main sequence were smaller, the further down the main sequence they were. This implied that stars lost a significant amount of mass as they cooled and progressed down this sequence, which seemed unlikely. This caused Russell to modify his theory in the following year, suggesting that only the most massive stars would reach the hottest temperatures on contraction, along the lines proposed by Ritter some years earlier. So only the most massive stars joined the main sequence at the top (from the right hand side), with the less massive stars joining the main sequence lower down (also from the right hand side), before following this sequence down as they contracted and cooled. Thus the average mass would decrease down the main sequence, as observed.

The stars are so far away that they were not resolved into discs in even the largest telescopes. In 1890, however, the American physicist A. A. Michelson had proposed that the size of stars could, in principle, be measured using an interferometer (see Page 306), in which two widely separated small mirrors simulate, from a resolution point-of-view, one large mirror of diameter equal to the separation between the small mirrors. The system was used to measure the diameters of the Galilean moons of Jupiter, but the idea was put "on hold", as far as measuring the diameters of stars was concerned, as it seemed that the mirror separation required would be too large for even the largest and closest stars to us. The idea was resurrected, however, when the existence of giant stars was established. In particular, George Ellery Hale decided, in 1917, that a Michelson interferometer should be built to operate with the newly-constructed 100 inch (2.5 m) Hooker telescope on Mount Wilson and, 3 years later, John A. Anderson and F. G. Pease succeeded in measuring stellar diameters with it. Mira gave a diameter of 0.056 arcsec, Betelgeux 0.047 arcsec and Antares 0.040 arcsec, which were equivalent to diameters of 400, 400 and 600 million km, or about 300, 300 and 450 times that of the Sun respectively. If the Sun were the size of Antares, for example, the orbit of Mars would be appreciably below the Sun's surface. This was the first direct evidence for, and was a crucial confirmation of, the existence of giant or supergiant stars that had been deduced earlier by Hertzsprung and Russell.

Walter Adams and Arnold Kohlschütter showed, at the Mount Wilson Observatory, that the intensity of certain lines in the spectra of stars correlated closely with the luminosity of those stars. They found, in 1914, that this correlation was accurate enough for F, G and K stars, allowing the absolute magnitude of these stars, and hence their distance, to be deduced from the intensity of these lines. This powerful new method for deducing stellar distance from the intensities of certain spectral lines, which was given the misleading name of "spectroscopic parallax", enabled the distances to be determined of stars that were too far away to produce a reliable ordinary parallax measurement. For stars of the Sun's absolute magnitude they could measure distances of up to 400 light years, but for the brightest supergiants distances of up to 40,000 light years could be determined.

Ionisation and the Abundance of Hydrogen in Stellar Atmospheres

In 1913, Niels Bohr proposed his theory of atomic structure (see Page 295), explaining how spectral lines were produced. In particular, he showed that the Pickering lines seen in some O stars were due to ionised helium, and as B stars exhibit the lines of un-ionised helium, then the O stars must be hotter than the B stars. Furthermore, the stellar spectra, that Lockyer had thought were due to proto-elements, were seen to be due to various ionised elements, including helium.

In 1920, Megh Nad Saha, an Indian physicist working in London, was able to show how the degree of ionisation of an atom depended on temperature and pressure. He also showed that atmospheres of the same composition, when heated to higher and higher temperatures, would exhibit spectra shown by the stellar sequence M, K, G, F, A, B and O, i.e. the sequence deduced by Antonia Maury in 1897 and confirmed by Annie Cannon. The hydrogen absorption lines, for example, become more evident for hotter stars because their gas is more ionised, not because there is more hydrogen present.

How could O-type stars of apparently the same colour and, therefore, temperature, have lines of different intensity, implying different degrees of ionisation? Megh Saha's theory showed that ionisation increased with decreasing pressure, and so the high intensity of the ionized lines of helium in very luminous O-type stars was due to very low pressure. The presence of Pickering lines of ionized helium in some O-type stars thus showed them to be low pressure blue-white giants. Similarly, analysis of spectral lines in high luminosity stars of other colours showed that some of these were also very low pressure giant stars, with some red giants having an average density of only one millionth that of the Sun.

Saha showed that differences in spectral type could be attributed to differences in temperature and pressure, but how could the abundances of the elements in the stellar atmospheres be determined from their spectra? In 1920, it was still generally assumed that the proportions of elements in stars were similar to those of the Earth, but was this correct?

In 1923, Arthur Milne and Ralph Fowler, who were working with Arthur Eddington at Cambridge University, developed equations from which the percentage of an element's atoms capable of absorbing radiation of a given wavelength could be calculated, from the temperature and pressure. Cecilia Payne was, at this time, also studying at Cambridge as an undergraduate student of Milne's but, later that year, she moved to Harvard, because she felt that she would have had limited prospects as a woman in British research. At Harvard she had access to an incredible collection of stellar spectra, which she analysed using Milne and Fowler's theory, making two assumptions, namely that:

(i) All stars have approximately the same proportion of elements.
(ii) All spectral lines at the limit of visibility are produced by the same number of atoms, given fixed observing conditions.

Using this analysis Cecilia Payne showed, in 1925, that hydrogen and helium are the most abundant elements in stellar atmospheres, but she then dismissed this finding as spurious because Henry Russell expressed strong doubts as to its validity. He still believed that stars had similar element abundances as the Earth.

Russell, helped by Charlotte Moore at Princeton University and Walter Adams at Mount Wilson, investigated the transition probabilities for two or more closely-spaced spectral lines produced by any one element, in so-called spectral multiplets. This analysis also enabled the relative abundances of the elements to be determined in the Sun and stars and, in 1927, it showed that there were large amounts of hydrogen present, but again Russell was unconvinced. In the following year, however, he met the German physicist Albrecht Unsöld at the Mount Wilson Observatory, who had independently shown that there were large amounts of hydrogen in the Sun by analysing the intensity profiles of the darkest lines. By 1929, not only had Russell accepted these findings for hydrogen's abundance, but he had also written a paper outlining the evidence, namely:

(i) Cecilia Payne's analysis of the relative abundances of elements in the stars and Sun.
(ii) Unsöld's analysis for the Sun which yielded results similar to those of Miss Payne for many elements.
(iii) Donald Menzel's analysis, just completed at the Lick Observatory, of the flash spectrum of the Sun, showing that the gas in the chromosphere has an average atomic weight of about 2, indicating that it must have a large amount of hydrogen (of atomic weight 1) to balance the smaller amount of much heavier elements (of atomic weights 3 and above).
(iv) The very dark hydrogen lines in the solar spectrum, seemingly caused by a relatively small *proportion* of the hydrogen atoms, indicating that the number of hydrogen atoms must be large.

Russell's paper on the abundance of hydrogen in the Sun's atmosphere was published in July 1929. Whether hydrogen was so prevalent in the interior of the Sun was left open to question, however.

Cecilia Payne had assumed that all stars have approximately the same proportion of elements in their atmospheres, and Megh Saha had shown that differences in spectral type from M to O could be attributable solely to differences in the temperature and pressure of the stellar atmospheres. So it was thought that maybe all stars have the same elements in approximately the same proportions in their atmospheres.

Doubts about this hypothesis were expressed in about 1930 by Ralph

Curtiss, of the University of Michigan, and Russell, for the cool K, M, R and N stars, as the K and M stars showed strong titanium oxide bands, while the R and N stars showed strong carbon and cyanogen (CN) bands. It was thought that the atmospheres of the K and M stars had, therefore, a higher oxygen to carbon ratio than those of the R and N stars.

Carlyle Beals of the Dominion Astrophysical Observatory near Victoria, Canada also found, in 1930, that Wolf–Rayet stars separated into two groups; those with intense nitrogen emission lines, and those with very strong carbon and oxygen and very weak nitrogen emission lines. He suggested that this showed real differences in chemical abundances but, in 1958, Anne Underhill, also of the Dominion Observatory, explained how they could be due to temperature differences.

The theory of supergiants was given an added impetus in 1956, when Armin Deutsch at Mount Wilson deduced that the M5 supergiant star α Herculis was surrounded by a huge chromosphere that was losing mass at the rate of about 10^{-8} solar masses per year. He speculated that maybe supergiant stars may be losing enough mass to replenish the interstellar medium, which would otherwise become exhausted because of the formation of new stars. Novae, supernovae and Wolf–Rayet stars were also known to be losing mass to the interstellar medium and, in the 1950s, suggestions were made that these may be the origin of the elements heavier than helium found in relatively young, or middle-age, stars like the Sun. It could also explain why the youngest stars in the Milky Way seemed to have more heavy elements than the older stars, as the proportion of heavy elements in the interstellar medium was increasing with time. An alternative proposal had been put forward in 1938, however, by Carl von Weizsäcker in Germany who had proposed that the heavier elements may have been produced (see Page 143) in the explosion that started the expansion of the universe.

The Surface Temperature of Stars

Wien's law for ideal radiating sources, or "black bodies" as they are called, enabled the surface temperature of stars to be determined from their colours; the redder the star, the lower the temperature. Julius Scheiner and J. Wilsing at Potsdam, between 1905 and 1910, and Hans Rosenberg in Tübingen in 1914, tried to use colour to estimate the surface temperature of about 100 stars, only to discover that their results were inconsistent because stars are not ideal black bodies, and so various corrections had to be made. After these corrections, the surface temperatures of A0 stars were seen to be about 10,500 K, and of M0 stars about 3,000 K.

It was difficult to measure the temperatures of the B and O stars in this way, as there were relatively few of them, and they were some distance away, causing their colours to be significantly modified by the intervening interstellar dust. A new technique was required and, in 1924, Cecilia Payne established the temperature of B5 stars as 15,000 K, B0 as 20,000 K, O9 as

25,000 K, and O8 as 30,000 K by measuring the intensities of the singly, doubly and trebly ionised silicon lines in their spectra.

Unfortunately, there was still a problem with the hotter stars as their radiation is mostly in the ultra-violet, which is absorbed by our atmosphere, and yet another method was required to estimate their surface temperatures. O stars were often found to be either surrounded by a planetary nebula, or embedded within a nebula, like the four "Trapezium" stars in the Orion nebula. The energy from these O-type stars excites the surrounding nebula (see Pages 228–229), and H. Zanstra and Louis Berman found, independently, that they could deduce the temperature of the O stars from the relative intensities of the hydrogen, oxygen and nitrogen lines of that nebula. This led to estimates ranging from 30,000 K for the trapezium stars in the Orion nebula, to more than 100,000 K for a few stars.

At the other end of the spectrum, the temperatures of many red M stars and the deep-red N stars could not be estimated by using simple colour measurements, because most of their energy is in the infra-red. Their temperatures, which were found to range from about 1,000 to 3,000 K, were deduced from measuring the intensities of their titanium dioxide bands, or the magnitude of their heat indices (see Page 305). The coolest of these stars are often so dim that they can only be seen on infra-red photographic plates, or be detected by infra-red sensors.

The Internal Structure of Stars

In the nineteenth century it was generally assumed that the primary heat transport mechanism in stars and the Sun was convection, but, in 1894, the British astronomer R. A. Sampson suggested that radiative transfer was more important in the solar atmosphere. In Germany, Karl Schwarzschild developed a model of the solar atmosphere, assuming a predominance of radiative transfer, and explained the limb darkening effect in 1906. Ten years later in England, Arthur Eddington extended Schwarzschild's theory to the interior of stars, where he assumed that the pressure of radiation travelling to the surface adds to the gas pressure in balancing the star's gravitational attraction. Eddington's initial work was on low-density giant stars, which he assumed were sufficiently diffuse to allow the use of the perfect gas laws, and for these he calculated a central temperature of 7 million K, assuming an average atomic weight of 54. Newall, Jeans and Lindemann independently pointed out, however, that, at such high temperatures, most atoms would be heavily ionised, which meant that there would be many free electrons, and so his assumption on average atomic weight was incorrect. Eddington, accordingly, modified his calculations, and produced a central temperature of 5 million K for a stellar giant.

Eddington now applied himself to the problem of calculating the luminosity of a star, knowing the temperature gradient throughout its interior, which he had just calculated. Photons of light that are emitted by atoms in the interior of a star are rapidly absorbed by other atoms or

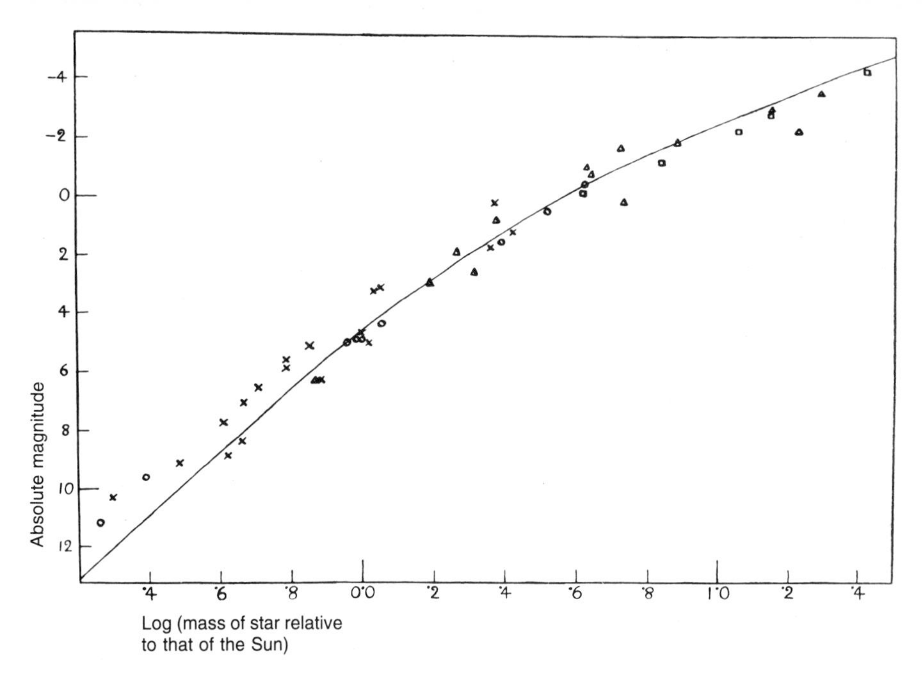

Figure 7.4 *The mass–luminosity diagram for stars as first produced by Arthur Eddington. (From* Monthly Notices of the Royal Astronomical Society, *Vol. 84 (1924), Plate 8; courtesy* The Royal Astronomical Society.*)*

electrons, long before they can reach the surface. These atoms and electrons re-emit photons and, eventually, after many more absorptions and re-emissions, the photons escape from the star. This "resistance" to the outward flow of energy from a star is called the opacity of a star, and the theory that enabled the opacity to be calculated theoretically was published by H. A. Kramers in 1923.

Eddington showed in his theory for giant stars, using the perfect gas laws, that the luminosity of a star is determined almost completely* by its mass (see Figure 7.4). He was surprised to find, however, that, when checking his theoretical predictions with real stars, main sequence† stars fitted on exactly the same mass–luminosity curve as the giant stars, clearly indicating that, contrary to his expectations, the gas in main sequence stars also behaves as a perfect gas, even though it is highly compressed. Eddington realised that this was because the high ionisation in stellar

*If two stars have the same mass, but different diameters, for example, the smaller star would emit slightly more light, as its brighter surface (caused by its higher density) more than compensates for its smaller size. This effect of diameter on luminosity is, however, trivial when compared with the effect of mass.

†Main sequence stars are often called dwarf stars, but this can be misleading as *white* dwarf stars are not main sequence stars. To avoid confusion, therefore, I have used the term "main sequence" rather than "dwarf" stars in this book.

interiors proposed by Megh Saha, James Jeans, Ralph Fowler, Arthur Milne and others allowed all stars, except white dwarfs, to behave as if they were made of perfect gases. In the interior of main sequence stars, the highly ionised atoms are much smaller than normal atoms, so they can move around the dense interiors of these stars, just as if they were normal atoms in a normal gas.

Eddington presented his analysis of stellar structures to the Royal Astronomical Society in March 1924, together with the resulting mass–luminosity correlation for both giant and main sequence stars. This correlation was consistent with the effect found observationally by Russell (see above), that the average mass of stars on the main sequence reduces with reducing luminosity. It meant, however, that Russell's modified evolutionary sequence was incorrect, as stars could not go down the main sequence unless they lost an appreciable amount of mass on the way, which seemed unlikely. Russell had also assumed that, as the stars went down the main sequence they contracted, and cooled, as they became too dense to follow the perfect gas laws. Eddington had shown that, at these orders of density, stars would still follow the perfect gas laws, and would thus heat up, rather than cool down, as they contracted. So Russell's theory, as it stood, was no longer viable.

According to Einstein's theory of relativity, matter can be annihilated to create energy, and so stars could progress down the main sequence and lose mass, creating energy in the process. Jeans calculated stellar lifetimes of a few thousand billion years if the energy was created by the annihilation of matter, resulting in an incredibly slow progression down the main sequence, although the mass loss would be appreciable (as Russell's theory required).

Russell adopted this concept in his modified theory of stellar evolution. In this, stars start life as red giants with central temperatures of one or two million degrees, but at this temperature the energy produced by the annihilation of matter is negligible. Radiation pressure is small at this stage, and so the giants contract rapidly under gravity, and heat up, as expected by the laws for a perfect gas. As the stars become smaller, the temperature in the centre becomes sufficient for the mass annihilation process to start to produce appreciable amounts of energy. As more energy is produced, the radiation pressure increases and gradually stops the rapid contraction phase. At this stage the star is a white giant at the top of the main sequence, having moved virtually horizontally from the top right to the top left of the H–R diagram.

The star now loses mass very slowly, as it gradually follows the main sequence down, gradually becoming less luminous, in accordance with the mass–luminosity law. During this phase the star is still gradually contracting and increasing in density. This higher density increases the opacity of the star, and so its surface temperature reduces.

In the alternative theory to Russell's, stars start their lives at various points on the right of the H–R diagram (as red stars of various masses), move to the left as they become more heated, until they meet the main sequence, where they spend most of their lifetime, before moving to the right of the diagram once again. Stars do not move down the main sequence, in this theory, and so there is no requirement for them to lose a significant amount of their mass during their lifetime. It, therefore, has the advantage over Russell's theory in that it applied no matter which* proposed method of energy generation is correct.

In 1927, Russell finally abandoned his concept of the main sequence being an evolutionary track, as the alternative theory seemed to be more convincing. The discovery, 2 years later, of the expansion of the universe virtually finished Russell's theory, as the universe is not old enough for stars to have moved any significant distance down the main sequence due to mass annihilation. As usual there were still a few die-hards who favoured the longer age of the universe and the concept of energy production by mass annihilation. They disputed the concept of the expansion of the universe, and pointed out that the age of the universe determined from that theory was less than the age of the Earth (see Page 253). They also pointed out that, in the theory of relativity the universe can either expand or contract, and maybe it could pulsate, alternating between expansion and contraction.

In 1924, Eddington had hit a problem in his theory which established the mass–luminosity relationship, as the luminosities of 45 out of the 46 stars that he had considered were all about ten times too high. He put this discrepancy down at the time to a fault in Kramers' opacity theory, and he decided to work backwards and calculate the value of the opacity required to make the calculated and observed luminosities of the star Capella the same. He then applied this correction to his calculated luminosities for all the other stars, and found a perfect fit between theory and observation for all but one star, the faint companion to Sirius, known as Sirius B.

We must now retrace our steps a little to 1910, when Russell was undertaking the background research for his work on stellar populations. He was intrigued by one star, σ^2 Eridani, that had two faint companions, one white and the other red. The white star, called 40 Eridani B, was strange, as it was not on either the main sequence or giant bands of his diagram, instead appearing as the lonely point in the bottom left hand part of the diagram (see Figure 7.3 earlier). It had a luminosity of only 1/400 times that of the Sun, while having a mass 0.4 times that of the Sun, so it had a very low luminosity for its mass, which was very unusual for a white star. In 1914 Walter Adams, at Mount Wilson, confirmed that it was a very low luminosity white star by measuring its spectrum as type A0.

Sirius B was known in the nineteenth century to be another star of low luminosity but normal mass (see Page 157), but its colour could not be determined because it was too close to its very bright Sirius A primary. Sirius B was assumed to be a low luminosity red M type star but, in 1914, Adams decided to try to measure its spectrum, spurred on by his success in measuring the spectrum of 40 Eridani B. In the following year, he found that

---

*Mass annihilation or hydrogen transmutation (see Page 141).

Sirius B was not a red M type star but was a white A0 type star like 40 Eridani B. It was known that A type stars have a very high surface brightness, so if the low luminosity of Sirius B and 40 Eridani B was simply due to their size, they had to be very small and very dense. Eddington estimated that Sirius B, for example, was only three times the diameter of the Earth, which implied an incredible density of about 50,000 times that of water. Clearly, the big question was why stars of normal mass have such low luminosities. Were they physically very small, in which case they must have an incredibly high density, or were they only radiating from a very small part of their surface for some reason? Either solution seemed unlikely in 1915, and astronomers began to wonder if there was something fundamentally wrong with their theories that had led them to these conclusions.

It was mentioned above, that Eddington found that Sirius B was the one star that did not fit his mass–luminosity law. Eddington had realised that the gas in main sequence stars behaved like a perfect gas because ionised atoms are much smaller than normal atoms. At the same time he realised that atoms that have been stripped of their electrons could be packed at the very high densities calculated for white dwarfs, so such very high densities were not impossible. Fortunately, there was one way of observationally checking whether these very high densities really existed in these stars that Eddington called "white dwarfs". If Sirius B and 40 Eridani B are very dense white dwarfs then, according to Einstein's theory of relativity, there should be a very small relativistic change in the frequency of their spectral lines, caused by the very large gravitational forces produced at the surface of these stars. Adams found this effect for Sirius B in 1925, confirming it to be of very high density, when he deduced that it had a red shift, unaccounted for by its binary nature, of 21 km/s, compared with Eddington's estimate based on relativity theory of 20 km/s.

In the mid 1920s, when Eddington was developing his theories, the range of stellar masses known was about 1/5 to 100 times that of the Sun. He reasoned that, as the radiation pressure inside a star is proportional to the fourth power of its temperature, there must be a temperature inside a star at which the radiation pressure exceeds the gravitational force holding the star together, and the star would then explode. Eddington concluded that this would happen if the star was about 100 times the mass of the Sun, thus explaining the observed upper limit on stellar masses. A few years later, in 1928, Milne showed theoretically that, under some conditions, stars would collapse, throwing off a shell of gas. This could explain why novae were often surrounded by expanding shells of gas shortly after they had exploded (see Pages 181–182), and why some dwarf O type stars were at the centre of a planetary nebula, the nebula being the remnant of an earlier explosion of the O type star. Thirty years later, in 1958, Martin Schwarzschild, the son of Karl Schwarzschild, and R. Härm showed that pulsational instability must disrupt any star whose mass is greater than about 65 times that of the Sun.

Russell and others had shown in the late 1920s that hydrogen was by far the most common element in the visible part of the Sun, and probably in the atmosphere of most stars, and yet Eddington's theory of stellar structures had been based on the assumption that the mix of elements in stars was similar to that of the Earth. Eddington was reluctant to concede that this may be wrong but, in 1932, he decided to see what would happen if he assumed that there was a high proportion of hydrogen in the interior of stars. Much to his surprise, the change in assumption improved his theory, solving the problem that he had had predicting the actual luminosities of stars. Kramers' opacity theory was correct after all, and Eddington no longer had to correct his opacity calculations, thus persuading Eddington that hydrogen was the most abundant element in the interior of stars, as well as in their atmospheres.

The Source of Energy in Stars

In the nineteenth century the best theory of the generation of heat in stars was the Waterston–Helmholtz contraction theory (see Page 6), but this could only explain lifetimes of the order of tens of millions of years, which even at that time appeared to be too short. In 1905, however, the brilliant young German physicist Albert Einstein gave the first clue of an alternative source of energy when he published his special theory of relativity in which he proposed the equivalence of mass and energy, i.e. that mass could be transformed into energy according to the relationship $E = mc^2$, where E is energy, m is mass and c is the velocity of light. It was immediately realised that, if the whole mass of a star could eventually be transformed into energy in this way, it would be sufficient to explain lifetimes running into a few thousand billion years.

In 1926, Eddington outlined an alternative to the theory of the complete annihilation of matter in stars, when he suggested that four hydrogen nuclei (i.e. protons) and two electrons could combine to form a helium nucleus. He calculated that the energy released in such a process, in which the star loses about 1% of its mass, would be enough to keep a star shining for a few tens of billions of years. Three years later, Edwin Hubble at Mount Wilson showed that most galaxies were receding from us at velocities proportional to their distance (see Page 236), and if these velocities had been constant with time, then the galaxies would all have been in the same place about 2 billion years ago. This timescale militated against the particle annihilation theory of stellar energy production, as that implied much longer stellar lifetimes, and so the building of helium from hydrogen (i.e. hydrogen transmutation) seemed the more likely energy source in stars.

Ernest Rutherford had shown experimentally in 1919 at the Cavendish Laboratory in Cambridge how particles can be knocked out of nuclei, but, unfortunately, the temperatures in the interiors of stars, even of the order of millions of kelvins, as calculated by Eddington, seemed to be too low to allow hydrogen transmutation to take place. Were Eddington's temperature calculations for the interiors of stars correct, or was there an alternative energy process to hydrogen transmutation?

In 1928, George Gamow, a young Soviet physicist working with Rutherford, started to investigate how alpha particles (helium nuclei) could be emitted during radioactive decay, and how they could enter other nuclei. When alpha particles approach nuclei, they meet a repulsive force which increases in magnitude the closer the alpha particles approach, because both the alpha particles and the nuclei have positive electrostatic charges which repel each other. If the alpha particles do not have enough energy to overcome this repulsive force then, according to classical mechanics, they cannot enter the nucleus, but Gamow showed that quantum mechanics allows a proportion of alpha particles to penetrate this repulsive barrier.

The English physicist Robert Atkinson decided to investigate how this principle could be applied in stars and, working with Fritz Houtermans in Berlin, suggested in 1929 that some protons (i.e. hydrogen nuclei) should be energetic enough to penetrate nuclei in the central part of stars, at temperatures of tens of millions of degrees calculated by Eddington. Atkinson moved to Rutgers University in the United States in 1931, followed by Gamow who defected from the Soviet Union in 1933, and Edward Teller a Hungarian physicist who left Germany 2 years later. These three émigrés and others attempted, over the next few years, to explain the details of the nuclear processes leading to the production of helium, as it was clear that the simple amalgamation of four protons and two electrons would not work. In 1938, Gamow arranged the Fourth Washington Conference on Theoretical Physics, with the stellar energy problem as the main topic on the agenda, and at this conference, Teller's student, Charles Critchfield, outlined his ideas on a chain of reactions starting with proton-proton collisions and ending with the synthesis of helium nuclei. Hans Bethe, an eminent theoretical physicist and German émigré to the United States, was fascinated by the problem and decided, before the end of the conference, to collaborate with Critchfield in an attempt to solve it.

Eddington had, in the meantime, reduced his estimate of the central temperature of the Sun from 39 million K (in 1926) to 19 million K.

Bethe analysed Critchfield's proton–proton chain and concluded, in 1938, that it would be the predominant source of energy for stars that were less massive, and thus cooler, than the Sun. He then proposed that a chain reaction using carbon as a catalyst would predominate in the Sun and more massive stars, and calculated a central temperature of the Sun of 18.5 million K, very close to Eddington's value of 19 million K deduced using gas dynamics. Bethe also estimated that, based on this cycle, the Sun would exist for another 12 billion years.

The two chain reactions worked as follows:

Proton–proton cycle

$$^{1}H + {}^{1}H \rightarrow {}^{2}H + e^{+} + \nu$$
$$^{2}H + {}^{1}H \rightarrow {}^{3}He + \gamma$$
$$^{3}He + {}^{3}He \rightarrow {}^{4}He + {}^{1}H + {}^{1}H$$

where ^{1}H = hydrogen nucleus (i.e. a proton)

^{2}H = deuterium nucleus
e^{+} = positron (or positive electron)
ν = neutrino
^{3}He = helium 3 nucleus
γ = γ-ray
^{4}He = helium 4 nucleus

Carbon-nitrogen-oxygen (CNO) or carbon cycle

$$^{12}C + {}^{1}H \rightarrow {}^{13}N + \gamma$$
$$^{13}N \rightarrow {}^{13}C + e^{+} + \nu$$
$$^{13}C + {}^{1}H \rightarrow {}^{14}N + \gamma$$
$$^{14}N + {}^{1}H \rightarrow {}^{15}O + \gamma$$
$$^{15}O \rightarrow {}^{15}N + e^{+} + \nu$$
$$^{15}N + {}^{1}H \rightarrow {}^{4}He + {}^{12}C$$

In the first cycle four hydrogen nuclei are transformed into a helium 4 nucleus, releasing two positrons, two neutrinos and two γ-rays in the process. In the second cycle, in the presence of carbon as a catalyst, four hydrogen nuclei are transformed into a helium 4 nucleus, releasing two positrons, two neutrinos and three γ-rays. Thus, either cycle can be used to explain how hydrogen is transformed into helium, with energy being released mainly in the form of γ-rays, but where did the elements in stars heavier than helium come from?

Carl von Weizsäcker, who independently discovered the carbon cycle in Germany in 1938, suggested that these middleweight and heavy elements, which I have shortened to "heavy elements" in the current discussion, could have been made in the gigantic explosion, the Big Bang, that had caused the recession of the galaxies. This idea was examined in more detail in the 1940s and, in 1948, Gamow and his students Ralph Alpher and Robert Herman proposed a theory in which the early universe, just after the Big Bang, consisted of a sea of neutrons which decayed to produce protons and electrons. Some of these protons captured neutrons to form heavier nuclei, and some of these nuclei were changed by the emission of an electron. These new nuclei then captured more neutrons to make even more complex nuclei. The nuclei of all the elements were produced in the first 2 hours after the Big Bang by this process and, as the universe continued to cool, the nuclei captured electrons to form atoms. Unfortunately, it was soon realised that the chain broke down at helium, as the nucleus produced after the capture of a neutron by a helium nucleus was unstable, and would immediately disintegrate.

Fred Hoyle had been working at Cambridge on the problem of the production of heavy elements since 1946. He was a proponent of the Steady State theory (see Page 253), in which hydrogen atoms are created spontaneously throughout the universe, and was very much opposed to the concept of the Big Bang*. It was essential, therefore, for him to find a way

---

*Hoyle was the first to use the term "Big Bang" to denote this theory, as he thought the theory was somewhat childish in concept.

for heavy elements to be formed, without relying on the conditions in the early universe after the Big Bang.

Other astronomers had also wondered if heavy elements are produced in stars and, although there had been an enormous amount of circumstantial evidence of this, the first direct proof was furnished in 1952 by Paul Merrill's discovery of technetium in red giants. An atom of technetium has 43 protons, compared, for example, with iron's 26, so it is quite a heavy element. The most important thing, however, was that technetium has a maximum half-life of a million years, so it could not have been present in red giants when they were formed billions of years earlier.

Triple-α process

$$^{4}He + {}^{4}He \rightarrow {}^{8}Be + \gamma$$
$$^{8}Be + {}^{4}He \rightarrow {}^{12}C + \gamma$$

Ernst Öpik and Edwin Salpeter suggested that the heavy elements are formed in the interiors of very hot stars, by the fusion of three helium 4 nuclei to form carbon 12, and by similar processes for heavier elements. The detailed process (called the triple-α process*) for the production of carbon 12 was worked out by Edwin Salpeter and Fred Hoyle, in 1956.

Salpeter showed that the formation of beryllium needs to take place at temperatures of 100 million K, much higher than those in the core of the Sun, otherwise too little beryllium would be produced for the further capture of a helium nucleus to take place with any sensible level of probability. So the triple-α process could only take place in the interiors of heavy stars, as only they had the required very high temperatures. Hoyle then showed that there must be an excited nuclear state of carbon possible, that had not previously been predicted, in order for the generation of carbon to occur in the way outlined above. The energy released by the triple-α process is 0.6 MeV per atomic mass unit, or about 10 times less than in the proton–proton cycle.

Other processes

Salpeter and Hoyle suggested that oxygen could also be produced by the fusion of carbon and helium as follows:

$$^{12}C + {}^{4}He \rightarrow {}^{16}O + \gamma$$

This reaction takes place at slightly higher temperatures than the tripleα process, and generates slightly less energy per atomic mass unit.

The classic paper, known as the B^2FH paper, on the production of the medium-weight and heavy elements in stars, was published in 1957 in *Reviews of Modern Physics* by Geoffrey and Margaret Burbidge, and Willy Fowler, all of Caltech, and Fred Hoyle. In it they explained that elements

---

*So-called as a helium nucleus is called an α particle.

could be produced in stars by a number of different processes, in addition to element fusion, as follows:

(i) The *e* process, in which *electrons* are absorbed and re-emitted.
(ii) The *s* process, in which neutrons are added *slowly* to middleweight nuclei, which have time to decay before the next neutron is added.
(iii) The *r* process, in which neutrons are added *rapidly* to middleweight nuclei, so that the nuclei do not have time to decay before another neutron is added.
(iv) The *p* process, in which *protons* are added to nuclei formed by the s and r processes.
(v) The *x* process, which was an unknown process which created stable lithium, beryllium and boron. At the time these were the problem elements, as a reasonable theory did not exist for their production. (It was later suggested that these elements are not created in stars, but in interstellar space by the collision of cosmic rays with interstellar hydrogen and helium. See Page 52.)

Modern theories of element production in stars, other than by element fusion, are based on those proposed in this B^2FH paper. The s and r processes are crucial, as they are the only way in which elements heavier than iron can be produced. The s process can operate in heavy stars, but the r process can only occur in supernova explosions where enormous amounts of neutrons are available. We are, unfortunately, now getting ahead of ourselves, and we must return to the subject of stellar classification.

The MKK and BCD Classification Systems

When the Harvard stellar classification system was first introduced in 1890 it was attacked as being too complicated, but, by the 1930s, developments in spectroscopic and stellar research had overtaken it, and astronomers were beginning to find it too simple.

Morgan, Keenan and Kellman

Hertzsprung had shown in 1905 that the stars with the sharpest lines (the c type stars of Antonia Maury) had a very high absolute brightness. Nine years later Adams and Kohlschütter had extended this approach to main sequence stars when they showed that the absolute magnitudes of F, G and K stars could be determined from the intensity of some of their spectral lines.

In 1929, Gwyn Williams, who was then at Cambridge University, showed that the strength of the hydrogen absorption lines in B0 to A5 stars was closely related to stellar luminosity. The most luminous stars were found to have relatively weak but sharp hydrogen lines in their spectra, while main sequence stars had wide and intense hydrogen lines. A number of astronomers carried out similar examinations into the correlation between spectral lines and the luminosities of stars in the 1920s and early 30s, and this enabled another new stellar classification system to be developed, the MKK system, named after William W. Morgan, Philip Keenan and Edith Kellman, its proposers.

Ralph Curtiss suggested in 1929 that the Harvard system needed modification as it was a one-dimensional system, and he thought that at least three dimensions or parameters were required to classify stars. Four years later, Otto Struve, a Ukrainian émigré and director of the Yerkes Observatory, voiced similar concerns in a paper written for the *Astrophysical Journal*. In fact, W. W. Morgan, also of Yerkes, had already started work to understand stellar spectra and, in 1938, he added another dimension to the Harvard system by introducing five luminosity classes, labelled I to V, to cover the range of luminosities from supergiants to main sequence stars, as follows:

Ia	Most luminous supergiants
Ib	Less luminous supergiants
II	Bright giants
III	Normal giants
IV	Subgiants
V	Main sequence stars

In 1943, the Yerkes astronomers Morgan, Keenan and Kellman published their *Atlas of Stellar Spectra* as a special publication of the *Astrophysical Journal*. In 55 plates, it presented examples of the spectra of various Harvard types and luminosity classes, and presented the classification criteria used for each spectral type and subtype. The criteria used for the luminosity classification were line intensity ratios, mainly of neutral to ionised atoms.

The relationship between luminosity type and absolute magnitude is somewhat complicated as, although all Ia stars have an absolute magnitude of about -7 (plotting horizontally at the top of the H–R diagram), category III stars have absolute magnitudes ranging from -6 to $+1$, and category V, the main sequence stars, from -5 to $+14$.

Barbier, Chalonge and Divan

Near the end of the nineteenth century, William Huggins had observed, on spectrograms of Vega and stars of a similar type, a sudden fall in the intensity of the continuous spectrum beyond the end of the series of Balmer lines in the ultraviolet (i.e. at wavelengths below about 365 nm). This so-called "Balmer jump" (see Figure 7.5), which showed that stars did not radiate as ideal black bodies, was also observed by W. H. Wright at the Lick Observatory in 1918, and Wright's student, Ch'ing Sung Yü, was the first to study it in detail in 1926. Yü found that the wavelength of the jump did not coincide exactly with the end of the Balmer series at 364.7 nm, and the magnitude of the jump varied from star to star.

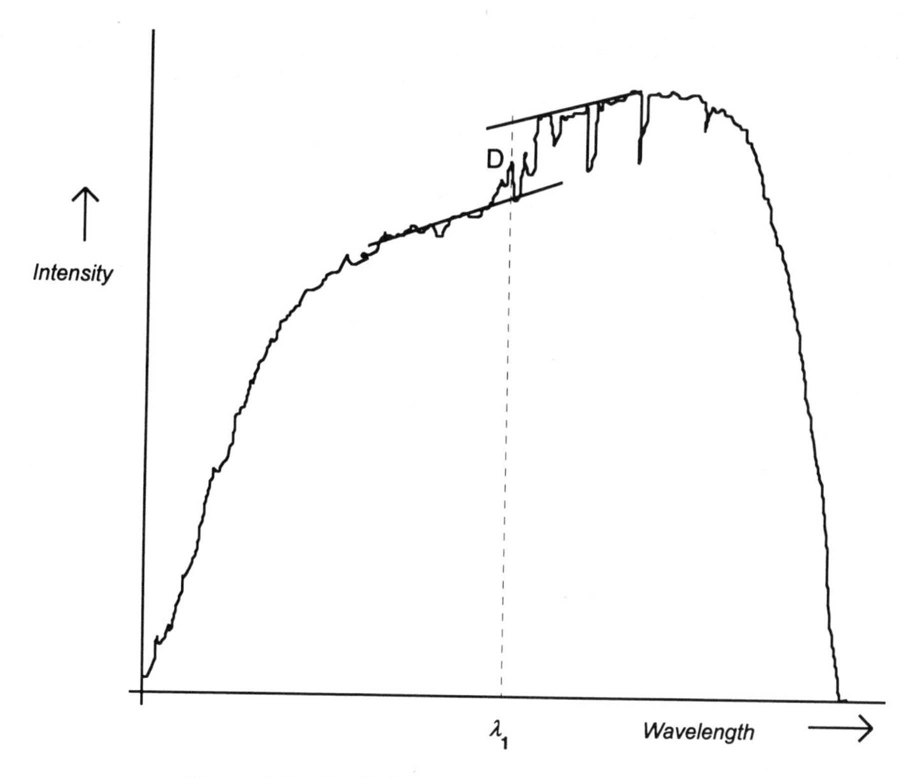

Figure 7.5 *The Balmer jump in the spectra of stars.*

In 1931, William McCrea in Edinburgh calculated the energy emitted by a star consisting of neutral hydrogen atoms and ionised hydrogen atoms (i.e. protons), and produced a spectral curve with the right value of Balmer jump for A0–type stars. For later type stars, McCrea's model badly overestimated the value of the Balmer jump, but this discrepancy was largely rectified by Subrahmanyan Chandrasekhar and Guido Münch at the Yerkes Observatory who added negatively charged hydrogen ions to the model in 1946.

The French astronomers Daniel Barbier and Daniel Chalonge started a detailed observational analysis of the Balmer jump in 1934, using a quartz objective prism telescope, at the newly-established high altitude observatory on the Jungfraujoch in Switzerland. They found that the magnitude of the jump, *D*, was a maximum for A0 stars because hydrogen absorption is strongest in these stars. As their work progressed further, they found that the wavelength of the mid-point of the jump, λ_1, and the magnitude *D*, correlated with the Harvard type. In addition, for A0 stars, λ_1 correlated well with absolute magnitude, and so, λ_1 and *D* provided the first* two-dimensional spectral classification, which was related to both absolute magnitudes and Harvard spectral type.

---

*As it pre-dated that of Morgan.

Unfortunately, after further work there appeared to be two problems with the Barbier–Chalonge system. Firstly, D was found to be relatively insensitive to spectral type around the A0 region of the Harvard classification. Secondly, stars on either side of about A0 could have the same Barbier and Chalonge co-ordinates. Stars with MKK classifications of B3V and F5III, for example, both had $D = 0.22$, $\lambda_1 = 376$ nm.

After the war, Chalonge installed a new instrument at Jungfraujoch and undertook further research. At first, the ambiguity problem for stars on either side of about A0 was solved by using two different (λ_1, D) diagrams, one for stars of A1 and earlier, and one for A3 and later. In 1953, however, Chalonge and Mlle. Lucienne Divan introduced a third parameter into their system, Φ_b, which was defined as the gradient of the intensity–wavelength curve between 380 nm and 460 nm. This three-dimensional scheme has since become known as the BCD (Barbier–Chalonge–Divan) system.

The BCD system can be used with low dispersion spectra (of about 25 nm/mm) and, in this respect, it is easier to use than the MKK system. It can only be used with stars where the Balmer jump is clear, however, and this limits its use to stars with spectral types earlier than about G0.

Given the luminosity and spectral types in the MKK system, both determined solely from the star's spectra, the position of any star can be plotted on the H–R diagram. Likewise, the (D, λ_1, Φ_b) co-ordinates in the BCD system, also determined solely from the star's spectra, can be transposed to the (Spectra, Luminosity) co-ordinates of the MKK system, and vice versa.

Various other systems were devised in the 1950s. In 1951, for example, Bengt Strömgren used a system based on D and the total intensity of the H_β line of hydrogen, as the latter is easier to estimate photoelectrically using interference filters, compared with estimating λ_1. Three years later Margherita Hack of the Milano-Merate Observatory used D and the central intensity of the H_δ line instead. Since then a number of modifications have been made to the MKK system but it, and the BCD system, are still the most widely used stellar classification schemes.

Later Evolutionary Ideas

In the early 1920s, Russell's theory of stellar evolution, that he had proposed in 1913, was generally thought to be correct. In this theory, stars evolve from the right hand side of the H–R diagram, and move rapidly to the left, until they meet the main sequence, which they then follow down to the bottom. The more massive stars join the main sequence near the top, and the less massive stars meet it lower down.

Robert J. Trumpler of the Lick Observatory began a detailed investigation of the distribution of stellar types in open clusters in the 1920s. He produced separate H–R diagrams for each cluster and noted that, although the results were consistent with the general shape of the H–R diagram with a Main Sequence and a Giant band, the population density of stars at various places on the Main Sequence and/or Giant branch was significantly different for the open clusters compared with that for stars much closer to the Sun. Some clusters, for example, had virtually no stars cooler than type F5, whereas others had virtually no stars hotter than F0 (see Figure 7.6). All of the open clusters had main sequence stars, but some had no giant stars.

In 1925, Trumpler assumed that all stars in a cluster were born at approximately the same time, and explained the different population density of stars on the H–R diagram as being due to some clusters having stars that were, on average, appreciably more massive than others. Most astronomers, while accepting his assumption, considered his conclusion somewhat speculative.

Unfortunately, in 1924, just before Trumpler published his results, Russell's theory of stellar evolution had been shown by Eddington to be incorrect (see Page 138), as stars could not move down the main sequence. This led Russell and others to make alternative proposals, but more observational data on stellar populations was required before further substantial progress could be made. It was also difficult to be altogether convincing with any theory, as long as the source of stellar energy was unclear (i.e. before Bethe and Critchfield's work of 1938).

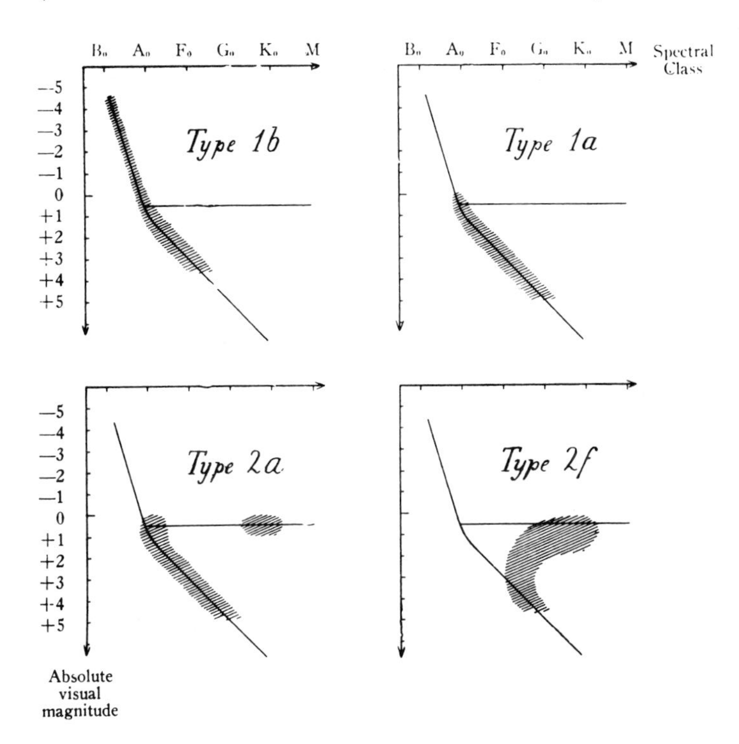

Figure 7.6 *The H–R diagrams of four stellar clusters by Robert Trumpler. The hatched areas show where the stars are located with reference to an idealised Main Sequence and horizontal Giant branch. (From* Publications of the Astronomical Society of the Pacific, *Vol. 37 (1925), p. 315; courtesy* The Astronomical Society of the Pacific. *Image supplied by the Royal Astronomical Society.)*

Although the theories that were put forward to replace Russell's theory differed in detail, they generally agreed that stars evolve by moving from the right of the H–R diagram to the left, until they meet the main sequence, where they stay for most of their lifetimes, before moving back to the right again. Strömgren, in particular, showed that a star on the main sequence would move towards the giant branch, in the top right of the diagram, as it gradually used up its hydrogen fuel.

Gerard Kuiper joined the Lick Observatory in 1933 having just obtained his doctorate working for Hertzsprung at Leiden in the Netherlands. Four years later, Kuiper produced his analysis of Trumpler's observational results of the early 1920s, assuming that all the stars in a cluster are of almost the same age. He was able to use the new theories of stellar evolution, however, that were not available to Trumpler 12 years earlier and, in particular, he had Strömgren's analysis at hand. Kuiper found that the lines on the H–R diagram for each cluster (see Figure 7.7) agreed approximately with Strömgren's lines of constant hydrogen content, and this led him to conclude that the hydrogen content for stars reduces progressively with time from the clusters called 12 Monocerotis and S Monocerotis (numbers 1 and 2 in the figure), which have hydrogen rich stars, to clusters like Coma Berenices and NGC 752 (numbers 13 and 14).

Very young clusters were analysed by Merle Walker at the Lick Observatory in the 1950s, who concluded that, although the hottest and most mature stars had reached the main sequence, the less-heavy cool stars, which were situated above the main sequence, had not. He concluded that these less massive stars were still in their Helmholtz contraction phase, so it was now possible to estimate the ages of these young clusters by observing how far away from the main sequence these young less-massive stars were.

It was clear in the early 1950s that the hot O and B stars could not have existed since the origin of the universe at least 4 billion years ago, as they burnt up their energy too fast. An O type star of mass 15 suns, for example, was estimated to have a total lifetime of only about 40 million years. The Soviet astronomer Viktor Ambartsumian, Otto Struve and others saw no problem with this, arguing that these O and B stars had formed only recently, but Hermann Bondi and Fred Hoyle, who favoured the Steady State theory of the universe, suggested that the O and B stars had been formed at the same time as the other less heavy stars, but that they were being kept burning by a steady accretion of material from surrounding space.

In the 1930s the Dutch–American astronomer Bart Bok had discovered a number of small dark patches in the Milky Way, which were later thought, in the 1950s, to be the birthplace of stars. It was suggested that the material in these Bok globules, as they became known, would gradually contract under its mutual gravitation, and would produce heat in the manner proposed by Helmholtz in 1854 (see Page 6). Eventually, when it became hot enough, it would become a star, producing energy by nuclear processes. In 1955 Louis Henyey, R. Le Levier and R. D. Levee of the Yerkes Observatory calculated the evolutionary tracks for this pre-main sequence

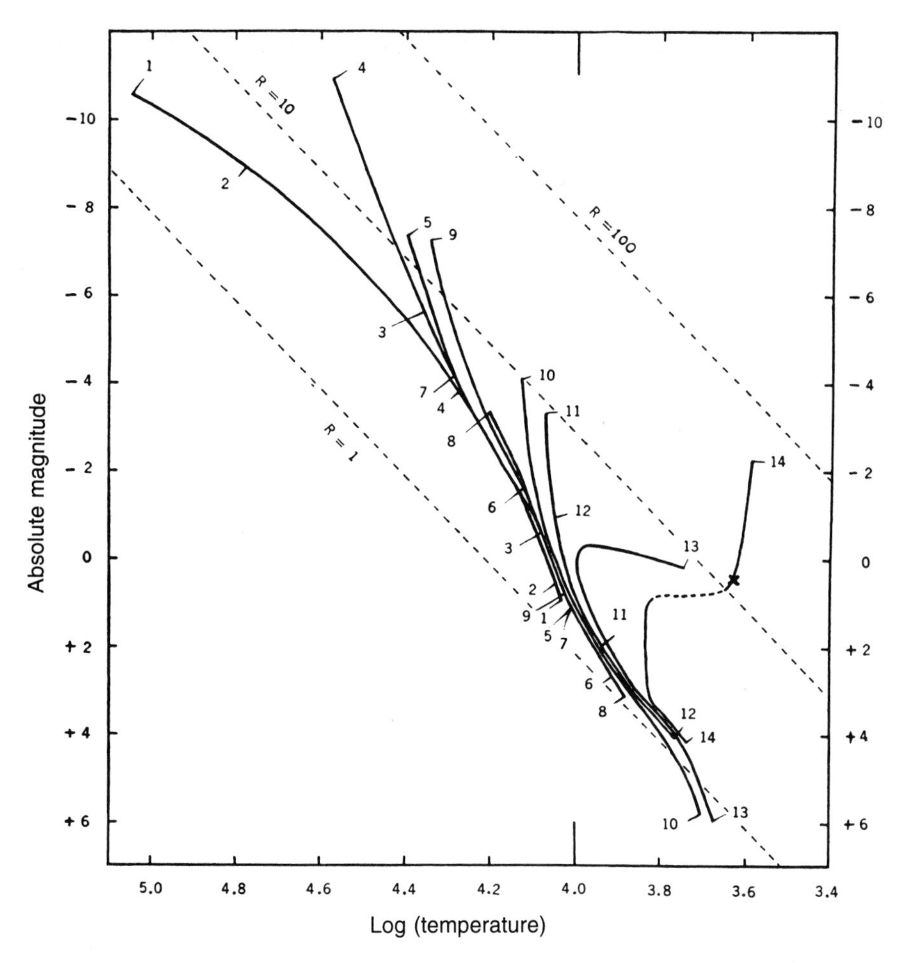

Figure 7.7 *The H–R diagrams of 14 stellar clusters by Gerard Kuiper. (From* The Astrophysical Journal, *Vol. 86 (1937), p. 185; reprinted with permission from* The Astrophysical Journal. *Image supplied by the Royal Astronomical Society.)*

stage, which were estimated to take, for example, from 300 million years for K7 stars of 0.5 solar masses, to only 30,000 years for B0 stars of 20 solar masses.

After this relatively rapid Helmholtz contraction phase, analysis showed that a star spends something like a thousand times as long on the main sequence, converting hydrogen to helium, before quickly moving to the top right of the H–R diagram, where it becomes a red giant or supergiant, with a core of about 50% helium. In this giant phase helium is converted into heavier elements like carbon and oxygen, but this happens much more rapidly than the production of helium on the main sequence, causing the star to rapidly run out of fuel. It becomes cooler, so that the pressure of radiation trying to leave the star can no longer counteract the star's gravity, causing the star to contract and move from the top right of the H–R diagram to become a white dwarf in the bottom left of the diagram. Stars that, at that stage, are heavier than about 1.4 times the mass of the Sun will contract even further to form neutron stars (see Page 187), rather than white dwarfs.

So it was clear in the 1950s that, as the heavy, brightest stars stayed on the main sequence for much less time than the dimmer stars, the age of a mature star cluster could be determined by observing which stars in the cluster had left the main sequence, assuming that all the stars in the cluster were formed at approximately the same time as each other. In 1957, Allan Sandage, working at the Mount Wilson and Palomar Observatories, presented a paper using this technique at the Vatican symposium on stellar populations, in which he deduced the ages of various clusters from the shape of their H–R diagrams. They ranged from less than 1 *million* years for cluster NGC 2362 to 5 *billion* years for cluster M67.

Work by Fred Hoyle in the late 1950s on stellar evolution caused the estimated age of M67 to be increased to 9.2 billion years, and analysis by Sandage for cluster NGC 188 produced an age of 24 billion years. This 24 billion years was uncomfortably more than the age of the universe of 11 billion years estimated in 1958. Sandage was doubtful about his calculation of the large age of NGC 188, however, and expected it to be revised downwards when nuclear processes in stars were better understood. Although we do have a better understanding of such processes today, there are still similar inconsistencies between the apparent ages of the oldest star clusters and the age of the universe (see Page 255).

Stellar Populations

In addition to the open star clusters discussed above, there are numerous tight groupings of stars, called globular clusters, scattered across the sky. Harlow Shapley examined these globular clusters in detail using the 60 inch (152 cm) Mount Wilson reflector, and concluded, in 1920, that they formed a halo around the Milky Way (see Page 215). He also found that the H–R diagrams for stars in globular clusters were quite different from the H–R diagram found by Russell for stars relatively close to the Sun. Shapley found that there seemed to be no main sequence of stars in globular clusters, and the red giants were much brighter than blue giants, whereas in Russell's diagram, for stars close to the Sun, the red giants were of almost the same intensity as the blue giants. These differences resulted in the absolute luminosity of stars in a globular cluster increasing from blue to red stars, whereas the reverse occurs for stars relatively near the Sun.

Trumpler's work in the 1920s showed that the H–R diagrams for stars in open clusters (like the Pleiades and the Hyades) were similar to that for stars in the Sun's neighbourhood, as they had main sequences located at the same place on the H–R diagram. Thus the star populations in globular clusters seemed to be different from both those in open clusters, and stars relatively close to the Sun. These differences were finally explained almost

30 years later by the German–American astronomer Walter Baade when he showed (see Pages 253–254) that stars fall into basically two populations; Population I being stars near the Sun, in open star clusters, and the arms of spiral galaxies, and Population II being stars in globular clusters, elliptical galaxies and the nuclei of spiral galaxies. As it was known that globular clusters are almost completely devoid of gas, Baade concluded that Population II stars are much older than Population I stars, the latter generally being in gaseous, star-forming regions.

8 Variable and Double Stars

Early Work

Eclipsing Binaries

The star Algol (β Persei) was known to ancient astronomers as the Demon Star, as it was the only star known to exhibit variations in brightness. Geminiano Montanari studied these variations at Bologna in 1667, but it was not until 1782 that their cause was correctly explained by the 18 year-old deaf-mute English astronomer John Goodricke. He noted that the brightness variations had a period of 2 d 21 h, and suggested that the bright visible star may be being partially eclipsed by a darker companion with this frequency, in what today we call an eclipsing binary pair.

The symmetrical shape and reproducibility of the light curve makes eclipsing binaries easy to identify. Algol, for example, shows a gradual brightness reduction over about 4 hours, stays constant for about 20 minutes, returns to its original brightness for another 4 hours, and then stays constant for the next 2 days and 13 hours, before reducing in brightness again. The minimum occurs when the darker star passes in front of the bright one, cutting off part of its light, but no minimum had been detected in 1890 when the bright star eclipsed its darker companion, indicating that the two stars must be of radically different luminosities.

In 1784 Goodricke discovered a variable star, β Lyrae, having two minima of different brightness. In this case it was presumed that there were two stars of different but relatively similar luminosities orbiting their common centre of mass, with one minimum being due to the eclipse of one star, and the other minimum being due to the eclipse of the other. β Lyrae was strange, however, as its spectrum observed in the late nineteenth century had sharp hydrogen and helium emission lines, indicating that it was not highly compressed* like a normal star, and its spectrum also showed periodic variations in its line structure. The cause of these variations was a mystery.

True Variable Stars

Omicron (o) Ceti was the first star, apart from Algol, found to vary in brightness. The Dutch minister David Fabricius saw the star disappear in October 1595, and his fellow-countryman Phocylides Holwarda observed its variability over the period 1638–39; the star then being called Mira, meaning "the wonderful", because of its strange behaviour. Ismaël Boulliaud in 1667 showed that it was varying with a period of about 334 days, and as such it was the first long period variable to be discovered.

The light curve of Mira was quite different from that of Algol in both its shape and the fact that its period and maximum brightness were not always the same; sometimes it reached magnitude 1.7 at maximum, but sometimes it did not even reach magnitude 5. It was clearly not an eclipsing binary, but was a star that was genuinely varying in luminosity, and by 1890 a further 95 long period variables had been discovered, all of them red in colour. In 1851, Rudolf Wolf at Berne saw similarities in the irregular behaviour of long period variables and the recently-discovered irregular sunspot cycle, and he suggested that the stellar variations may be due to star spots.

Brightness variations over much shorter periods were also seen in stars, for example, the English astronomer Edward Pigott discovered in September 1784 that η Aquilae was varying in brightness with a period of 7 days, and that it increased in brightness more rapidly than it decreased. A few weeks later, his friend John Goodricke found that δ Cephei was exhibiting similar fluctuations with a period of 5 days. Over the next century many more of these short period variables were discovered. Some astronomers thought that these stars were eclipsing binaries with highly eccentric orbits, but the shape of their light curves was very difficult to explain, and then in 1873 the German physicist Arthur Ritter came up with the radical suggestion that they could be stars oscillating in size. Astronomers of the time had difficulty with this oscillating concept, however, and tended to favour the binary theory in spite of its difficulties.

Double Stars (Optical Doubles and Visible Binaries)

Two stars, which are at very different distances from us, can appear to be close together simply because they are on almost the same line of sight from the Earth. These stars are called optical doubles because they are only apparently close together. Some double stars really can be close together, however, and can be interacting gravitationally in what is now called a

---

*The pressure near the surface of a normal star causes the spectral lines to be slightly unsharp, due to what is called "pressure broadening".

binary pair, with each star rotating about their common centre of gravity. For example, Goodricke had suggested that Algol was really a binary star although, in this case, only one star could be seen. In the eighteenth century, many astronomers wondered if some of the apparent double stars were really binary stars. If so, how many were there, and did they follow Newton's laws of motion? The answer came when William Herschel was carrying out a different investigation.

When Herschel tried to measure stellar parallaxes in 1779, he decided to use a method first proposed by Galileo in the previous century, namely to observe double stars, as if one of the two stars is much further away from Earth than the other, it will show much less parallax and so the distance between the two stars would vary cyclically over a year. It is also much easier to measure the relative rather than the absolute positions of stars.

Herschel had originally thought that the vast majority of double stars were optical doubles (i.e. due to the line-of-sight effect) and were not binaries, but as his researches continued he found that there were far more star pairs than could be justified statistically as line-of-sight doubles.

Herschel's equipment was not sufficiently sensitive to measure any parallaxes but, in 1802, he noticed that some double stars were changing their separations with periods that were not annual, indicating that these stars were really binary stars moving around their common centres of gravity. In 1804 he published a paper describing the relative movement of 50 pairs, but it was not until 1827 that there were sufficient data for the orbit of such a binary pair to be accurately calculated. In that year Felix Savary calculated the orbit of the binary pair ξ Ursae Majoris, showing that both stars were orbiting in ellipses about their common centre of gravity with a period of 58 years.

Astrometric Binaries

The Königsberg astronomer Friedrich Bessel studied the proper motion (or true motion) of many stars, and in 1844 he found that both Sirius and Procyon had a secondary oscillatory motion superimposed on their linear proper motions. He concluded that this was caused by the visible star being perturbed by an unseen companion in both cases, and estimated the periods of both binary pairs at about 50 years. These were the first examples of astrometric binaries, or binary stars discovered from their non-linear proper motions.

Sirius only remained an astrometric binary for 18 years, as in January 1862 Graham Clark and his father Alvan accidentally saw the unseen companion (called Sirius B) from their workshop in Cambridge, Massachusetts when they were testing a new telescope that they were building for the University of Mississippi. Observation of the two stars over the next few years confirmed the 50 year period estimated by Bessel.

By 1890 the distance of Sirius from the Earth had been measured by parallax and so the luminosities of the two stars and the distance between them could be calculated. This allowed the masses of Sirius A and B to be deduced from the period of their orbits around their common centre of mass, using Newton's laws. The result was quite unexpected as, although Sirius B appeared to be only slightly less massive than Sirius A, its luminosity was considerably less (0.01 suns, compared with 50 suns), and both Sirius A and B were appreciably heavier than the Sun.

So, Sirius B was seen to be very dim for its mass, and this caused astronomers to realise that there may be many objects relatively close to Earth that could not be seen because they were too dim. In other words, what we see in the universe may give a very misleading picture of what is there, even for regions very close to us.

Although the companion of Sirius had been observed, the companion of Procyon had still not been seen by 1890, so Procyon was still an astrometric binary at that time.

Spectroscopic Binaries

A new method of detecting binary stars was discovered by Edward Pickering of the Harvard College Observatory in 1889, when he found that the spectral lines of Mizar A were not only split in two, but the separation between the two components was varying cyclically. He attributed the two sets of lines to two different stars, and the variation in the line separations he attributed to changes in the Doppler shifts, as the two stars orbited their common centre of gravity in a binary pair. Mizar A was described as a "spectroscopic binary" in recognition of its mode of discovery, and shortly afterwards his assistant Antonia Maury found that β Aurigae was a spectroscopic binary also.

In the same year, Hermann Vogel at Potsdam found that the spectral lines of Algol oscillated slightly in frequency with the same period as the variation in its light curve, so Algol was also a spectroscopic binary, although in its case only the spectrum of one of the stars was bright enough to be detected. Vogel had finally proved the eclipsing binary nature of Algol that had been first hypothesised by Goodricke in 1782 to explain its brightness variations. The parameters of this binary, calculated from its Doppler shifts and light curve, were a big surprise. The diameter of the orbit was estimated at only 3 million km (only 2% of the distance of the Earth from the Sun), and the component stars were estimated to have diameters of 1.6 and 1.3 million km, which clearly meant that their surfaces must be very close together. Assuming both stars were of the same density, Vogel calculated their masses to be 45% and 22% of the mass of the Sun, and their density was only 0.4 g/cm^3, or less than half that of water. Their surface gravities must be rather low and, coupled with their closeness to each other, this meant that the two stars must be somewhat distorted from the usual spherical shape.

Irregular Variables

Occasionally a star shows brightness variations which are unlike any other, so it cannot be fitted into any of the above categories. Such a star was η Carinae, previously known as η Argus, which had been observed to show large irregular brightness variations since 1677, when it had first been studied by Edmond Halley.

A number of other irregular variables were known in 1890, none of which could be explained. For example:

(i) Gamma (γ) Cassiopeiae which, like β Lyrae, showed sharp hydrogen and helium emission lines in its spectrum, but which, unlike β Lyrae, had a spectrum that varied quite randomly. It was thought that both β Lyrae and γ Cassiopeiae were stars of low density, being in transition between the nebula phase (where bright emission lines were normal) and the stellar phase (where dark absorption lines were the norm). In this transition, γ Cassiopeiae was showing some instability.

(ii) Epsilon (ε) Aurigae which was found by F. W. A. Argelander, the director of the Bonn Observatory, and Eduard Heis at Münster to have decreased from its normal third magnitude to fourth magnitude in 1848, but then in the following year it recovered its previous brightness. It was also seen to be of fourth magnitude for a few months in 1875, but it again retained its original brightness. These variations seemed to be very irregular.

(iii) R Coronae Borealis which was observed by Pigott in 1795 to suddenly reduce in brightness and then recover again. Over the following century, R Coronae Borealis was seen to exhibit similar sudden large reductions in brightness, but there seemed to be no periodicity. It spent most of its time at maximum.

(iv) U Geminorum which was observed to dim from magnitude 9 to 12 from 15th December 1855 to 10th January 1856, and then was observed by Eduard Schönfeld at the Bonn Observatory to increase by 3 magnitudes in 24 hours in 1869. Its variations in brightness were apparently irregular, spending most of the time at minimum, unlike R Coronae Borealis which spent most of its time at maximum.

So by 1890 the following types of variable and double stars were known:

Variable Stars

Eclipsing binaries as per Algol	(1782, J. Goodricke)
Long period variables as per Mira	(1638, P. Holwarda)
Short period variables as per δ Cephei	(1784, J. Goodricke)
Irregular variables as per η Carinae, γ Cassiopeiae, R Coronae Borealis, U Geminorum	

Double Stars

Optical doubles as per Mizar	(1651, G. Riccioli)
Eclipsing binaries as per Algol	(1782, J. Goodricke)
Visible binaries	(1802 W. Herschel)
Astrometric binaries as per Sirius and Procyon	(1844, F. Bessel)
Spectroscopic binaries as per Mizar A and Algol	(1889, Pickering and Vogel)

By 1890 over 200 variable stars were known, three of which were in globular clusters.

As more and more variables were discovered, the method of designating them had to be expanded. In 1844, Argelander had introduced a nomenclature for variables that ran from R to Z per constellation. By 1881, however, as the number of known variables increased, this scheme proved inadequate, and a double lettering system was introduced by Ernst Hartwig of the Dorpat Observatory in Russia, running from RR to RZ, SS to SZ and so on. This was later extended to include AA to AZ, BB to BZ, etc., using all the letters except J, and thus allowing 334 variables to be listed per constellation.

Short Period Variables

Eta (η) Aquilae and δ Cephei had been discovered in 1784 to be variable stars with periods of about 6 days, whose brightness increased more quickly than it decreased. Most astronomers in 1890 thought that they were binary stars, but Arthur Ritter thought that they were pulsating stars. In 1894 Aristarkh Belopol'skii at the St Petersburg Pulkovo Observatory measured a velocity range of from −19 to +24 km/s for δ Cephei, indicating that it may be a spectroscopic binary, but the variation of velocity with time was peculiar. Not only did the minimum brightness occur a day earlier than it should, but the maximum brightness occurred when the star was apparently approaching us, but not when it was apparently receding.

In 1899 Karl Schwarzschild found that the photographic intensity range was 50% greater than the visual range for Cepheids, indicating that they were bluer, and therefore hotter, at maximum brightness than at minimum. Eight years later, Sebastian Albrecht analysed their spectra at the Lick Observatory, confirming that the variations in luminosity of Cepheids appeared to be caused by variations in temperature. These temperature changes were attributed to some sort of tidal effect on the Cepheid caused by a dark binary companion, but when the tidal behaviour of a star in such a system was analysed, the results did not correlate with the observations.

An alternative theory was suggested by Robert Emden of Munich in 1906 and Forest Moulton of Chicago in 1909, in which the stars oscillated in shape about their natural spherical shape, rather like a large soap bubble floating in the air. Unfortunately this theory produced a radial velocity period that was twice that of the luminosity period, whereas they were both observed to be the same. So, at this stage, all the theories about the cause of the Cepheids' variability had failed in some way or other.

Meanwhile, in 1896 the 24 inch (61 cm) Bruce refractor had been installed at the southern Harvard College Observatory at Arequipa, Peru where it was used to photograph the complete southern sky down to magnitude 14. This enabled Solon Bailey of Harvard to find, by 1902, over 500 short-period variables in 17 globular star clusters, all having periods of less than 1 day (now called the cluster variables). Williamina Fleming, also of Harvard, discovered that the star RR Lyrae in the Milky Way also had a period of less than 1 day. Carl C. Kiess at the Lick Observatory then studied the brightness and velocity curves of the so-called cluster variables, of RR Lyrae, and of a number of similar stars in the Milky Way with periods of less than 1 day, concluding in 1912 that they were all part of the Cepheid variable family.

The Magellanic Clouds, which are seen as two large clouds of stars in the southern sky, were surveyed in the early years of the twentieth century by Henrietta Leavitt of the Harvard College Observatory from plates taken from the Arequipa station in Peru. In 1907 she published a list of 1,777 short period (mostly Cepheid) variables in the two clouds, and, in the following year, she was able to measure the periods of 16 of the Cepheid variables in the Small Magellanic Cloud (SMC), finding that the brightest stars had the longest periods. By 1912 she had increased the number of measured Cepheid variables in the SMC to 25 with the same result (see Figure 8.1). Miss Leavitt concluded that, as the SMC was very distant, its stars were all at very similar distances from us, and so differences in apparent magnitude were virtually the same as differences in absolute magnitude. This meant that the periods of the Cepheid variables correlated with their absolute magnitudes.

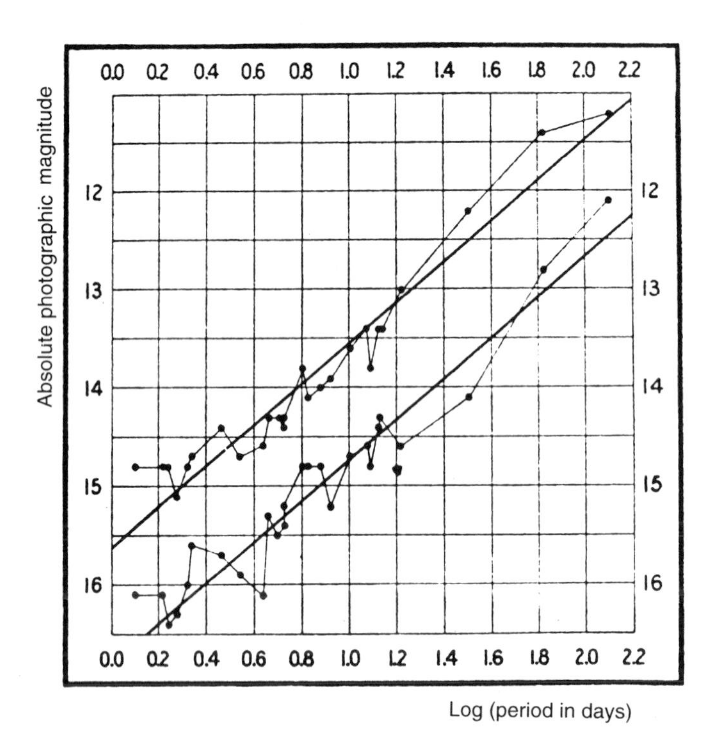

Figure 8.1 *The period–luminosity diagram as first produced by Henrietta Leavitt. The two lines indicate maximum and minimum photographic magnitudes. (From* Harvard College Circular, *No. 173 (1912); courtesy* Harvard College Observatory.*)*

Ejnar Hertzsprung had also been working on Cepheid variables, and he realised that once the absolute magnitudes of the closest Cepheids had been measured, Miss Leavitt's period–luminosity relationship could be used to determine the distance of any Cepheid from its measured period. Hertzsprung determined the small proper motions of 13 relatively close Cepheids in 1913, and found that a Cepheid of 6.6 day period, for example, had a median absolute visual magnitude* of −2.3, which is very bright, putting it near the top of the H–R diagram (see Figure 7.3). Miss Leavitt had shown that Cepheids with such a period in the SMC had apparent visual magnitudes of 13.0†, or 15.3 less than Hertzsprung's absolute magnitude of –2.3, and this enabled Hertzsprung to deduce that the SMC is about 37,000 light years‡ away.

Hertzsprung had established the distances, and hence the absolute magnitudes of Cepheids, using the proper motions of just 13 stars, but this could only yield an order-of-magnitude calibration, because of the random nature of the proper motions. (As stars do not all move in the same direction in space, the fact that one has a larger proper motion than another, only indicates that, on average, it is closer, not that it is so in each and every case.)

Harlow Shapley re-calibrated the absolute magnitude scale of Cepheids between 1916 and 1918, using a much larger statistical sample of stars than Hertzsprung, so making it much more accurate. With this new calibration he established, in 1918, that the distance of the SMC needed revising from 37,000 to 95,000 light years.

Shapley had taken photographs of 86 globular clusters in 1916–17, and found that he could detect Cepheids in only the largest and, presumably, the closest clusters. In these clusters he discovered that the average luminosity of the 20 or so brightest non-variable stars was about ten times that of the average luminosity of the Cepheids in the same cluster. Assuming this to be so for all globular clusters, enabled him to estimate the distance of additional clusters which were too far away for their Cepheids to be detected, by measuring the apparent magnitudes of their brightest stars. When he had done this, Shapley found that all the globular clusters whose distance he had estimated, using either method, were of about the same *absolute* size and intensity. This enabled him to estimate the distances of the furthest clusters from their apparent sizes and intensities, and gave him confidence in his assumption that the Cepheids in all the clusters behave in basically the same way.

So measuring the periods of Cepheid variables and their apparent magnitudes had enabled the distance of the globular clusters to be determined. In 1924, Edwin Hubble was able to identify Cepheid variables

---

*The median is the average of the maximum and minimum luminosities.

†Magnitude 14.5 photographically (see Figure 8.1). Log 6.6 = 0.82.

‡For the mathematically minded, this distance = $32.6 \times \sqrt{10^{15.3 \times 0.4}} \approx 37{,}000$ light years, where both 32.6 and 0.4 are constants.

in the Andromeda and M33 nebulae (see Page 233), and this enabled him to estimate their distances for the first time, proving that they were galaxies of stars outside of the Milky Way, thus settling one of the major controversies of the late nineteenth and early twentieth centuries.

In parallel with these dramatic developments, theories were being developed to explain the behaviour of Cepheid variables. Shapley reviewed all the theories available in 1914 and concluded that Cepheids were the pulsating stars suggested by Ritter, but it was not until Arthur Eddington published his theory of pulsating stars in 1918 that most astronomers accepted this theory. Unfortunately, however, Eddington's theory left one key fact unexplained, namely, why did the maximum brightness and temperature occur when the star was rapidly expanding, and not when it was at greatest compression a little earlier in its period?

Eddington was able to estimate the diameters of Cepheids using his new theory, and show that they were larger than the orbit that a companion star would have to have around it, if the light variation was due to them being binary stars. The orbit of the secondary would thus have to be within the surface of the Cepheid,* and this seemed to put the final nail in the coffin of the binary theory of Cepheids.

Then, in 1938, Martin Schwarzschild at Harvard was able to account for the phase lag between temperature and expansion that Eddington's theory did not explain. He suggested that the maximum temperature is reached at maximum density in the centre of the star, but that the whole star does not vibrate in phase. In particular, the maximum compression and temperature in the outer layers occur a little later in the cycle than they do in the centre, thus producing the observed effect.

Long Period Variables

Mira (o Ceti) was the first long period variable discovered, and by 1890 almost 100 such stars were known, having periods of from 90 to just over 600 days. They were all cool red stars having spectra similar to stars of Secchi type III (Harvard class M). Bright hydrogen lines near maximum brightness, which were found to be characteristic of these variables, were first discovered in 1869 by Angelo Secchi for the star R Geminorum, and in 1904 Alfred Fowler showed that the dark bands in their spectra were generally due to titanium or titanium oxide.

Long period variables have irregular periods, and their maximum and minimum magnitudes vary with time. In the middle of the nineteenth century, Rudolf Wolf in Berne had suggested that these effects were caused

---

*Shapley had come to a similar conclusion in 1914, but his analysis was based on a somewhat speculative theory of stellar structures devised by Emden, which assumed convective rather than radiative heating.

by star spots, but this explanation became unlikely when Henry Russell showed in 1918 that Mira stars are red giants of low density. A. Brester suggested in 1908 that the light variations are caused by the periodic appearance and disappearance of holes in a cool shell of gas around the star, and in 1916 Paul Merrill of the University of Michigan suggested that there is a shell of cool condensed gases in the upper atmosphere of the star itself. He suggested that the shell is relatively opaque and absorbs heat, until it becomes too hot and evaporates, after which it gradually forms again.

Robert Aitkin of the Lick Observatory discovered in 1923 that Mira, which was then known to be a red giant of about 400 million km in diameter, had a faint companion which was a white dwarf. Whether the white dwarf was, in some way, responsible for Mira's variability was unclear, however. Then, in the following year, a new line of investigation was begun when Hans Ludendorff of Potsdam showed that the redder, long period variables have the longer periods and greater light fluctuations. As the redder stars were thought to be younger, this observation indicated that the variability was gradually reducing as the star developed. Boris Gerasimovic of Harvard found, in 1928, that the long period variables with shorter periods (of about 100 to 200 days) are considerably brighter than those with longer periods and, in the 1930s, Seth B. Nicholson and E. Pettit at Mount Wilson measured the temperatures of the 200 day variables at about 3,100 K, and the 400 day variables at about 2,600 K. These findings indicated that the stars become brighter and hotter as they evolve and reduce their variability, en route to the main sequence.

Irregular Variables

Gamma Cassiopeiae

Gamma (γ) Cassiopeiae was known in the nineteenth century to have apparently random variations in its spectrum. In the period from 1863 to 1894 astronomers sometimes observed hydrogen and helium emission lines superimposed on the normal absorption spectrum of a B type star. The magnitude of the star hardly changed during this period, however, with a maximum variation of only 0.2 magnitude being reported.

The absorption lines of γ Cassiopeiae were generally broader than for normal B stars, and this was presumed to be due to the rapid rotation of the star. In addition, Otto Struve of the Yerkes Observatory suggested in 1931 that the bright emission lines in this so-called Be type star, which were generally double and centred on the absorption lines, were due to a ring of material around the star.

In 1929, the relative intensities of the two components of the bright emission lines, called the Violet (V) and Red (R) components, were observed to start varying in intensity, such that the intensity ratio V/R reached 1.5 after just 6 months. The V/R ratio continued to vary in a random fashion and in 1931 it reduced to 0.65. In 1932 a more rapid variation phase set in and V/R reached 2.5 in 1933, but reducing to 0.4 in 1934. The variations continued until 1942 when the star gradually became stable once more.

The variations of γ Cassiopeiae were attributed to the star emitting shells of gas that gradually dispersed. There are other stars that exhibit similar changes, but these changes tend to be less spectacular and are often periodic.

R Coronae Borealis Type Stars

R Coronae Borealis had been discovered in 1795 to exhibit a sudden reduction in brightness and then recover again, and its light curve had been recorded consistently since 1843. It showed brightness reductions of up to nine magnitudes, with slower recovery to normal, but with no periodicity, and in 1922 Alfred Joy and Milton Humason at Mount Wilson found bright titanium emission lines when the star was at minimum.

By the 1920s, a number of stars had been discovered that exhibited sudden brightness reductions like R Coronae Borealis. These stars were generally located in the plane of the Milky Way, causing Ludendorff to attribute the brightness reductions to irregular dust clouds in that plane (see Page 225). But, in 1935, Louis Berman found that carbon was by far the most abundant element in their spectra, indicating that the obscuration was caused by a build-up of soot in the star's atmosphere. Then in 1986 Gillett, Backman, Beichman and Neugebauer detected a vast dust shell, with a temperature of only 30 K, surrounding R Coronae Borealis itself, using the IRAS satellite. The shell is 40 light years across and consists mainly of silicon and carbon particles.

Dwarf Novae

U Geminorum had been observed in the second half of the nineteenth century to spend most of the time at its minimum brightness, but to quickly increase and decrease in brightness at irregular times, as if it were showing large-scale eruptions. In 1912, its spectrum was found to be of class F with broad calcium absorption lines, and its hydrogen lines were found to vary in intensity, often appearing in emission.

Starting in 1896, a gradually increasing number of similar stars have been found, so that by today over 600 are known. They are called recurrent or dwarf novae.

Flare Stars

Ejnar Hertzsprung noticed in 1924 that a faint star, now called DH Carinae, suddenly increased in brightness by two magnitudes. Then, 15 years later, A. A. Wachmann noticed a marked change in the spectrum of one of the faint stars (now called V371 Orionis) in the Orion nebula, together with a simultaneous decrease in its brightness. In the same year, the Dutch–American astronomer Adriaan Van Maanen noticed that the star Lalande 21258 (now called WX Ursae Majoris) was two magnitudes brighter than normal, and later found that the star Ross 882 (now called YZ Canis Minoris) was 1.5 magnitudes brighter than normal.

The first real advance in the study of these stars, now called flare stars, occurred accidentally while Carpenter was taking a number of short exposures of the star Luyten L726–8 in December 1947. He found that the star had increased in brightness by almost three magnitudes over a 3 minute period, and had then faded more gradually to its previous brightness. William Luyten and Hodge concluded that a violent flare had occurred on the star, similar in type but of much greater magnitude than those seen on the Sun. It was eventually found that Luyten L726–8 is a binary star, with components Gliese 65A and 65B, and it was 65B, now called UV Ceti, that exhibited the flare.

Two years later, on 30th April 1949, Gerald Kron and Gordon were measuring the light curve of the star BD +20°2465 (later called AD Leonis) with a photoelectric cell, when the meter deflection went off the end of the scale for 5 minutes. A similar effect was observed when Roques was measuring the magnitude of UV Ceti with a photoelectric cell, and then Luyten measured a six-magnitude increase in the intensity of UV Ceti photographically. In January 1969 a major flare was observed for YZ Canis Minoris which lasted for 4 hours, having an amplitude of 1.7 magnitudes visually, but over eight magnitudes in the ultraviolet.

It is now known that these flare stars are red dwarfs with surface temperatures of 3,000 K, and with masses of only 10% that of the Sun. They are generally dwarf M-type stars lying near to the bottom of the main sequence on the H–R diagram, with the usual titanium oxide bands, but with hydrogen and calcium emission lines also. As such they are classed as dMe stars.

Eclipsing Binaries

Goodricke had suggested, as long ago as 1782, that Algol was an eclipsing binary, and Vogel had detected the Doppler shift caused by its orbital motion in 1889, but the companion star was too faint to be seen either visually or by means of its spectrum. Vogel had calculated the density of the two Algol stars as only 0.4 g/cm^3, but in 1899 Russell reduced the density estimate to only 0.14 g/cm^3. Russell also estimated the density of 16 other Algol-type stars as ranging from 0.73 down to 0.06 g/cm^3.

The brightness reduction seen in Algol was due to the eclipse of the brighter star by the dimmer one, but the reduction caused by the eclipse of the dimmer star by the brighter (i.e. the secondary minimum) was undetected for many years. Finally, in 1910, Joel Stebbins of the University of Illinois, using an early prototype of his selenium photometer, detected this secondary minimum as a 6% dip in the light curve of Algol (see Figure

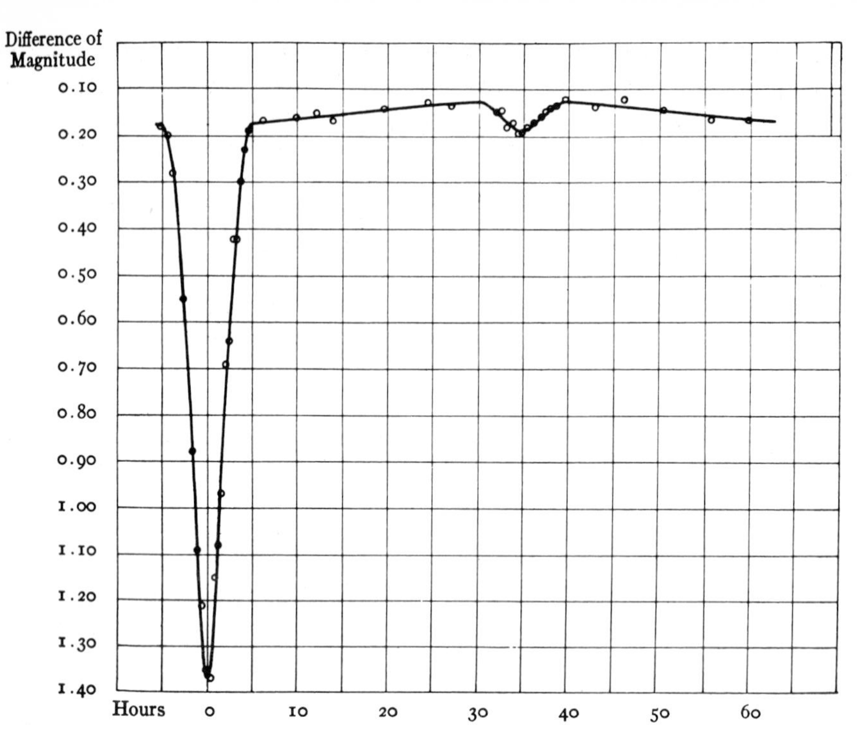

Figure 8.2 *The light curve of Algol measured by Joel Stebbins, using his prototype selenium photometer, showing both the primary and secondary minima. (From* The Astrophysical Journal, *Vol. 32 (1910), p. 199; reprinted with permission from* The Astrophysical Journal. *Image supplied by the Royal Astronomical Society.)*

8.2). In the meantime, in 1906 Belopol'skii noticed a long-period oscillation in the radial velocity of Algol caused by a third star of the system.

In 1901 the star ε Aurigae was found to decrease in brightness from its normal third magnitude to fourth magnitude, as it had also done in 1875 and 1848, and in 1903 Ludendorff showed that it was an eclipsing binary with a period of 27 years. During the eclipsing sequence the star decreased in brightness for 7 months, remained constant for 10 months, and then increased back to its normal level over the next 7 months. The reason for the very long eclipse period of 2 years was due to the size of the stars, both of which were about 1,000 times the diameter of the Sun.

Eclipsing binaries will usually have two minima of unequal brightness; at one minimum we see star A plus star B eclipsed, and at the other we see star B plus star A eclipsed. Unless both stars have exactly the same brightness per unit area, these two minima will be different (see Figure 8.2).

For two stars at some distance from each other, the total brightness of the pair out of eclipse will be the same at all times, provided that they are of constant luminosity as they rotate on their axes (or unless a reduction in the luminosity of one is exactly counterbalanced by an increase in the other). In 1912, Russell pointed out that the stars, like the Sun, should appear to be brighter at the centre than near the edge. He also suggested that, if one star is very much brighter than the other, it will illuminate the other sufficiently for the latter's luminosity to vary, being a maximum seen by us when it almost behind the bright star, and a minimum when it virtually in front of it. If the stars are very massive or very close, however, one or both of the stars can be distorted with their bulges permanently facing each other, and the maximum area presented to us will generally be when they are at maximum separation as seen from Earth.

The star β Lyrae was first seen to be variable by Goodricke in 1784 and, over the years, the brightness of this binary was found not only to have two different minima, but also to be continuously variable out of eclipse. Its variations in radial velocity were measured by Belopol'skii in 1892 and Norman Lockyer in 1893, and its orbital elements were computed by G. W. Myers in 1896 and J. W. Stein in 1907. These showed that it was an extreme case of an eclipsing binary star system, as the two stars were so close together and distorted that they appeared to be almost touching.

In 1882, the B8 spectrum of β Lyrae had been found to vary by the Hungarian, Eugen Von Gothard, and in 1891 Williamina Fleming showed that this variation was periodic, having the same periodicity of 12 d 22 h seen in its light output. This indicated that the two stars in the binary had very different spectra.

Gerard Kuiper and Otto Struve calculated the masses and luminosities of the two β Lyrae stars, assuming that Eddington's mass–luminosity relationship holds (see Page 137), concluding, in 1941, that the brighter star has a mass of 75 times that of the Sun, and an absolute visual magnitude of −7, which would put it right off the top of the H–R diagram (see Figure 7.3). Helmut Abt then discovered a companion star of β Lyrae, which enabled a much better estimate of its distance from Earth to be made. With this revised estimate (of 860 light years) the brighter star of the binary was found by Abt, Jeffers, Gibson and Sandage to have an absolute magnitude of −3.9, but the dimmer star was found to be heavier than its very bright companion, thus violating the mass–luminosity relationship.

In the eighteenth century, the French mathematician Joseph Lagrange developed a theory for the behaviour of a small particle near to two bodies in space, and showed the existence of a critical surface around each body, which touched at the L_1 Lagrangian point (see Figure 8.3). These surfaces,

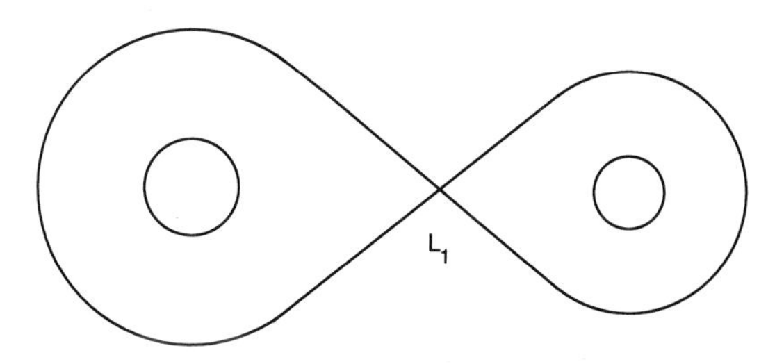

Figure 8.3 *The Roche lobes around two stars.*

called the Roche lobes, define the extent of the gravitational attraction of the two bodies (stars) individually. Inside one or other of these lobes, the particle is dominated by the gravitational attraction of one of the two stars, and is retained by that star, but outside of these lobes, the particle is subjected to the gravitational attraction of both stars as a system, and is not dominated by either.

If one of the stars ejects a stream of gas, for example, the gas can either (i) pass from one lobe to the other, via the L_1 Lagrangian point, and be captured by the other star, or (ii) be lost to either star individually, but be retained by the system, or (iii) be lost by the system completely. If one of the stars expands during its evolution to fill its Roche lobe, material can transfer from that star to the other through the L_1 Lagrangian point.

In the 1940s, the period of the binary β Lyrae was known to be increasing at the rate of 9 seconds per year, and its spectrum indicated that it was ejecting large amounts of gas. The primary (i.e. brighter) β Lyrae star was thought to be a giant B8 star that filled its Roche lobe. Some of the gas was thought to flow from one star to the other through the L_1 Lagrangian point, at one point in their orbit around each other, and return at another point in their orbit. Some of the gas was also lost to both stars individually, but retained by the system as a whole, as a steadily expanding shell which rotates around the binary system. This loss of mass explained the increase in the period of the binary.

Frank Schlesinger, the director of the Allegheny Observatory, found that the Doppler shifts observed in 1909 for the two eclipsing binaries δ Librae and λ Tauri were not symmetrical about the mid-eclipse. He explained this as being due to the Doppler shift caused by the stars rotating on their axes, and he was the first to measure the axial rotation of a star. Adams and Joy measured a similar effect in 1919, when observing the short-period spectroscopic binary W Ursae Majoris, deducing an equatorial velocity of the limb of 120 km/s (compared with 2 km/s for the Sun).

Otto Struve of Yerkes and G. A. Schajn of the Simeis Observatory in the Crimea concluded, in a joint analysis in 1929, that many short-period spectroscopic binaries had fast axial rotation rates. Then 2 years later, Struve and Christian Elvey produced a statistical analysis of axial rotation rates for both single and binary stars. They found that the O and B-type stars, with emission lines, generally had equatorial rotational velocities of 250 to 500 km/s, but the velocities decreased substantially in traversing the spectral sequence from O to M-type stars, so that the G, K and M-type stars all had rotational velocities of less than 50 km/s.

In 1938 Gerald Kron at the Lick Observatory measured the light curve of the eclipsing binary 21 Cassiopeiae with an accuracy of 0.002 of a magnitude, and this enabled him to measure the limb darkening of a star for the first time (from the detailed shape of the light curve as the companion star crossed its disc).

Non-Eclipsing Binaries

William Herschel discovered binary* stars in 1802, and Felix Savary of Paris calculated the orbit of the binary pair ξ Ursae Majoris in 1827. Initially it was thought that binary stars were unusual, but by 1900 over 1,000 were known, and by 1950 over 30,000, most of them non-eclipsing. In the third edition of the Yale Parallax Catalogue, published in 1952, there were 48 stars within 16 light years of the Sun, and of these, 23 (including the Sun) were single stars, but the other 25 formed 11 binaries and one triple. Between 16 and 32 light years there were 167 stars, of which 94 were single, and 73 formed 27 binaries, five triples, and one quadruple (ξ Ursae Majoris). So it was clear that about half of the stars in the neighbourhood of the Sun were constituents of multiple systems, and there was no reason to suppose that this should not be generally true for more distant stars.

It was realised that if a large proportion of stars are binaries, this could cause accuracy problems with plotting their positions on the H–R diagram if they are unresolved binaries. In the worst case, if what is thought to be a single star is, in fact, really a pair of identical stars, then the "star" will appear twice as bright as it actually is (assuming one star is not in eclipse from the other star) and it will plot 0.75 magnitudes too high in the H–R diagram. Moreover, if the two stars of a binary are too close, as in the case of β Lyrae, the evolutionary history of both stars may be greatly affected by the transfer of matter between them.

Between 1914 and 1919 Ejnar Hertzsprung at Potsdam applied photography to the measurement of double stars and obtained separation accuracies of better than 0.01 arcsec, an order of magnitude better than previous visual measurements. His measurements of the binary star 61 Cygni (the star whose parallax Bessel had measured in 1838) showed that it had an invisible companion with a minimum mass of 0.02 times the mass of the Sun. Further work on 61 Cygni, undertaken by Strand and Armin Deutsch in the late 1940s, produced a more accurate minimum estimate of 0.008 times the mass of the Sun, indicating that this companion could, therefore, be the first planet to have been discovered outside of the solar system.

The most massive star known for a long time,† called HD 47129 or Plaskett's star, was found in 1922 by J. S. Plaskett, the director of the Dominion Astrophysical Observatory at Victoria, Canada. It consists of two O7-type supergiants in a binary pair with a period of 14 days. Plaskett deduced that the masses of each of the stars is about 90 suns, although this was reduced in 1961 by J. Sahade of the La Plata Observatory in Argentina, who estimated that the brighter (primary) star has a mass of 40 suns, and

---

*Binary stars are double stars which are gravitationally connected to each other, both stars orbiting their common centre of mass.
†η Carinae is the most massive star known at present (see Pages 171–172).

the dimmer (secondary) star has a mass of 60 suns. This is another case, like that of β Lyrae, where the dimmer star appears to be the heavier, again violating the mass-luminosity relationship.

The two stars in HD 47129 were found to yield different values for the radial (i.e. line-of-sight) velocity of the centre of mass of the system, and Abhyankar found, in 1959, that the velocity curve of one of the stars was distorted. This was attributed to one of the stars shedding gas, like β Lyrae, producing erroneous radial velocity measurements. The distance of HD 47129 was estimated by Morgan and Albert Hiltner, in 1965, as about 2,700 light years, giving a very bright absolute magnitude of −4.0, like that of β Lyrae.

9 Young Stars, Old Stars and Stellar Explosions

Young Stars

The KL Nebula

Douglas Kleinmann and Frank Low discovered the KL Nebula in 1967, when they were observing the Orion Nebula at a wavelength of 22 μm (microns) in the infrared. KL is near the small Trapezium group of stars in the centre of the Orion Nebula, and was called the Kleinmann Low, or KL, Nebula after its discoverers. In the same year Eric Becklin and Gerry Neugebauer discovered the Becklin Neugebauer or BN object, having an apparent temperature of only 600 K, in the KL Nebula. Astronomers thought, at first, that the BN object was a protostar (i.e. a condensing cloud of gas that has not yet become hot and dense enough to start shining as a star), but radiation was soon found coming from hydrogen atoms at a temperature of 10,000 K, indicating that the BN object is a cloud of warm dust surrounding a star. The star is thought to be only a few thousand years old, as the dust cloud remaining after the formation of the star has not yet dispersed .

Further analysis showed that an object known as IRc2, which is also in the KL Nebula, is five to ten times brighter than the star at the centre of the BN object. IRc2 appears to shine as brightly as 100,000 suns, but it is surrounded by a large cloud of dust, which has greatly reduced its apparent brightness.

The Carina Nebula

In the 1920s, Robert Trumpler catalogued two star clusters called Trumpler 14 and 16 in the Carina Nebula which appear to be mainly responsible for heating the nebula. The star η Carinae, which is in Trumpler 16, had been observed in the nineteenth century to vary in intensity by a considerable amount over a number of years. It was clearly not a nova, as it took years to reach its peak intensity, so what was it?

The first clue was found in 1914 when the itinerant Scottish astronomer Robert Innes discovered that η Carinae was surrounded by a shell of gas and dust. It took about another 50 years, however, until the true nature of the star η Carinae was revealed. Firstly, in 1968 Gerry Neugebauer and Jim Westphal at Caltech found, much to their surprise, that η Carinae is the brightest star in the sky at an infrared wavelength of 20 μm. This infrared radiation is, in fact, being generated by a surrounding dust cloud which the star has heated up to a temperature of 250 K. Then, in 1979, the Einstein Observatory spacecraft found η Carinae to be a strong source of X-rays. Detailed analysis showed that η Carinae is the most massive star known, being about 120 times as heavy as the Sun, and shining as brightly as five million Suns.

Robert Gehrz, of the University of Wyoming, and Ed Ney measured the size of the dust and gas cloud surrounding η Carinae, from records made by many observers since its discovery by Innes. They showed that the cloud was ejected by the star during its brightening in the 1840s, and that the cloud is still gradually expanding. Kris Davidson recorded the ultraviolet spectrum of the gas cloud in 1981 at the University of Minnesota, and found that it was typical of one ejected by a star near to the end of its lifetime.

The central star HD 93129A of the other nebula, Trumpler 14, is also very massive, having about the same mass and energy output as η Carinae, but it does not have a dust and gas cloud surrounding it. It radiates mostly in the ultraviolet, as it has a surface temperature of 52,000 K.

A number of other stars in the Carina Nebula were found by the Einstein Observatory to be intense emitters of X-rays. The star HD 93162, for example, is only one tenth as massive as η Carinae, but it is emitting almost as much X-ray energy.

Because η Carinae is so massive, it will have a stellar lifetime of only a million years. It thus appears that, about 1 or 2 million years ago, the Carina Nebula was a large gas cloud, which has gradually contracted over the intervening period to form many stars. The most massive ones, η Carinae and HD 93129A, are now near the end of their lifetimes, whereas the smaller, more normal stars are near the beginning of their lives. The Carina Nebula is thus a gigantic geriatric ward and nursery, all in one.

Rho Ophiuchi

In the mid 1980s, the IRAS satellite discovered a cool infrared object (designated IRAS 16293–2422) embedded in an interstellar dust cloud surrounding Rho (ρ) Ophiuchi, about 500 light years from Earth. Charles Lada, Philip Maloney, Christopher Walker and Erick Young of the University of Arizona, assisted by Bruce Wilking of the University of Missouri, examined this IRAS object using the 12 m radio telescope on Kitt Peak, and found in 1986 that it is a cloud of dust, part of which appears to be collapsing onto an embryonic stellar core, possibly only 30,000 years old. Further studies, particularly at long wavelengths, have shown that the ρ Ophiuchi cloud contains many other stars in their early stages of development.

Pulsars

The Discovery

Quasars had been found in 1963 to be almost point-like radio objects at large distances (see Page 238) and, shortly after their discovery, Anthony Hewish at Cambridge set out to estimate their angular diameter by measuring the degree of scintillation, or twinkling, of their radio emission.

Light from ordinary stars twinkles at visible wavelengths because of thermal currents in the Earth's atmosphere, but planets don't twinkle because they are not point sources. In the mid 1960s it was found that exactly the same thing happened at radio wavelengths with quasars, with the degree of twinkling, or scintillation of their radio waves, being a strong indication of their size. The quasars which subtend the largest angle at the Earth did not scintillate at all, but the smaller the angular size, the more their radio signals seemed to vary, the scintillation being caused by variations in the solar wind.

It was noticed in the 1960s that all the scintillating radio sources were quasars, so Hewish's first task was to survey the sky for scintillating objects, and to do this he needed a radio telescope that could detect rapid fluctuations in faint signals. His design consisted of an array of permanently fixed wires connected to over 1,000 posts, which were spread over an area of 4½ acres (2 hectares). It could measure variations in very faint signals over periods as short as one tenth of a second. Clearly the telescope could not be moved to physically track sources as they moved across the sky, but this was a price worth paying to get such a large collecting area.

Shortly after they had started work with this new telescope, Jocelyn Bell, one of Hewish's research students, noticed an unusual signal coming from source CP 1919 on the tape made on 6th August 1967. After a short diversion to other work, she decided to try a faster sampling rate and found, on 28th November, much to her surprise, that the source showed a regular pulse with a frequency of 1.34 seconds.* The source passed through the field of view of the telescope every sidereal day of 23 h 56 min, and so it was unlikely to be man-made, as man-made devices tend to operate on a 24 hour clock. As astronomers work in sidereal time, however, a check was made that no radio telescopes were transmitting these signals. They were not.

If the pulses were not man-made what were they? They were clearly

*Pulsars often have two pulses of different amplitudes and/or shapes. The pulse rate, or pulsation frequency, refers to the time between two similar pulses.

coming from outside the Earth, but the pulses were so short and the pulsation frequency so short and consistent* that Hewish could not conceive of them coming from a natural source. Maybe they were coming from some extra-terrestrial intelligence? The source was accordingly designated LGM 1, standing for Little Green Man 1, but Jocelyn Bell found another pulsing source in December, and two more in January, making this theory implausible.

The announcement of the discovery of the first pulsing source was made in February 1968 and, during that year, other radio observatories found even more sources. It was suggested that maybe these pulsars, as they became known, were rapidly rotating white dwarf stars, but white dwarf stars are too large to spin at about one revolution per second. Thomas Gold of Cornell University and Franco Pacini then suggested that they were rapidly rotating neutron stars, with typical diameters of only 10 to 20 km, which were emitting beams of synchrotron radiation. These beams would be picked up on Earth as pulses from the rotating star, in the same way that the rotating light of a lighthouse is seen to flash by a stationary observer. Some years earlier it had been suggested that neutron stars were produced by supernova explosions, so astronomers speculated that these pulsars may be the stellar remnants of such explosions.

The Crab Pulsar

In November 1968, a pulsar was discovered in the Crab Nebula, which was known to be a supernova remnant (see Pages 196–198). This pulsar had a period of only 33 milliseconds, the shortest period that had been observed up to that date. The other short-period pulsar found in 1968 was also associated with a suspected supernova remnant, the Vela remnant, and this led to the suggestion that pulsars slowed down with age, eventually outliving the visible nebula that had been produced by the explosion. Within a month John Comela, using the 1,000 ft (300 m) radio telescope at Arecibo, Puerto Rico, confirmed this theory when he showed that the pulse rate of the Crab pulsar had lengthened by 10^{-6} seconds. This deceleration rate also enabled an estimate to be made of the age of the Crab pulsar, which turned out to be about 1,000 years, which was consistent with it being the residual star of the supernova explosion that had been observed by Chinese and Japanese astronomers in the year 1054.

In September 1969, less than a year after its discovery, the period of the Crab pulsar suddenly decreased by 3×10^{-10} second and, since then, similar "glitches" have been observed for other pulsars. In particular, the period of the Vela pulsar has been observed to change suddenly by 2×10^{-7} second. The surface crust of a neutron star is oblate, because of centrifugal force, but

*Each pulse lasted for only 0.016 seconds. The pulsation frequency was accurate to one part in 10 million. (*Note*. The pulse duration meant that the source emitting the pulses could not be larger than about 0.016 light seconds, or 5000 km, in diameter.)

this oblateness has to reduce as the star spins down and the centrifugal force reduces. The glitches are thought to be due to the surface crust of the neutron star suddenly readjusting itself to the reduced centrifugal force by periodic fracture.

William Cocke, Michael Disney and Donald Taylor were the first to observe optical pulses emitted by a pulsar when, in 1969, they discovered optical flashes coming from the Crab pulsar with exactly the same frequency as the radio pulses. They observed these pulses using the 36 inch (90 cm) telescope at the Steward Observatory in Tucson, Arizona. In March 1969 the Crab pulsar (now called NP 0532) was also observed to be pulsing in X-rays, and 7 years later it was found to be pulsing in γ-rays.

The second optical identification of a pulsar was made by astronomers working at Siding Spring, Australia, in 1977, who found that the Vela pulsar, PSR 0833–45, which has a period of 0.089 seconds, is an object of mean magnitude 24 in the Vela supernova remnant (see Pages 201–203). The pulses were very difficult to detect as, with such a faint object, they were at the limit of sensitivity of even the 153 inch (3.9 m) f/3.3 Anglo-Australian telescope, using a state-of-the-art image intensifier.

Cygnus X-3

Cygnus X-3 was discovered as an X-ray source from a rocket flight in 1966, and it was later found to be the source of very high energy γ-rays. Cygnus X-3 was also detected at infrared wavelengths, but it could not be seen in visible light owing to absorption by galactic dust. Then, in 1971, measurements made with the Uhuru satellite showed that the X-ray intensity varied with a period of 4.8 hours, indicating that Cygnus X-3 was a rapidly orbiting binary star, although no such variations have been observed at radio wavelengths.

On 2nd September 1972, Philip Gregory at the Algonquin Radio Observatory in Ontario noticed that the radio emissions from Cygnus X-3 had brightened over 1,000 times to make it one of the strongest radio sources in the sky. A second radio burst was observed in 1982, and a third in 1983.

In 1934, the Soviet physicist Pavel Cerenkov had discovered flashes of bluish light in the Earth's atmosphere. This Cerenkov radiation, as it is called, is caused by high energy particles travelling through the atmosphere at a speed greater than the velocity of light in the atmosphere (which is less than its velocity in vacuum). These particles can be produced, for example, when a high energy γ-ray (of energy $> 10^{11}$ electron volts) hits the atmosphere, producing a cascade of electrons and positrons (positively charged electrons).

Following the radio outburst of Cygnus X-3 in 1972, the Crimean Cerenkov group reported detecting Cerenkov radiation in the Earth's atmosphere, with a period of 4.8 hours, and γ-ray energies in excess of 10^{12} electron volts. The Whipple Observatory Group confirmed the period of this Cerenkov radiation but, since 1980, this radiation has generally not shown any variation.

The SAS-2 γ-ray group reported detecting γ-rays from Cygnus X-3 with a 4.8 hour period in 1977, but the Cos-B satellite was unable to find any such γ-ray variations during its 7 years of observation starting in 1975, in spite of an initial report that it had found a 4.8 hour period. So doubt exists as to whether the 4.8 hour variation exists at γ-ray frequencies. In August 1985, the Durham group at Dugway announced the discovery of a completely different sort of variation in the γ-ray output, when they found that it was pulsing with period of 12.5908 seconds. This pulsation was in γ-rays with energies greater than 10^{12} electron volts, but no such pulses have yet been found at any other frequency. Cygnus X-3 was the first pulsar to be discovered at γ-ray wavelengths.

Binary and Millisecond Pulsars

In 1971, the X-ray satellite Uhuru detected that the source Centaurus X-3 was emitting pulses of X-rays with a period of 4.84 seconds. In addition, every 2.1 days the source disappeared for about 12 hours, indicating that this X-ray pulsar was part of an eclipsing binary system. This was proved by accurately measuring the time of arrival of the pulses, that were also seen to vary with a period of 2.1 days, as the X-ray pulsar orbited the common centre of gravity of the binary system.

Joe Taylor and Russell Hulse of Princeton University discovered the first binary pulsar (PSR 1913+16) at radio wavelengths in 1974, using the Arecibo radio telescope. The pulsar was found to spin at a rate of 17 times per second, and its intensity was also found to be varying with a period of 7 h 45 min. Analysis showed that the pulsar is a 1.4 solar mass neutron star in orbit around the common centre of mass of an invisible star of slightly less mass, possibly a white dwarf. This binary pair provided an excellent check of Einstein's theory of relativity as, not only are the stars reasonably heavy, they are orbiting very close to each other. As a result the periastron of the binary pulsar's orbit was found to precess at 4.2° per year (which is 35,000 times faster than the rate of precession of Mercury's perihelion), in good agreement with theory.

In 1982, Ali Alpar of the Scientific Research Institute in Turkey, Jacob Shaham of the Institute of Physics in Jerusalem, and Andrew Cheng and M. A. Ruderman in the United States suggested that some fast pulsars, like PSR 1913+16, could be old pulsars that have been spun up as the result of material being attracted onto the pulsar from the other star in a binary pair. The Hercules X-1 pulsar was observed to be speeding up, giving added credibility to the theory.

Don Backer of the University of California and Bill Erickson of the University of Maryland became intrigued in the late 1970s by the scintillation of the radio source 4C 21.53 that had been discovered in the 1960s. This source lay in the plane of the Milky Way, which was unusual as the interstellar gas should smear out such scintillations. A number of astronomers thought that 4C 21.53 may be a pulsar, but the search for pulses that began in 1979 failed to detect any. The search did discover, however, that the source was not one radio source but three, one of which was a point source. Further work with radio telescopes at Culgoora (Australia), Bonn (West Germany) and Westerbork (the Netherlands) disentangled these sources. One was found to be a radio galaxy, one a nebula which is accidentally in our line of sight, and one was the point source.

In September 1982 Shrinivas Kulkarni found evidence for pulses coming from the point source using the Arecibo radio telescope. Then in November the team of Kulkarni, Don Backer and Carl Heiles, all of the University of California, Mike Davis of the Arecibo Observatory, and Miller Goss from Groningen in the Netherlands hit the jackpot. They found that the object was a pulsar, but it was pulsing at the amazing rate of once every 1.56 milliseconds. This first millisecond pulsar to be discovered, now called PSR 1937+214, was found to be about 7,000 light years from Earth. It was thought to have been spun up like the binary pulsar PSR 1913+16, by taking material from a binary companion, although 1937+214 did not seem to have such a companion.

The binary theory for millisecond pulsars led astronomers to search for more millisecond pulsars in globular clusters, which were known to have many X-ray binary stars. T. T. Hamilton, David Helfand and Robert Becker used the Very Large Array radio telescope in 1983, and found a point-like radio source in the globular cluster M28, but they could find no evidence of its pulsations. Bill Erickson found that its radio spectrum was similar to that of other short-period pulsars, however, and this convinced Backer and Kulkarni from the University of California at Berkeley that the source was a short-period pulsar, but try as they might they could not find any pulsations. Backer and Kulkarni, at the suggestion of Trevor Clifton, a visiting astronomer from Jodrell Bank, asked Jodrell Bank to undertake a new set of observations for them. Andrew Lyne and Andrew Brinklow undertook the observations in the UK, sending the raw data to Backer, who had it processed by John Middleditch of the Los Alamos National Laboratory using the Cray supercomputer. In May 1987, after 5 hours of computer processing time, the millisecond pulsar PSR 1821−24 was found with a period of 3.054 milliseconds.

As mentioned above, the search for millisecond pulsars had concentrated on globular clusters because of the large number of binaries in them, and binaries were thought to be likely candidates for millisecond pulsars. Ironically, the pulsation rate of PSR 1821−24, like that of PSR 1937+214 before it, was found to be very stable, so neither of these two millisecond pulsars appeared to be part of a binary. On 13th October 1987 a second millisecond pulsar, PSR 1620−26, was discovered in a globular cluster, this time in M4 in Scorpio. Its period of 11.076 milliseconds was not constant, however, indicating that it may have a binary companion.

The discovery of the strange pulsar PSR 1957+20, with a period of 1.6074 milliseconds, was announced by Andrew Fruchter, Daniel Stinebring and Joseph Taylor of Princeton University in May 1988. Its intensity was found to vary slightly over a period of 9 h 10 min, which indicates that it is one member of an eclipsing binary pair. Assuming that the pulsar is a neutron star of mass 1.4 suns, the mass of the companion can be calculated as being only 0.022 suns, which is surprisingly low, even for a white dwarf. The strange thing is that, from intensity measurements taken during the eclipse of the binary, the "white dwarf" appeared to have a diameter of 0.75 times that of the Sun, which is much too large, so it was suggested that the white dwarf is surrounded by a large gaseous envelope. Such a gaseous envelope would not be stable, however, and it should be driven away from the white dwarf by the elementary particles and γ-rays emitted by the pulsar. This led astronomers to suggest that the pulsar, now called the Black Widow pulsar, is evaporating the white dwarf, hence its low mass, and this could explain why some millisecond pulsars are not now in binary pairs, as they have completely evaporated their binary companion. Less than a year later, a photograph taken by Jeff Hester, of Caltech, showed the shock wave produced by the pulsar PSR 1957+20 as it blasts its white dwarf companion.

Finally, a sobering story.

In 1985, Trevor Clifton and Andrew Lyne discovered an apparently ordinary pulsar, PSR 1829−10, near the centre of the Milky Way, about 30,000 light years away. Six years later, in May 1991, Setnam Shemar was analysing the pulsation periods of various pulsars at Jodrell Bank, when he noticed something strange in the behaviour of PSR 1829−10. After further work by Shemar, Andrew Lyne (his supervisor) and Matthew Bailes, they announced in *Nature* on 25th July that they had found evidence for a planet, at least ten times the mass of the Earth, orbiting the pulsar. Much effort was spent by a number of astronomers trying to explain how a planet could have survived the supernova explosion that produced the pulsar and then, 6 months later, Andrew Lyne announced that the data analysis used to discover the planet had been in error. There was no planet around pulsar 1829−10.

Novae and Supernovae

Early Work

About a dozen novae, or "new stars", had been observed to reach naked-eye visibility between 1570 and 1890, the brightest of which were called Tycho's and Kepler's stars, after the two famous astronomers who studied them. Tycho's star had emerged from obscurity to become the brightest star in the sky in November 1572, before rapidly losing intensity, and Kepler's star had behaved in a similar way in 1604.

More recently, Nova T Coronae Borealis was discovered by John Birmingham of Millbrook, Ireland just before midnight on 12th May 1866, when it reached magnitude 2. Four hours earlier, Julius Schmidt of Athens had been surveying that part of the sky, and he was sure that the nova was not visible to the naked eye at that time (so it was fainter than magnitude 6). This nova was important as it was the first to be studied spectroscopically. William

Huggins examined its spectrum on 16th May, finding it to be that of a typical Secchi type III star but with bright hydrogen emission lines added. Nine days after its discovery the star was again invisible to the naked eye.

Nova Cygni was discovered by Schmidt at Athens on 24th November 1876 when it reached third magnitude, and over the next two weeks it was seen to decrease quickly in intensity to sixth magnitude. The exact date of its outburst was not known, but the star had not been recorded as late as 20th November, and so it must have been very faint at that stage. Spectroscopic analysis of the star in its nova state showed a number of bright emission lines due to hydrogen, and probably helium, superimposed on a continuous spectrum with strong absorption. As the star decreased in brightness, the continuous spectrum reduced in intensity and the number of bright lines in its spectrum also reduced. The star was lost to view in March 1877, but when it reappeared in September its spectrum had changed completely to resemble that of a planetary nebula with a single bright green emission line.

So astronomers were faced with explaining how a star could increase dramatically in intensity in a few days, or maybe hours, and then decrease by three or four magnitudes in just 2 weeks, with its spectrum changing months after the event. Two alternative explanations were considered plausible, either novae were due to the sudden explosion of a star, or they were due to a collision, but the rapid reduction in intensity observed seemed to be too fast for either theory.

Novae 1890–1945

The Rev. T. D. Anderson of Edinburgh discovered Nova T Aurigae, and announced it on 1st February 1892 by sending an anonymous postcard to Ralf Copeland, the Astronomer Royal for Scotland. Subsequent searches showed that the star was fainter than magnitude 13 as late as 8th December 1891, but it had been first recorded photographically, in its nova state, by the Harvard Observatory on 10th December 1891 at magnitude 5.4, reaching a peak of magnitude 4.2 on 17th December.

Nova T Aurigae was the first nova to have its spectrum photographed, the best early spectrum being that produced on 22nd February by Sir William and Lady Huggins. It showed bright hydrogen, helium and sodium lines, together with a continuous spectrum and dark absorption lines. The Doppler shift of the bright and dark lines indicated that the emitting gases were receding from the Earth at 300 km/s, while the absorbing gases were approaching the Earth at 500 km/s, suggesting that the nova had been caused by the close approach of two stars, one with an emission line spectrum, and the other with an absorption line spectrum. It seemed unlikely, however, that the two stars that appeared to be involved should have identical line spectra, although one had bright lines and the other dark ones.

Nova T Aurigae was last recorded as magnitude 16 on 26th April before disappearing, but on 17th August it reappeared at magnitude 10, and

4 days later its spectrum was seen to consist of a series of emission lines with the strong green emission line of planetary nebulae clearly evident (like Nova Cygni in 1877). These observations caused the two star theory to be abandoned, and in 1892 Hugo von Seeliger of Munich suggested that novae are caused when stars are heated when they enter a dense cloud of gas. Three of the five dimmer novae discovered in the remainder of the nineteenth century showed the same spectral development.

The Rev. Anderson discovered the most intense nova since 1604 when he saw a very bright new star near Algol in the early hours of 22nd February 1901. Nova Persei, as it was called, went from magnitude 12.8 on 20th February, to magnitude 2.7 on 22nd February, to magnitude 0.1 on 23rd February, before starting to lose intensity, so a year later it was at magnitude 7. Its spectrum on discovery was that of a B type star with hydrogen and helium absorption lines but, by the following day, when it was at maximum intensity, it had changed to an A type spectrum, indicating that a reduction in its surface temperature had taken place. By the following day, 24th February, the familiar hydrogen emission lines had appeared, showing, for the first time, that the emission line spectrum of a nova does not appear when the star starts its rapid increase in intensity, but only when the decrease in intensity has begun.

The bright hydrogen lines, although very broad, were, on average, stationary with respect to the Earth, but the material causing the dark lines was approaching the Earth with a velocity of 1,200 km/s. About 3 weeks after discovery, these dark lines were found to be split into many fine dark lines, indicating that the various layers of gas were approaching us with different velocities. A little later, Nova Persei started oscillating in intensity with an amplitude of over one magnitude, with corresponding changes in its spectrum. When the nova brightened, the continuous spectrum became brighter, the absorption lines became darker, and the emission lines faded, and when the nova faded all these changes reversed. By July the nova was showing the usual emission line spectrum of a nebula.

On 19th August 1901, 6 months after the outburst of Nova Persei, Camille Flammarion and E. M. Antoniadi working in France found that the star was partially surrounded by a faint nebulous arc of about 6 arcmin in diameter. Max Wolf at the Königstuhl Observatory of Heidelberg independently discovered this arc 3 days later, and G. W. Ritchey found at Yerkes on 20th September that it was a complete circle centred on the nova. On 7th November, Charles Perrine at the Lick Observatory found that the circle had become larger and, from the estimated distance of the nova, calculated the expansion rate as about that of the velocity of light. Such high expansion rates are impossible for matter, but J. C. Kapteyn of Groningen and Seeliger suggested that we were seeing the light from the original outburst being scattered by a pre-existing dust cloud. Confirmation was achieved using an exposure totalling 35 hours over several nights, when the spectrum of the brightest part of the ring was found to be like that of the nova at maximum in February 1901, rather than the emission line spectrum that had first been seen from the nova in July 1901. Later photographs showed that the diameter of the ring was expanding at the rate of 11 arcmin per year, giving a distance of the nova of 660 light years, assuming that the ring was expanding at the velocity of light.

Fifteen years later, in September 1916, the British astronomer W. H. Steavenson detected a very small faint ring around Nova Persei which, by August 1919, was about 10 arcsec in diameter, with an expansion rate of 0.5 arcsec per year. The velocity of the gas was known from its Doppler shift, so the expansion of this ring enabled the distance of the nova to be estimated as 1,500 light years. This distance also suggested that the light ring, that had been observed previously, was apparently expanding at faster than the velocity of light, but this apparent superluminal velocity was the result of a trick of perspective. The light was really moving at its normal speed, but it was being reflected off a layer of gas between us and the nova, not, as originally assumed, off gas in the same plane as the nova.

The spectral development of Nova Geminorum in 1912 was similar to that of Nova Persei. The earliest spectrum that was taken just as Nova Geminorum started to decline showed dark absorption lines, mainly of hydrogen, displaced towards the blue, indicating that the absorbing gas was moving towards the Earth, and faint bright lines that were undisplaced (like Nova Persei). A few days later, both the dark and bright lines had become more evident, and a second set of absorption lines had appeared, even more displaced to the blue than the first set. The displacement of both sets of absorption lines gradually reduced, and then, a week after maximum, a third set of dark lines appeared, more displaced to the blue than either of the others. The first two sets of absorption lines were like those of an A type star, but the third set was like those of an earlier B type star.

Nova Geminorum started oscillating in magnitude about 3 weeks after discovery, with the corresponding changes in its spectrum of the type seen earlier for Nova Persei. Eventually, the spectrum changed to the typical bright line spectrum of a nebula.

Nova V603 Aquilae, discovered by Luyten at Utrecht on 6th June 1918 at magnitude 5.8, reached a spectacular −1.1 magnitude 3 days later, but its most important claim to fame was that it was the first nova whose spectrum had been recorded in a pre-nova state. The best spectrum had been taken at Harvard in 1899, showing a bluish-white class A star, of magnitude 11, with a continuous spectrum and hydrogen absorption lines. As the star brightened in June 1918, the lines became very narrow, like the c lines of supergiants, and in its nova state the Doppler shifts of the absorption lines indicated gas velocities of up to 3,400 km/s. This velocity was so large that some astronomers wondered if their interpretation of the lines as being Doppler-shifted was correct. These doubts were generally satisfied, however, when the gaseous expansion ring, of the sort detected for Nova Persei, was detected for Nova Aquilae in 1918, enabling its distance of 1,200 light years, and absolute magnitude of −8 (at maximum), to be determined. It was later found that the star had still been of magnitude 10.5 on a plate taken at Heidelberg on 5th June, and so it had increased in intensity by 11.5 magnitudes in just 4 days.

Because Nova Aquilae was so bright, the effect of the changes of its spectral emission lines on its colour were clearly seen. At maximum intensity the nova was white, but as it lost its intensity it changed successively to yellow, pink and cerise. The change from yellow to pink was due to the red emission line of hydrogen becoming more and more prominent in its spectrum, and the change to cerise was due to the addition of a number of blue emission lines. Finally the nova changed to green when its spectrum changed to the emission line spectrum of a nebula, but at this stage it was below naked-eye visibility.

Based on these and other novae, it became progressively clear, in the first two decades of the century, that a nova was a star that threw off two or three successive shells of gas over a few days, starting when the nova was at maximum intensity. These shells produce absorption lines in the light coming from the central star, and emit light of their own in bright emission lines. The centres of these bright lines were generally undisplaced, but the lines were broad because the gas that was emitting them was in a spherical expanding shell, moving in all directions relative to the nova-Earth line. The dark absorption lines were displaced, however, because only that part of the shell that was directly between the central star and the Earth was absorbing the light that we saw.

The next major step in understanding novae was taken by analysing the behaviour of Nova Pictoris, which was discovered by R. Watson in South Africa on the morning of 25th May 1925. It took the unprecedented period of 15 days to reach maximum intensity and, during this time, it was studied intensively by astronomers at the Cape Observatory. They found that its spectrum remained essentially unchanged during its 15 day rise to maximum, thus indicating that its temperature was constant, and that its increase in luminosity must be due to a large increase in size. This was quantitatively confirmed by the displacement of the absorption lines that showed an expansion rate of 115 km/s. In parallel, J. F. Hartmann, who was director of the National Observatory in Buenos Aires, had come to the same conclusion, that the star was getting brighter because it was expanding. Harold Spencer Jones at the Cape estimated that, when the star was at its maximum size and intensity, it had a diameter similar to that of a supergiant, that is about 400 times that of the Sun.

After maximum, the spectrum of Nova Pictoris developed in the normal way. There were two sets of absorption lines which appeared one after the other, with displacements indicating expansion gas velocities of 130 and 320 km/s. From a careful study of the fading of the continuous spectrum, Spencer Jones calculated that the central star was shrinking at the rate of 100 km/s. The two expanding shells of gas were first observed directly by van den Bos and W. S. Finsen at the Union Observatory, Johannesburg in 1928, 3 years after maximum.

So Nova Pictoris was caused by a star that expanded and burst for some reason. The star, when it burst, was seen to throw off expanding shells of gas, while the original star shrank back in size.

At 4.30am on the morning of 13th December 1934, Prentice, who was director of the Meteor Section of the British Astronomical Association, a society for amateur astronomers, discovered the next major nova, namely Nova Herculis. He immediately reported his discovery to the Royal Observatory at Greenwich, which enabled them to take a spectrogram of the nova that same morning. The nova then decreased in intensity for 4 days, before increasing in intensity again by about two magnitudes. It reached its maximum magnitude of 1.4 on 22nd December, and gradually reduced to about magnitude 4 in the following April, when it dropped to magnitude 13, over a period of only 1 month. After a short interval, the nova started increasing in intensity again and, 5 weeks later, it reached magnitude 7, where it remained for over a year before fading once more. On discovery the spectrum was that of a B type star, but this changed to an A type as maximum approached, indicating a reduction in its surface temperature. When it faded out of naked-eye visibility in April it was cerise in colour, but when it reappeared in July it was a beautiful emerald green colour, caused by the nebula emission lines.

So by 1935 the spectra of novae were beginning to be understood. During the rise to maximum, which had been seen to take from one to 15 days, the star swells in size. Its spectrum is largely unchanged in many cases during this rise, although, in the case of both Nova Persei and Nova Herculis, the star changed from type B to type A, indicating a reduction in its surface temperature. At maximum the star bursts, expelling a shell of gas and, over the next few days, one or two further shells of gas are expelled as the star reduces in size. The shells of expanding gas can temporarily reduce the intensity of the central star, as in the case of Nova Persei, but as the shells expand, the central star becomes visible again. The shells generate a spectrum of their own with broad emission lines and, at the same time, produce absorption lines in the spectrum of the central star. The absorption line spectra generated by successive shells of gas tend to become those of a slightly earlier spectral type, indicating a reduction in density. Eventually, after a few months, all that can be detected is the bright green spectrum of a tenuous nebula.

Supernovae 1890–1945

In 1911 Edward Pickering had differentiated, in his classification of variable stars, between normal novae and novae seen in nebulae. The first of the latter was Nova S Andromedae, which had been discovered in the Andromeda nebula in August 1885 by Ernst Hartwig, of the Dorpat Observatory, and Ludovic Gully of Rouen. Nova S Andromedae was of 9th magnitude on 20th August, but it had brightened to 7th magnitude by 31st August, from which it declined to 16th magnitude by 7th February 1886 and thence into obscurity. The nova showed the continuous spectrum of a normal star, but the bright emission lines detected in previous nova were only seen with difficulty. Hermann Vogel thought that the spectrum was continuous, for example, but the Hungarian N. von Konkoly detected four broad emission bands on 4th September from the Vienna Observatory.

Williamina Fleming discovered the next nova in a nebula on 12th December 1895 when she was examining a plate taken with an objective prism (see Page 292) in the previous July. Nova Z Centauri, as it was called, was found in the nebula catalogued as NGC 5253. The nova's maximum intensity was of 7th magnitude, and Annie Cannon found that the spectrum was unlike any other nova except Nova S Andromedae*. Twenty-two years later, in July 1917, G. W. Ritchey discovered a nova of magnitude 14.6 in the nebula NGC 6946, and this led him to examine the Mount Wilson plate collection, where he found six relatively faint novae in nebulae, two of which were in the Andromeda nebula, but they were both very much fainter than S Andromedae. In parallel, Heber D. Curtis examined plates taken with the Crossley reflector at the Lick Observatory, and found one nova in NGC 4527, and two in NGC 4321.

One of the great outstanding issues at the start of the twentieth century was whether all nebulae were associated with the Milky Way, and were thus relatively close, or whether some of them were island universes of stars at great distances (see Pages 209–211). If some of them were the latter, the novae seen in them must be incredibly bright, and in 1917 Harlow Shapley calculated in that case that the absolute magnitude of Nova S Andromedae must be about −16, or about ten magnitudes brighter than normal novae seen in the Milky Way. This seemed ridiculous, so Shapley suggested that the nebulae were much closer, as essentially part of the Milky Way. Ten novae were then found, however, in the Andromeda nebula over the next 2 years that were ten magnitudes fainter than S Andromedae, leading Curtis to suggest that the Andromeda nebula was an island universe of stars, or galaxy, at some distance from the Milky Way, there being two different types of novae, normal novae and supernovae† in it and other galaxies. Tycho's nova of 1572 and Kepler's of 1604 were very bright novae in the Milky Way, so maybe they were supernovae also.

Edwin Hubble settled the matter in 1924, when he discovered Cepheid variables in the Andromeda nebula and other spiral nebulae (see Page 233). This enabled him to estimate their distance, which clearly showed that they were far away from the Milky Way.

Milton Humason and Walter Baade were the first to photograph a supernova with a slit spectrograph when they investigated supernova 1936a with the 100 inch (254 cm) Hooker telescope. They found bright emission lines like those of an ordinary nova, but they were much broader, indicating that the star was expanding at about 6,000 km/s. In the same year, W. A. Johnson and Cecilia Payne-Gaposchkin‡ re-examined the original spectrum of the supernova Z Centauri, taken in 1895, and found the same great broadening of the emission lines.

---

*The spectrum of Nova S Andomedae was only observed visually. The first spectrum of a nova in a nebula to be photographed was that of Nova Z Centauri.

†The term "super-novae" was first used (by Fritz Zwicky) in 1931. The hyphen was omitted a few years later.

‡Cecilia Payne had married Sergei Gaposchkin, another astronomer, in 1934.

In 1937 a supernova was discovered by Walter Baade and Fritz Zwicky of Caltech to be 100 times brighter than its parent galaxy IC 4182. The light curve of the supernova showed a quick decline in the first month, followed by a more gradual reduction of about one magnitude every 2 months for the next 2 years. The discovery of the supernova in IC 4182 was no accident, as Zwicky and J. J. Johnson had started a systematic search in 1936 of nearby galaxies to find supernovae, using the newly constructed 18 inch (45 cm) Palomar Schmidt telescope. Over a three year period they took 1625 photographs of 175 regions of the sky, and found 12 supernovae.

In 1941 Rudolph Minkowski of the Mount Wilson Observatory analysed the first results of the Palomar Supernova Search, and concluded that supernovae were of two types; Type I being of outstanding luminosity and having broad emission bands, and Type II having spectra resembling the spectra of normal novae with hydrogen emission lines. Type I had no such hydrogen lines. He further concluded that Type I supernovae reached a typical absolute magnitude of −16, whereas Type IIs reached "only" −14.* The Type Is, of which the supernova in IC 4182 was typical, decreased in magnitude after maximum initially much faster than Type IIs, however.

Baade analysed the historical records of the intensities of Tycho's and Kepler's supernovae of 1572 and 1604, and concluded in 1943 that they were both Type Is. The historical data on the supernova of 1054 were too sketchy, however, to decide what type it was.

Early Theories

In the 1920s, the collision and explosion theories for the production of novae were still competing with each other. A. W. Bickerton of New Zealand had suggested that the collision was between two stars, but the probability of that happening was considered by most astronomers to be too small, so William Pickering of the Harvard College Observatory proposed that the collision was between a star and a planet, assuming that there were more planets than stars. Hugo von Seeliger suggested that the star was in collision with a dense cloud of gas that sometimes varied in density, so causing the oscillations in magnitude observed for some novae. He also pointed out that the pre-nova dust cloud had been detected in 1901 for Nova Persei.

All of these collision theories had one major problem, however. Why was it that, of the 67 novae that had been found in the Andromeda nebula by 1926, all but one of them (Nova S Andromedae) had approximately the same magnitude as each other? It also appeared that, of the novae observed in our galaxy, most of them had increased in intensity from their pre-nova to their nova state by about the same amount (11 to 13 magnitudes). These facts were difficult to square with any of the collision theories, which should

*Modern values are −19.5 and −17.5, respectively. At the time of Minkowski's analysis the distances of the galaxies had been considerably underestimated.

produce a greater range of luminosities. The explosion theory, on the other hand, may not have these problems, if the stars that exploded were all of similar type and size, but what could cause a star to suddenly swell up and explode?

Arthur Eddington had explained that in a star the radiation pressure and gas pressure balance the gravitational attraction. He suggested, however, that, as the radiation pressure is proportional to the fourth power of the temperature, a star could explode if the temperature became too high. In a parallel development, he had also explained in 1924 that white dwarfs, like Sirius B, are made of atoms that are almost completely ionised, which enabled the atoms to be packed very close together into what was called degenerate matter. In the following year, Nova Pictoris was seen to be caused by a star swelling up and exploding.

Ralph Fowler of Cambridge developed a theory of degenerate matter in 1926, in which he confirmed that a white dwarf is largely composed of degenerate matter, with only its outer layer remaining gaseous. As the white dwarf cools near the end of its life, more and more of this gaseous layer becomes degenerate and, eventually, the whole star consists of degenerate matter at a temperature of absolute zero. The star is then a black dwarf radiating no energy.

Two years later, Arthur Milne of Oxford examined the properties of stars, not just white dwarfs, assuming that there was some degenerate matter in the centre of most of them. He showed in 1930 that, if a small part of the mass of a star is degenerate matter, the star's luminosity would be slightly higher than if a star of the same mass had no degenerate matter. The effect for most stars is small, so it does not significantly affect the mass–luminosity relationship discovered by Eddington. He found, however, that if the star has a small degenerate core, and if, for some reason, it started to lose intensity (because it was getting old, for example, and running out of fuel) then the degenerate core would vaporise. As the star continued to get dimmer, the radiation pressure from the interior could no longer support the weight of the star, and the whole star would implode and condense rapidly into the degenerate state as a white dwarf. The energy released in this sudden collapse would be enormous and would blow off the outer layer of the star. He concluded that this expanding layer is what is seen as the expanding star prior to the maximum of the nova. When the outer layer gets too large, however, the continuity of the surface is destroyed, but it continues to expand as a shell of cooling gas, while the star collapses on itself as a white dwarf.

Subrahmanyan Chandrasekhar developed the theory of white dwarfs in the early 1930s, and showed in 1931 that white dwarfs could not weigh more than about 1.4 times the mass of the Sun. In such stars, the electron degeneracy* pressure, that supports a star against gravitational collapse

---

*According to Pauli's exclusion principle, which he proposed in 1925, no two electrons can be in exactly the same quantum state. So the electrons in a white dwarf

beyond the white dwarf state, is insufficient to do so, and further collapse is inevitable. This Chandrasekhar limit, as it is called, for the maximum mass of white dwarfs varies between 1.44 and 1.76 solar masses, depending on the type of nuclei in the star. If they are all helium nuclei the maximum mass figure is 1.44, and if they are all iron the maximum figure is 1.76.

Chandrasekhar was studying at Cambridge University at the time, where Eddington, the foremost authority on stellar interiors, was Plumian professor of astronomy. Eddington fundamentally disagreed with Chandrasekhar's conclusion and, 2 years later, Lev Landau, a Russian Nobel Laureate in physics, who had independently reached the same conclusion as Chandrasekhar, then decided to reject that conclusion as it was "ridiculous" (his word).

In 1933, Baade and Zwicky showed that, if a star is appreciably heavier than 1.4 suns, it will not produce a nova when it explodes, but a supernova. They explained that the amount of energy released in a supernova explosion is appreciably higher than that released in a nova, as the final star is not a white dwarf, but a much more compact neutron* star. J. Robert Oppenheimer of Caltech and George Volkoff published their theory of neutron stars in 1939, showing that the mass of a neutron star cannot be greater than 3.2 suns. They concluded that if a central object is produced by a supernova explosion with a mass greater than this limit, the neutron degeneracy pressure is not sufficient to prevent further collapse, and a black hole will be produced whose gravity is so strong that light cannot escape. So, in order of increasing density, stars can end their lives as white dwarfs (of maximum mass about 1.4 suns), neutron stars (mass about 1.4 to 3.2 suns), or black holes (mass $>$ 3.2 suns).

Later Novae

T Coronae Borealis had been observed as a nova in 1866 reaching magnitude 2, but decreasing to magnitude 6 within 9 days. It repeated its performance in 1946 as what is now called a recurrent nova, showing that at least some novae are not one-off events. This caused a problem with the stellar explosion hypothesis of novae, as a star could hardly blow up twice.

In 1955, Merle F. Walker analysed the light curve of Nova Herculis, and showed that it is part of an eclipsing binary, one component of which is pulsating with a period of about 71 seconds. Ahnert also showed that the 4.5 hour period of the binary pair has increased by $3\frac{1}{2}$ minutes since the nova erupted, due to the loss of mass caused by the nova explosion. A few

---

resist compaction to greater than a certain density, producing what is described as electron degeneracy pressure.

*A neutron star consists of neutrons, which are the neutral particles that exist with positively charged protons in an atomic nucleus in normal conditions. A neutron star is prevented from further collapse by neutron degeneracy pressure, which is a quantum effect, analogous to electron degeneracy pressure in a white dwarf.

years later, Robert Kraft of the University of California found that most novae occur in binary star systems consisting of a white dwarf, which became the nova, and a main sequence star.

This was a crucial discovery, as novae were clearly not caused by isolated stars exploding at the end of their lifetimes, but appeared, instead, to be caused by mass leaving a main sequence star, and falling directly onto the surface of its white dwarf companion, which then explodes. A few years later, however, this theory had to be modified when it was noticed that white dwarfs flickered over a period of a few minutes, as if they were surrounded by a luminous disc of gas in what is now called an accretion disc. The gas appears to be being pulled off the main sequence star into the swirling accretion disc that reaches right down to the surface of the white dwarf, so the gas does not fall directly onto the white dwarf, but via the accretion disc.

X-ray astronomy was developed using sounding rockets in the 1960s (see Page 345), and the first comprehensive satellite observations at X-ray wavelengths were made in 1970. Then, in August 1975, the British satellite Ariel V discovered an X-ray source called A 0620-00 (see Page 205) that became, for a short time, the brightest X-ray source in the sky. This stellar source was also found to have suddenly brightened in visible light, and further examination of old photographic plates showed that it had previously brightened in 1917, thus showing it to be a recurrent nova. The X-ray output was found to be flickering at the time of its 1975 outburst, and this indicated that the X-rays were being generated as matter fell onto an accretion disc around the star, and thence onto the surface of the dense star itself. The temperature of the accretion disc was estimated at about 1 million K.

Nova Cygni 1975 (now called V1500 Cygni) appeared on 28th August 1975 and peaked at magnitude 1.7 two days later but, within a week or so, it was back below naked-eye visibility. Analysis showed it to be 3,500 light years away, with a peak absolute magnitude of −10. In September 1975 Nova Cygni was found to vary in brightness with a period of 3 h 20 min but, within a year, this period had reduced by 5 minutes. A year later the pulsation pattern had changed again.

The cause of these changes in pulsation behaviour remained a mystery for some time until, in 1987, Peter Stockman, Gary Schmidt and Don Lamb examined the white dwarf that had produced Nova Cygni 1975 and found, from the polarisation of its light, that the white dwarf was highly magnetised. This led them to conclude that the white dwarf collected material from an adjacent binary star at its magnetic poles and, in the explosion seen as Nova Cygni, the material was ejected from the white dwarf along the field lines at its two poles. The spin axis and magnetic axis of the white dwarf were not the same, and so these two beams of material appeared to an observer like a flashing light as the star rotated. This was the cause of the 3 h 20 min pulsations. The emitted gas then formed a gradually expanding nebula around the white dwarf, but this nebula, and through it the white dwarf, was spun up by the orbiting binary star. Hence the

5 minute reduction in spin period. Some of the ejected material left the binary system, some was captured by the binary companion, but some fell back onto the white dwarf, causing its rotation velocity to change once more.

If accretion discs are involved in novae, the temperature of the nova should continue to increase as more and more gas is lost from the main sequence star to the disc, and then from the disc to the surface of the white dwarf. If, on the other hand, a nova is caused by matter falling directly onto the white dwarf from the main sequence star, the maximum temperature should be reached almost instantaneously. Most novae show a gradual rise to maximum over a few days, thus indicating that accretion discs are probably present in most cases.

Later Supernovae

The war, and other research programmes, virtually stopped the Palomar Supernova Search for almost 15 years but, in 1954, it was resumed, first with the 18 inch Schmidt and then, in 1958, with the 49 inch (1.2 m) Palomar Schmidt telescope. By 1960, Zwicky had discovered a total of 44 supernovae, ranging in absolute magnitude from about −13 to −18, and by 1964 a total of 152 supernovae had been discovered world-wide since S Andromedae had been seen in 1885. When the Palomar search was officially closed in 1975 it had resulted in the discovery of 281 supernovae.

In the 1930s Chandrasekhar had defined the maximum mass of about 1.4 solar masses for a white dwarf, and Oppenheimer and Volkoff had defined the maximum mass of about 3.2 solar masses for a neutron star, and in the middle of the century the nuclear processes in stars were calculated for stars of different masses. It only became evident gradually, however, that stars could lose a significant amount of mass in the red giant phase, in between the main sequence and white dwarf phases. This allowed main sequence stars with masses significantly above the Chandrasekhar limit to become white dwarfs.

The red giant mass loss was first examined by Armin Deutsch in 1956. Further work by Ed Ney, Neville Woolf and Andrew Bernat showed that the gas haloes surrounding red giants are extensive, being up to 1,000 times the diameter of the solar system, and massive, weighing up to 2 solar masses. It began to appear as though stars of up to 8 solar masses could lose enough matter to become white dwarfs below the 1.4 solar mass Chandrasekhar limit without exploding, sometimes producing planetary nebulae in the process. More massive stars could not lose enough matter, however, and they implode to produce a Type II supernova and a neutron star.

• *Type I*

Type I supernovae were, like novae, found to occur only in binary systems consisting of a white dwarf and a normal main sequence star. The theory of their behaviour was gradually developed in the second half of the twentieth century, but even today there is not a consensus on exactly what happens.

The sequence of events for the production of a Type I supernova starts with two main sequence stars in a binary pair. The heavier star evolves the more quickly, becoming a red giant when the hydrogen burning in its core is exhausted and helium burning commences. The red giant continues to burn helium in its core to produce carbon and oxygen, but loses its envelope of unburnt hydrogen to its companion. So the red giant is transformed into a white dwarf made of carbon and oxygen, and its companion becomes the heavier star. The companion now becomes a red giant itself, losing some of its outer envelope to the white dwarf, via an accretion disc. If this takes the mass of the white dwarf over the Chandrasekhar limit, the whole white dwarf explodes in a Type I supernova. If the extra mass is not sufficient to take the white dwarf over that limit, however, the captured material, which is mostly hydrogen, is compressed on the surface of the white dwarf, where it gets hotter. When the temperature is high enough, nuclear fusion starts on the surface of the white dwarf, producing an explosion which is limited to the surface. This nova explosion, intense as it is, leaves the white dwarf and its companion virtually unscathed, so the whole process can be repeated to produce recurrent nova explosions at intervals of from tens to thousands of years.

In summary, if the mass of the white dwarf, after it has received gas from its companion, is greater than the Chandrasekhar limit, the white dwarf explodes completely producing a Type I supernova. If it is below that limit, the explosion is limited to the surface and the explosion appears as a nova.

There are a number of possible processes, when a white dwarf explodes in a Type I supernova. In one, the extra mass collected by the white dwarf causes all the carbon to detonate in less than a second to produce an isotope of nickel, called nickel 56. The outer layers of the star are ejected, and a small white dwarf of nickel 56 is left behind. The nickel 56 decays radioactively to cobalt 56 which, in turn, decays to iron 56, to produce an iron white dwarf.

In another theory of Type I supernovae, the explosion of the white dwarf takes about three seconds, with only about one half of the carbon being converted to nickel 56, the remainder being converted to calcium, sulphur and magnesium. This explosion leaves no stellar remnant, but the nickel 56 also decays to iron 56, via cobalt.

In either case, as the white dwarf has virtually no hydrogen, there are no hydrogen lines found when it explodes as a Type I supernova.

• *Type II*

A Type II supernova is produced when a star with a main sequence mass in excess of eight solar masses gets to the end of its lifetime.

When hydrogen in the core of a heavy star has been converted into helium, the core collapses and the temperature becomes high enough for helium to fuse into carbon and oxygen. This produces even more heat, and when the helium has been used up in the core, the carbon is converted to neon, magnesium and sodium. Successive processes continue, producing ever higher temperatures and heavier elements, until iron is produced in the core. Then there are no further processes possible, and the heat energy is abruptly cut off. At this stage the star consists of shells of gas, with the lightest elements hydrogen and helium in the outer shell, with shells of ever heavier elements going towards a core of iron (see Figure 9.1).

Once the star is no longer producing heat and radiation, nothing can stop it collapsing and the shells rapidly collapse onto the iron core. There is a limit to this contraction, however, and when the core of the star has been compressed to the density of neutrons in an atomic nucleus, it cannot usually be compressed any more, and rebound occurs, setting up a shock wave which progresses outwards through the star to its surface. Neutrinos are produced when the core collapses, and they move towards the surface

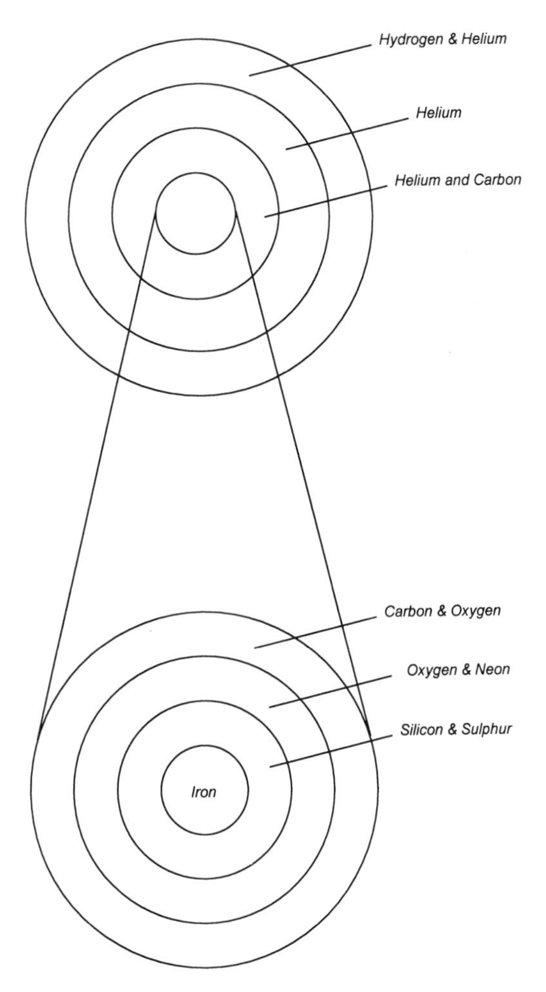

Figure 9.1 *A schematic cross-section through a star just before it explodes to produce a Type II Supernova.*

much faster than the shock wave. When the shock wave reaches the surface, we see the sudden increase in light which is the visible signal of a Type II supernova. As the neutrinos arrive at the surface of the star first, however, and as the neutrinos and light both travel with the speed of light outside of the star, the time difference in receiving their signals on Earth tells us how much longer it took the shock wave to get to the surface of the star from the core. This gives us some idea of the size and condition of the progenitor star.

The theory further predicted that, after the explosion of a Type II supernova, there would remain a central very dense neutron star, or possibly a black hole (if the star had been massive enough to overcome the neutron degeneracy pressure during its collapse). Surrounding this neutron star, or black hole, is a rapidly expanding shell of gas which cools at it expands, blown off by the shock wave.

When the iron core collapses, an enormous number of neutrons are released, and these are captured by atomic nuclei in the so-called "r-process", as described in the B^2FH paper of 1957. This key process enables elements heavier than iron to be produced.

• *Supernova 1987A*

The discovery and investigation of Supernova 1987A is not only interesting, but it was crucial in testing the predictions of the theory of supernovae explosions, outlined above. Supernova 1987A confirmed the major predictions of the theory, although it also produced a number of surprises which required the theory to be modified in detail.

In the early hours of 24th February 1987, Ian Shelton, a Canadian astronomer working at the Las Campanas Observatory in Chile, developed a 3 hour exposure of the Large Magellanic Cloud (LMC) that he had just taken with a 25 cm (10 inch) telescope. He immediately noticed what appeared to be a new bright star, which was not on the plate that he had exposed the night before. He went outside to check, and was astonished to see a new star that could be easily observed with the naked eye. By 0800 hours UT (Universal Time) on 24th February it had brightened to visual magnitude 4.5, making it an easy object to see with the naked eye.

Oscar Duhalde, an assistant at the Las Campanas Observatory, had been the first to notice something unusual about the LMC on 24th February when he had been walking outside at about 0530 UT. He had made his visual sighting when Ian Shelton was in the middle of his 3 hour exposure, and so Oscar had been the first to observe the supernova visually, although he had not fully appreciated what he had seen.

Thousands of miles to the west in New Zealand, Albert Jones, an amateur astronomer, independently discovered the supernova at about 0900 UT, when he was setting up his telescope for an evening's observation. He immediately telephoned Robert McNaught at the Siding Spring Observatory in Australia, who confirmed the discovery at 1055 UT. They had all seen the nearest supernova to be discovered since Kepler's star appeared in

1604, and within hours all the main observatories in the southern hemisphere had been informed. The world astronomical community was buzzing with excitement at the prospect of the first close-up look at a supernova (see Figure 9.2), if 160,000 light years can be said to be "close-up".

By good fortune, Robert McNaught had taken some photographs of the LMC on the previous night, but had not developed them. When he did so, he found that his photograph taken at 1040 UT on 23rd February recorded the supernova (now called supernova 1987A) at magnitude 6.0, or just on the limit of naked-eye visibility. Ian Shelton's first photograph, that had not shown the supernova, had been taken at 0230 UT on 23rd February. Albert Jones had observed the same area with his finderscope at 0920 UT on 23rd and had not seen anything unusual. So the timing of the supernova's rapid increase in intensity could be deduced as sometime between 0920 and 1040 UT on 23rd February.

Figure 9.2 *The star Sanduleak −69°202 before (top) and after (bottom) the explosion that produced Supernova 1987A. (Courtesy* Anglo-Australian Observatory, *(1987), photography by David Malin.)*

Theories of supernova explosions predicted the generation of neutrinos of very high energy a few hours before maximum light output, as they travel through the collapsing star faster than the shock wave which produces the maximum intensity of the supernova when it reaches the surface. Astronomers did not realise it at the time, but they had recorded the neutrinos from the supernova on 23rd February. Eleven neutrinos had been detected by the Kamiokande II detector in Japan over a period of 10 seconds at 0736 UT on 23rd February, about 3 hours before the supernova had been detected optically, and eight further neutrinos had been detected, simultaneously with the Japanese neutrinos, by the IMB detector in the USA, over a period of 8 seconds.

These neutrinos were the first to be detected as the result of a supernova explosion, giving a powerful confirmation of the theory. As neutrinos have no electric charge, and little or no mass, they can travel straight through the Earth with little reduction in intensity, which is why they could be detected in Japan and the United States, when the supernova was below the horizon. The neutrinos that had been detected had come right through the Earth!

After its initial rapid rise in intensity, supernova 1987A continued to gradually increase in intensity from a visual magnitude of 4.5 on 24th February to magnitude 2.8 in mid May, before it began a rapid decline. Astronomers had, in the meantime, examined old photographs of the area of the supernova, and McNaught had found that the position of a blue supergiant called Sanduleak −69°202 (see Figure 9.2), of magnitude 12.2, coincided with that of SN 1987A. Sanduleak −69°202 was of spectral type B3, with a surface temperature of 20,000 K and a mass of 20 suns. This was the first time that the progenitor (i.e. pre-explosion) star had been found for a supernova, but it presented something of a problem as it had been thought that red supergiants, rather than blue ones, developed into supernovae.

For SN 1987A, infrared measurements showed a gas temperature of 5,700 K on 1st March, with a cloud expanding at 18,000 km/s. By the end of March the gas temperature had fallen to 4,700 K, and the expansion rate had reduced to 10,000 km/s.

SN 1987A was thought to be a Type II supernova but, at first, astronomers were puzzled as to why its peak intensity appeared to be two or three magnitudes fainter than it should have been. Then they realised that it was because it was a blue supergiant, rather than a red supergiant, that had exploded. Blue stars are hotter than red stars, and so blue supergiants are smaller than red supergiants of the same intensity, and with a smaller star the explosion is smaller. Interestingly, the time delay on Earth between the arrival of the neutrinos and the visible light of the explosion was only about 2 or 3 hours, confirming that it was a blue supergiant that had exploded, as the time delay for the shock wave to get to the surface of a larger red supergiant would have been about a day. It was also found that the number of neutrinos detected on Earth was consistent with the theory of the formation of a neutron star in the explosion. All of this could be deduced from the detection on Earth of just 19 neutrinos.

Theory also predicted that, not only are neutrinos produced when the core of the supergiant star collapses, to briefly produce a core temperature of 10,000 million K, but a large amount of nickel 56 is also produced. This nickel 56 is radioactive and decays spontaneously to cobalt 56, which in turn is radioactive, decaying spontaneously to iron 56. In radioactivity not all of the atoms decay at the same time, and the time taken for half of the atoms to decay is called the half-life of the process. The half lives of the nickel–cobalt and cobalt–iron processes are 6 and 77 days respectively. These radioactive processes heat up the core of the supernova, which is why the light output of the supernova peaked at about 88 days after its initial outburst, again confirming the theory of supernovae.

The decay of cobalt 56 to iron 56 was also expected to produce a great number of very energetic γ-rays, some of which would be converted in collisions in the supernova to less energetic X-rays. It was expected that these γ-rays and X-rays would be detected as the gas shell surrounding the central neutron star got thinner, allowing them to escape into space. The theory was vindicated when, in August 1987, γ-rays were first detected by the Solar Maximum Satellite and, at about the same time, X-rays were discovered by the Japanese GINGA satellite. The X-ray intensity peaked in October 1987.

A double ring structure, caused by light from the supernova explosion being reflected off an intervening gas cloud, was discovered by Arlin Crotts of the University of Texas on 3rd March 1988, using the 40 inch (1.0 m) telescope at Las Campanas. Subsequent investigations showed that these rings had first been recorded on 13th February 1988 by Michael Rosa using the European Southern Observatory 3.6 m telescope. The two rings indicated from their size that there were sheets of material 400 and 1,100 light years in front of the supernova.

The luminous ring surrounding 1987A enabled a more accurate estimate to be made of the distance to the LMC, which had been previously estimated as 161,000 $\pm$ 18,000 light years. The non-circularity of the ring enabled its inclination to our line-of-sight to be determined, and analysing the data from IUE spacecraft gave the difference in time taken for the light to reach the Earth from the nearer and further edges. Given the geometry, the distance to the LMC was then calculated as 169,000 $\pm$ 8,000 light years.

The remnant of the supernova explosion should be a neutron star which would appear as a pulsar, if the spin axis orientation causes the beams of radiation emitted by the neutron star to be detected on the Earth. It was expected that the best chance of detecting the possible pulsar through the clouds of gas surrounding it, would be to observe it at far infrared wavelengths, where dust and gas scattering is a minimum. Would the neutron star appear as a pulsar as the gas clouds dispersed? The anticipation was intense as the search for the pulsar got underway.

The pulsar was first apparently detected by John Middleditch on 22nd January 1989, when he was analysing data produced 4 days earlier by Tim Sasseen and himself with the 4 m infrared/visible telescope at Cerro Tololo in Chile. The big surprise was that the pulsar appeared to have a pulsation frequency of 0.508 milliseconds, implying that it was rotating at about 2,000 times per second. If the neutron star was rotating at this rate, its surface velocity must be very close to the velocity of light, even if the neutron star was only 10 km or so in diameter. The fact that the surface appeared to be rotating at such an incredibly high speed, left many astronomers feeling somewhat uncomfortable. What was just as bad, however, was that all previous millisecond pulsars appeared to be old objects, which had probably been spun up by accreting material from a companion star over hundreds or thousands of years, so finding that the fastest pulsar yet discovered was the youngest caused considerable interest. Careful analysis showed that the pulsation rate appeared to vary regularly with a period of 8 hours, and many theories were advanced to explain this.

On 31st January 1989, attempts to measure the pulse rate failed when the pulsar could no longer be detected. Repeated attempts over the next few months also failed, but this was put down to obscuring material from the explosion temporarily blocking the light. Then, in February 1990, it was discovered that the so-called pulsar signal was actually due to radio interference from a television camera used on the telescope. The real pulsar is still awaited.

Progress in understanding some of the best-known and best-researched remnants from supernova explosions is outlined in the next few pages. At present, about 200 supernova remnants have been discovered in the Milky Way, and new ones are being found every year.

Supernova Remnants

• *The Crab Nebula*

The English doctor and amateur astronomer John Bevis first recorded a dim, nebulous object in the constellation of Taurus in 1731, and the French astronomer Charles Messier saw it in 1758, and recorded it as the first object in his catalogue of nebulae, published in 1781. It was given its name of the Crab Nebula by the third Earl of Rosse, who made a detailed study of nebulae in the mid nineteenth century from his observatory at Birr Castle in Ireland.

Joseph Winlock, the director of the Harvard College Observatory, and Edward Pickering had observed the Crab Nebula with their spectroscope in 1868, finding the green nebula line which showed that the nebula was gaseous. Forty-five years later, when Vesto Slipher photographed its spectrum with a high-dispersion spectrograph at the Lowell Observatory, he saw that all the spectral lines were double. The double lines were not straight, however, but lenticular (lens-like), joining at top and bottom, when the slit of the spectrograph was placed across the middle of the nebula. Slipher realised that the doubling* was due to the Doppler effect,

---

*It was deduced that the lenticular shape was caused by gas in the centre of the nebula moving straight towards and away from us, while gas at the outer part of the

with the line separation indicating a nebula expansion rate of about 1,000 km/s.

John Duncan measured the Crab Nebula on photographs taken a few years apart with the 60 inch (1.5 m) Mount Wilson reflector and showed, in 1921, that the nebula was clearly expanding. The angular expansion rate measured on the photographs, and the Doppler velocity measured spectroscopically enabled the distance of the nebula to be estimated at 6,300 light years. In the same year the Swedish astronomer Knut Lundmark published a list of novae that had been observed by the Chinese, and pointed out that a supernova observed in 1054 was very near to the position of the Crab nebula in the sky. Could they be the same?

Hubble compared the angular expansion rate of the nebula with the Doppler expansion rate, and estimated in 1928 that the expansion had been underway for about 800 or 900 years. This fitted the timescale of the 1054 supernova remarkably well, so he suggested that the Crab nebula was the remnant of that supernova. For some reason, however, Hubble did not publish his conclusion in the academic press, and so it was not generally known. Jan Oort and J. J. L. Duyvendak independently undertook a similar analysis during the Second World War in German-occupied Holland with the same result, although Oort pointed out that the position of the supernova, as recorded by the Chinese, appeared to be over 1° away from the position of the Crab Nebula. As both the timescale and position were close, however, it was generally accepted that the supernova recorded by the Chinese was the precursor of the Crab Nebula, with the Chinese position estimate assumed to be in error.

Later research showed that the supernova of 1054 had also been seen by Japanese astronomers, but they did not record its position accurately, unfortunately. The Japanese recorded it in early June 1054 as bright as Jupiter, which pre-dated the Chinese discovery on 4th July, by which time it had brightened to be as bright as Venus. The Chinese records showed that it was so bright as to be visible in daylight for about 3 weeks, but after 2 years it had dimmed so rapidly that it could not even be seen at night. These records show that the supernova must have had an absolute intensity of about magnitude −17 at maximum, which is equivalent to that of 400 million suns, assuming a distance of 6,300 light years.

Walter Baade and Rudolph Minkowski published a paper on the Crab Nebula in 1942, in which they analysed the movement of the gas and stars in the nebula since the explosion in 1054. There are now two stars equidistant from the apparent centre of the nebula's expansion. The northeast star, as it is known, shows no proper motion, but the southwest star is clearly moving, and when Baade and Minkowski took this movement into account, they were able to show that it was originally closer than the northeast star to

---

nebula moved across our line of sight. The Doppler shift is therefore a maximum at the centre of the nebula, which is why the spectral lines have their widest separation there, and zero at the edge of the nebula, where the lines are no longer double.

the centre of the explosion. This led them to conclude that the southwest star is probably the remains of the original supernova.

The data were all beginning to tie quite neatly together, except for the 1° discrepancy in position of the Crab compared with the Chinese records, until it was discovered, in the 1940s, that the date of 1054 didn't quite match the expansion rate either; the most accurate estimate of the expansion rate giving an explosion date of 1140 ± 10. The data could only match if the rate of expansion had increased with time, which seemed unlikely, unless there was a powerful energy source in the nebula causing such an acceleration. Baade thought that he had detected an acceleration, using photographs taken over the preceding 30 years, but the data were not clear.

In 1949, John Bolton, G. J. Stanley and O. B. Slee discovered that the Crab Nebula was a strong source of radio waves (called Taurus A), using a radio telescope in both Australia and New Zealand. It had been known for some time that the nebula has much less structure in blue light than red light (see Figure 9.3), and that the inner part of the nebula gives a continuous spectrum, whereas the outer part gives an emission-line spectrum similar to that of a planetary nebula. In 1953 the Russian astrophysicist Iosif Shklovskii suggested that the radio emission from the Crab Nebula was caused by

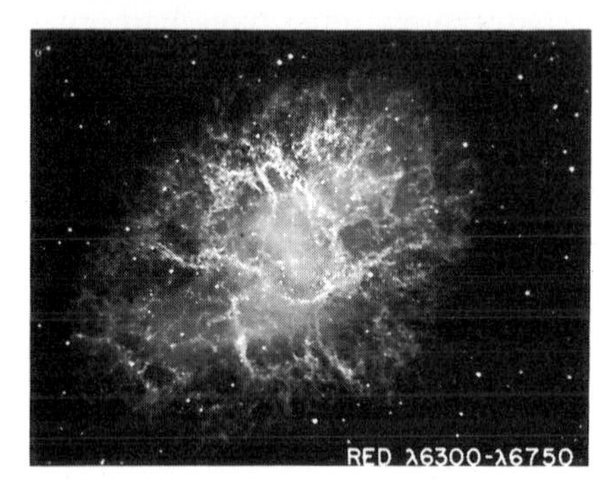

Figure 9.3 *The Crab Nebula photographed in blue light (top) and red light (bottom), using the Mount Wilson 100 inch (2.5 m) telescope. The wavelengths (λ) are quoted in Ångström units, where 10Å ≡ 1 nm. (Courtesy* The Observatories of the Carnegie Institution of Washington.*)*

a process known as synchrotron radiation, which is produced by electrons spiralling in a strong magnetic field, and that the bluish-white light with the continuous spectrum, which he thought was also caused by synchrotron radiation, should thus be polarised. The idea of large amounts of synchrotron radiation coming from the Crab Nebula was treated with a great deal of scepticism, until the Soviet astronomers M. A. Vashakidze and V. A. Dombrovsky showed that the light from the Crab Nebula is strongly polarised. Then, in 1957, Mayer, McCullough and Sloanmaker found that the radio signals were also polarised, further confirming Shklovskii's theory of synchrotron radiation.*

An experiment launched by an Aerobee sounding rocket in April 1963 by Herbert Friedman's group at the Naval Research Laboratory, discovered that the Crab Nebula was a strong X-ray source. Friedman speculated that the source of the X-rays was a neutron star, and he and his team took advantage of the occultation of the nebula by the Moon on 7th July 1964 to measure the size of the source. Unfortunately, they found it to be 1 arcmin, or about 2 light years, in diameter, clearly too large to be a stellar object.

Five years later, David Staelin and Edward Reifenstein decided to search for a pulsar in the Crab Nebula, following Bell and Hewish's discovery of the first pulsar the previous year. They used the radio telescope at the National Radio Astronomy Observatory in Green Bank, West Virginia and, in November 1968, they discovered the Crab pulsar with a period of 33 milliseconds. At this time there had been no clear identification of a pulsar with an optical source but, in early 1969, William Cocke, Michael Disney and Donald Taylor, of the Steward Observatory in Tucson, Arizona looked for optical flashes from the centre of the Crab Nebula. They were almost immediately successful, but they were not sure which of the two faint central stars was the source of the 33 millisecond flashes. The southwest star, which is of magnitude 17, was then shown to be the source by Joe Miller and Joe Wampler, of the Lick Observatory, who used a stroboscopic technique, with a rotating shutter and TV camera attached to the 120 inch (3 m) reflector. So Baade and Minkowski had chosen the right star in 1942 as the stellar remnant of the supernova explosion.

The Crab pulsar had been found to pulse at both radio and optical frequencies, but did it pulse at X-rays?

Friedman, Byram, Chubb and Fritz launched an Aerobee rocket in March 1969 to search for X-ray pulses from the Crab, and found the same pulse pattern that had been seen at both radio and optical frequencies. The X-ray pulsar was seen to be very powerful, generating about 20,000 times as much

---

*Synchrotron radiation is now known to be a common feature of *supernova* remnants. In the 1980s Roger Chevalier and Steve Reynolds found, using the Very Large Array radio telescope, that the remnant of *Nova* Persei was also a synchrotron emitter. Seaquist, Evans, and Bode then observed 26 other *nova* remnants, using the VLA, but found that none were synchrotron sources, so Nova Persei appears to be a unique nova in this respect.

power in X-rays as in radio waves, and about 20 times as much power in X-rays as in visible light. They had found the source of energy that was accelerating the expansion of the nebula, so making the explosion year of 1054 credible.

In 1976, the Cos-B spacecraft found that the Crab pulsar was the third most powerful source of γ-rays in the sky, and 3 years later, the Einstein Observatory imaged both the pulsar and the nebula in X-rays. So, whether we look at the Crab pulsar in γ-rays, X-rays or radio waves, we are shown the same intense object, a neutron star with a surface temperature of 2 million K. Its pulses, or spin rate, are slowing down at the rate of 10^{-5} s/year, and it is therefore losing rotational energy at a rate equivalent to the radiation from 100,000 suns. It is this energy that accelerates and lights up the nebula.

• *Tycho's Supernova Remnant*

The discovery, in the early decades of the twentieth century, of the link between the Crab nebula and the supernova seen by the Chinese in 1054, naturally caused astronomers to search for the remnants of the Tycho and Kepler supernovae seen in 1572 and 1604, respectively. Baade used the 100 inch (2.5 m) telescope on Mount Wilson in 1947 to search for the Kepler supernova remnant, but all he could find was a few filaments of gas. Looking for the Tycho remnant he could find nothing of interest, then in 1952 a radio source was discovered near the correct place by Robert Hanbury Brown and Cyril Hazard, using the 218 ft (65 m) transit radio telescope at Jodrell Bank. In the 1970s the Tycho supernova remnant (SNR), as it is now called, was also discovered to be a source of X-rays. The Einstein Observatory X-ray image taken in 1980, showing the shell of hot gas, and the radio image taken by the Cambridge Five Kilometre Telescope, showing the synchrotron radiation, both show an almost-circular hollow shell of gas, 8 arcmins in diameter, which is brightest at the edge. This is quite unlike the Crab Nebula which is brightest near the centre. There also seems to be no central star in the Tycho SNR, which should be visible in X-rays, even if its pulsar beam missed the Earth.

Tycho's SNR is, at 7,500 light years, a little further away from the Earth than the Crab Nebula, and the outer shell is expanding at 6,000 km/s, compared with the outer part of the Crab which is expanding at 2,500 km/s. At optical wavelengths only a few extremely faint filaments of gas can be seen at the position of the Tycho SNR.

• *Cassiopeia A*

In 1942, the American pioneering radio astronomer, Grote Reber, found a region of the sky in Cassiopeia to be a source of radio waves (see Page 311). This source, now known as Cassiopeia A, was rediscovered by Martin Ryle and F. Graham Smith at Cambridge in 1948. Three years later, Graham

Smith obtained an accurate estimate of its position using a radio interferometer, enabling Baade and Minkowski at Mount Wilson to optically identify Cassiopeia A, which is the brightest radio source in the sky, with a very faint nebulosity of about 4 arcsec in extent. F. Graham Smith at Cambridge, and Hanbury Brown, Roger Jennison and Das Gupta at Jodrell Bank helped to confirm this visual correlation, when they showed that the radio source has a diameter of about the same size. The radio source was not dispersed like the optical source, however, but seemed to have a circular shape.

Minkowski measured the movement of the nebulosity on photographs taken at optical wavelengths over a period of time, and calculated, in 1958, that the supernova that had caused it had exploded about the year 1700. Later measurements by Sidney van den Bergh and Karl Kamper of the David Dunlap Observatory in Canada yielded a date of 1660. Minkowski also estimated that the maximum gas velocity was 7,400 km/s which, given the angular expansion rate that he had measured, enabled him to deduce a distance for Cassiopeia A of about 11,000 light years. At this distance the preceding supernova should have been seen as one of the brightest stars in the sky, but no such star had been recorded, so Minkowski suggested that the supernova had been dimmed by a considerable amount of interstellar dust in our line of sight. The only candidate supernova discovered is a sixth magnitude star recorded in 1680 by the first Astronomer Royal John Flamsteed about 3 arcmin from the position of Cassiopeia A. This is larger than Flamsteed's normal error of about 1 to 2 arcmin, but one or two stars in his catalogue had errors of more than 10 arcmin and, as 3 Cassiopeia can no longer be seen, it could well have been the supernova.

More recently, Cassiopeia A has been imaged in X-rays but, like the Tycho SNR, no central stellar source has yet been found at any wavelength.

• *The Vela Supernova Remnant*

Radio astronomers discovered, in the 1950s, a strong source of radio waves in the southern constellation of Vela, which was flanked by two other radio sources. Further work showed that these three sources, known as Vela X, Y and Z, are all part of a very extended source about 4°, or 100 light years, in diameter. It is near the centre of the spectacular nebula, called the Gum Nebula (after its discoverer), the central part of which is shown in Figure 9.4.

In 1968, a radio pulsar, PSR 0833-45, was discovered near the centre of the bright radio source Vela X by Large, Mills and Vaughan in Australia, with a period of 89 milliseconds. Astronomers had long suspected that the extended Vela radio source was a supernova remnant, and the discovery of the pulsar confirmed it. Seven years later, the pulsar was found by the SAS-2 satellite to pulse in γ-rays and, in 1977, a team of Australian and British radio and optical astronomers detected optical pulses after 10 hours integrating the signal with the Anglo-Australian telescope. The optical pulsar has an average magnitude of only 24, or about 1/1,000th that of the Crab pulsar. The optical, radio and γ-ray pulses were found to be out of phase

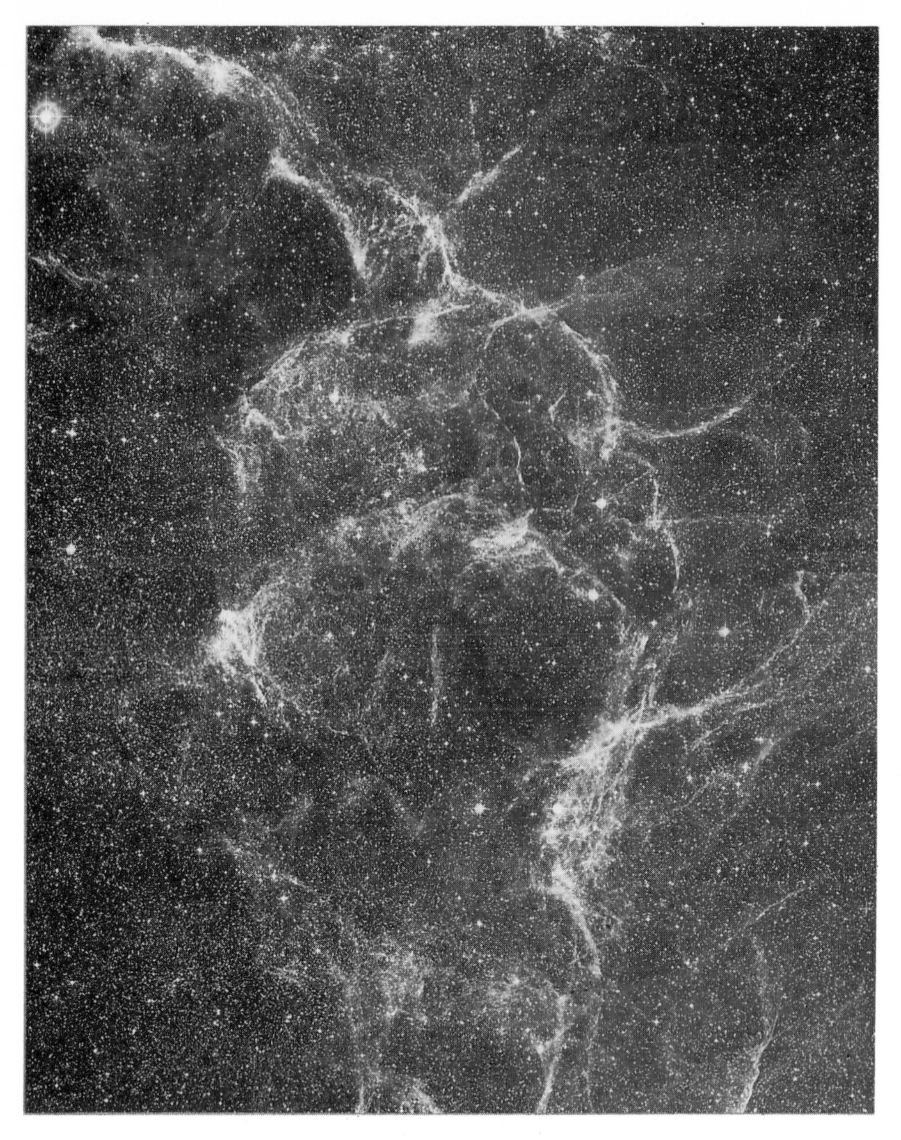

Figure 9.4 *Part of the Vela supernova remnant. (Courtesy* ROE/Anglo-Australian Observatory *(1979), photography by David Malin.)*

with each other (i.e. the pulse does not occur at precisely the same time at optical, radio and γ-ray wavelengths) and, although the Einstein Observatory spacecraft recorded the X-ray image of the Vela pulsar in 1979, it was found that the pulsar does not pulse in X-rays. The surface temperature of the pulsar, deduced from its X-ray and optical intensities, appears to be 1,500,000 K.

The period of the Vela pulsar is lengthening by 4×10^{-6} s/year and, working backwards, astronomers have estimated that, what is now the Vela

Supernova Remnant is the scattered remains of an explosion 11,000 years ago. It is one of the closest SNRs, the centre being only 1,500 light years away.

• *SS 433*

Gart Westerhout, a Dutch radio astronomer, discovered a large ($2° \times \frac{1}{2}°$) radio source in the 1950's, now catalogued as W50, which was found some years later to have a point radio source at the centre. In the early 1970s the American Uhuru spacecraft showed that W50 was also a source of X-rays, and then Fred Seward, of the Mullard Space Science Laboratory, discovered a point X-ray source, using the UK Ariel V spacecraft, at the position of the point radio source.

David Clark and Jim Caswell had tried for some time to correlate radio stars and supernova remnants. Now that a point X-ray source had been found, Clark asked Caswell for a precise position fix on the coincident radio source (whose position could be determined more accurately than that of the X-ray source), to enable a search to be made for the optical counterpart. Clark and Paul Murdin used the Anglo-Australian telescope and found three stars in the right position, two of which were ordinary stars, but the third was a binary with strong hydrogen emission lines, and a period of 13 days. They had, in fact, re-discovered a star that had been observed and catalogued over 10 years earlier.

The American astronomers Bruce Stephenson and Nicholas Sanduleak had discovered this star in the 1960s when they were looking for stars with emission line spectra, and had catalogued it as SS 433, a 14th magnitude star, with bright hydrogen emission lines. In 1978, Ernie Seaquist, a Canadian astronomer, found SS 433 to be a radio source, but he did not realise that it was surrounded by the large radio source W50.

David Clark and Paul Murdin examined the optical spectrum of SS 433 in detail, and found that some of the emission lines did not correlate with those of any known element, then Bruce Margon of the University of Washington discovered that these lines were changing dramatically in frequency over a period of just over 5 months. There were three sets of hydrogen emission lines, one set of which remained constant in frequency, whereas the other two sets moved to and fro on either side of the central set, indicating Doppler velocities ranging from 81,000 km/s to zero over a period of 164 days. It appeared that SS 433 is emitting hydrogen, with a velocity of 81,000 km/s (or a quarter of the speed of light), in two oppositely-directed beams which are rotating with a period of 164 days. The existence of the beams was revealed directly, in 1979, by the Einstein Observatory which found X-ray sources extending up to $\frac{1}{2}°$, or 100 light years, on either side of the main X-ray source, and by the Very Large Array, in 1981, which showed the change in structure of the jets much closer to the star, over the 164 day beam rotation period.

The binary SS 433, which is the source of these two beams, is thought to consist of a small dense star (probably a neutron star) which is dragging gas from its normal stellar companion. This gas forms an accretion disc around the small dense star, where it heats up and emits X-rays. It is thought that the dense star may be the remains of a supernova explosion whose shell is the much larger radio source W50. The mass of the dense star was estimated in 1991 by D'Odorico, Oosterloo, Zwitter and Calvani as 0.8 solar masses, using the 3.6 m New Technology Telescope at the European Southern Observatory, which is consistent with it being a neutron star of 10 km diameter. The mass of its normal stellar companion was calculated as 3.2 solar masses.

Black Holes

The English astronomer John Mitchell had suggested in 1783 that, if stars were heavy enough, light would be prevented by gravity from leaving the surface, and Pierre Laplace had made a similar suggestion 13 years later, but both proposals had been largely forgotten. The subject was resurrected in 1916 by Karl Schwarzschild who solved Einstein's field equations to show that light cannot escape from within what is now called the Schwarzschild radius around an extremely dense object. In 1939, J. Robert Oppenheimer and Hartland Snyder, a graduate student at the University of California, showed that an object produced in the supernova explosion of a star which has a mass greater than 3.2 suns would be dense enough to prevent light from escaping.

In 1967 Iosif Shklovskii suggested that a black hole could be detected if it were in a binary pair with a supergiant. He reasoned that the black hole would drag the outer shell of gas off the supergiant, and would compress the gas, causing it to emit X-rays. So Shklovskii suggested that X-ray sources should be examined to see if they indicated that a supergiant star was in orbit around a compact object whose mass was greater than the maximum of 3.2 solar masses allowed for a neutron star.

Cygnus X-1

The X-ray source Cygnus X-1 was discovered by astronomers at the United States Naval Research Laboratory (NRL) in 1964 and, when observed in the following year, its X-ray intensity was found to have reduced by 75%. In December 1970 the Uhuru spacecraft found that Cygnus X-1 showed variations in its X-ray intensity over many different timescales down to a few milliseconds which, it was thought, could be caused by a black hole sucking in material from another star.

Cygnus X-1's average X-ray intensity decreased abruptly in March 1971, and a radio source appeared in its place, which was fortunate because the position of radio sources could be measured far more accurately than that of X-ray sources. The radio source was found to be coincident with a 9th magnitude O-type blue supergiant, called HDE 226868, that was a member of a binary with a period of 5.6 days and, although the other member of the binary was invisible, its mass could be deduced from the Doppler shift of the spectrum of the visible star. Assuming that the O-type star had a normal mass for this type of object of 30 solar masses, the mass of the invisible X-ray star was found to be at least 6 solar masses. This is much greater than the theoretical maximum mass of 3.2 solar masses for a neutron star, and Cygnus X-1 is the best candidate yet for a black hole. In 1979, γ-rays were found by the HEOS 3 satellite coming from Cygnus X-1 with an energy of about 10^6 electron volts, adding to evidence for the black hole theory.

A0620–00

The British satellite Ariel V detected a strong source of X-rays in the constellation of Monoceros on 3rd August 1975 that was rapidly increasing its X-ray intensity, so that 5 days later it was stronger than Scorpius X-1, and was thus the strongest X-ray source in the sky. It reached maximum X-ray intensity on 15th August, at five times that of Scorpius X-1, before starting a gradual decline. This X-ray nova, called A0620–00, was found to coincide with a blue 11th magnitude star in visible light.

A search of the Palomar Observatory plates that had been taken in 1955 indicated that the 11th magnitude star was normally reddish and of 18th magnitude, so the nova A0620–00, now called V616 Monocerotis, was also detectable at optical wavelengths. A further search showed that the star had also exploded in November 1917, when it had reached 12th magnitude.

After the 1975 outburst had faded, A0620–00 became an 18th magnitude orange dwarf star of spectral type K5. In 1983, Jeff McClintock of the Harvard–Smithsonian Center for Astrophysics announced that the orange dwarf's intensity was varying regularly with a period of 7.75 hours, indicating that it was a binary. The shape of the intensity curve also indicated that the orange dwarf was not spherical, as it was apparently being distorted by its invisible companion. McClintock and Ronald Remillard of MIT measured the Doppler shift in the spectral lines of the orange dwarf and, in 1986, concluded that its dark companion's mass must be at least 3.2 solar masses, making it a candidate for a black hole. The system appeared to be about 3,300 light years from Earth.

Further work by a number of different research groups showed, in 1988, that the X-ray spectrum of A0620–00 is very similar to that of Cygnus X-1, thus adding to the evidence that A0620–00 may contain a black hole, like Cygnus X-1. Then, in August 1990, Carole Haswell of the University of Texas and Allen Shafter of San Diego State University were able to measure the Doppler shift of the light and the dark companion, concluding that the dark companion had a mass of 10.6 times that of the orange dwarf. Independent evidence showed that the mass of the orange dwarf was at least 0.36 solar masses (and maybe twice that), giving a minimum mass for the dark companion of 3.8 solar masses. Haswell and Shafter suggested that the dark companion was not a star but an accretion disc around a black hole, as the mass of the accretion disc plus that of the central object (the black hole candidate) was clearly in excess of the maximum mass of 3.2 suns allowed for a neutron star.

ASM 2000 + 25

On 26th April 1988 the Japanese GINGA satellite detected an extremely bright X-ray source in the constellation of Vulpecula, which had not been present 4 days earlier. At the time of its discovery, the nova (called ASM 2000 + 25) was one of the most intense X-ray sources in the sky. Okamura and Noguchi found two possible optical counterparts which they called A and B. Star A had an optical spectrum typical of a late M-type star, but star B's spectrum, taken with the William Herschel Telescope on 7th June 1988, strongly resembled the optical spectrum of A0620–00 in its nova state in 1975. The Very Large Array radio telescope found a radio source very close to the position of star B on 26th May 1988, that had not been present on 3rd May, so star B appeared to be the X-ray nova ASM 2000 + 25. Measurements of its X-ray spectrum by the Soviet Space Station MIR showed that it was similar to that of A0620–00 and Cygnus X-1, thus indicating that ASM 2000 + 25 may also be associated with a black hole.

V404 Cygni

On 22nd May 1989, an X-ray nova was discovered in the constellation of Cygnus, which was identified with a nova of magnitude 12.8 photographed in visible light 4 days later. Further photographic investigations showed that the nova had brightened from about magnitude 20, and was in the same position as Nova Cygni 1938 (now called V404 Cygni). Then in February 1992 Jorge Casares, Phil Charles and Tim Naylor discovered periodic variations in the spectrum of V404, using the William Herschel telescope on La Palma. They found that a G or K type star, slightly cooler and less massive than the Sun, was orbiting a dark companion every 6.5 days, and calculated that the dark companion had a mass of at least 6.3 solar masses, making it also a black hole candidate.

Non-Stellar Black Holes

Black holes are not restricted theoretically to stellar objects, as they can occur whenever the escape velocity from an object is equal to, or greater than, the velocity of light. Astronomers had speculated for some time that enormous black holes may exist in the centre of galaxies and, in 1978, Peter Young and Wallace Sargent found evidence of a massive black hole at the centre of the galaxy M87. Then, in 1987, John Kormendy of the Institute for Astronomy in Hawaii measured the velocity of stars 1 arcsec from the nucleus of the Andomeda galaxy (M31). He concluded that their high velocity, together with the abrupt fall-off in velocity with distance from the nucleus, showed that the centre of M31 contains a mass of between 10 million and 1,000 million suns. Such a large mass concentration in such a relatively small volume appeared to indicate the presence of a massive black hole, and then, in 1992, a radio source was discovered at the centre of M31, supporting the black hole theory.

In 1987 the existence of a black hole of 3 million solar masses was also suspected at the centre of the galaxy M32, because of the abrupt increase in the orbital velocities of stars near the centre. Four years later, Tod Lauer of the National Optical Astronomy Observatories, Sandra Faber of the University of California and colleagues found further evidence for a black hole at the centre of M32, when high resolution images taken with the Hubble Space Telescope (HST) showed a rapid increase in brightness at the centre of the galaxy. Holland Ford of the Johns Hopkins University found similar evidence in 1992, using HST data, for a similar-sized black hole at the centre of the M51 galaxy.

Joss Bland-Hawthorn of Rice University, Andrew Wilson of the University of Maryland and Brent Tully of the University of Hawaii found evidence in 1991 for an even more massive black hole in the galaxy NGC 6240, which the IRAS satellite had shown in its 1983 survey was a very powerful infrared source. Bland-Hawthorn, Wilson and Tully used an interferometer on the University of Hawaii's 88 inch (2.2 m) telescope on Mauna Kea, to investigate NGC 6240 further, and found that the galaxy has two spinning gas discs. The motion of one of these discs indicated that there was a mass of about 100 billion suns at the centre of NGC 6240, that is equivalent to the mass of the whole of the Milky Way galaxy being contained in a volume of only 1×10^{-4} that of the Milky Way. This high density implied that there was a black hole with a mass of 100 billion suns at the centre of NGC 6240.

The HST imaged the nucleus of the elliptical active galaxy NGC 4261 in 1992, and found a dark disc of dust and gas surrounding the centre. The outside of the dark disc is about 350 light years in diameter, and it extends to within less than a light day of the nucleus, which is seen as a bright point. Radio telescopes had previously shown that NGC 4261 has a pair of jets, spanning a distance of 100,000 light years, coming from the nucleus. These radio jets are aligned along the spin axis of the gas disc, indicating the presence of a very hot accretion disc within the bright core of the galaxy, the accretion disc surrounding a black hole. Some astronomers have suggested that the bright ring of light that surrounds the gas disc is the accretion disc, but if this is so it is very much larger than theory predicts. Either way, the visible structure seen around the centre of this galaxy, and the radio jets coming from the nucleus region, provide strong circumstantial evidence that there is a black hole at the centre of this active galaxy.

10 The Milky Way

Early Work

The Movement of the Sun

William Herschel discovered in 1783 that the Sun and solar system were moving towards the star λ Herculis (with celestial coordinates of 17 h 31 min, or 263°, right ascension and +26° declination), by analysing the proper motions of 13 stars. During the nineteenth century, a number of other astronomers made further measurements of stellar proper motions, but all produced very similar estimates to that of Herschel of the apex, or target area, of the Sun's motion.

Stellar proper motions show the movement of stars across our line of sight, but, to get a better assessment of the movement of the Sun relative to the stars, it was also necessary to measure the relative motion of stars along our line of sight. This motion, called radial motion, could, in principle, be detected by measuring the Doppler shift of stellar spectral lines, but this was more difficult than it seemed in the nineteenth century.

In 1842 Christian Doppler, who was professor of mathematics in Prague, had proposed his theory that if a light (or sound) source is moving either towards or away from an observer, then the frequency or wavelength of the source will appear to the observer to differ from its true value. If the light source is moving towards the observer, the frequency will appear to increase (and the wavelength decrease), to make the source appear bluer, and if the source is receding, it will appear redder.

Doppler thought that the velocity of the stars relative to the solar system would be so great that it would affect their apparent colour and intensity as seen from Earth, and this idea caused great controversy at the time. It was known by the early 1860s, however, that if the wavelength of the spectral absorption lines and the colour of the stars were changed as a result of their velocity relative to the Earth, the effect must be small. Then, in 1868,

William Huggins sent a paper to the Royal Society announcing the first measured shift in the Fraunhofer absorption lines of a star, when he deduced a recession velocity for Sirius of 47 km/s, by observing the slight displacement of the hydrogen F line. Over the next 4 years, Huggins observed the deviation of the lines in 30 stars, and further work was carried out, in particular, by Walter Maunder and W. H. M. Christie at Greenwich. By the mid 1880s, however, the results of Huggins, Maunder, Christie and others were so inconsistent, and the errors so large, as to render these line-of-sight velocity measurements almost meaningless.

Huggins, Maunder and Christie had made their measurements of the displacements of the stellar absorption lines visually, but Hermann Vogel decided to measure them photographically using the 11 inch (28 cm) Potsdam refractor. This enabled Vogel and Julius Scheiner to make the first reasonably accurate stellar radial velocity measurements in 1888, when he announced results for Sirius, Procyon, Rigel and Arcturus with an accuracy of 5–10 km/s. This was almost an order of magnitude better than that achieved by Huggins but, in 1890, James Keeler was able to make visual measurements with an accuracy of 1 km/s, almost an order of magnitude better than Vogel's figures, using the powerful new 36 inch (91 cm) refractor at the Lick Observatory, which augured very well for the future. These, and other data, would be used in the last decade of the century to help determine the apex of the Sun's motion.

Nebulae

Diffuse objects, or nebulae, had been observed in the sky since the invention of the telescope in the early seventeenth century, and in 1781 Charles Messier produced a catalogue of the brighter nebulae to avoid them being confused with comets. Two years later William Herschel decided to list all the non-stellar objects that could be seen from his observatory in Slough and, over the next 20 years, he catalogued over 2,500 of them.

The German philosopher Immanuel Kant had suggested in 1755 that nebulae were systems of stars like the Milky Way, but which were so far away that their individual stars could not be resolved. Originally, Herschel believed Kant's theory, but his discovery of a planetary nebula in 1790, where the nebula surrounds a central star, caused him to change his mind and conclude that at least some nebulae consist of luminous non-stellar material which would eventually condense to form stars. In particular, Herschel mentioned the Orion nebula as an example of self-luminous, non-stellar material.

Herschel found that he could resolve some nebulae into stars, and when the Third Earl of Rosse used his much larger 72 inch reflector in the mid nineteenth century, he found that he could resolve even more nebulae into stars. This led some astronomers to believe that if one had a large enough telescope, all nebulae would be resolved into star systems like the Milky Way, with the exception of planetary nebulae, which appeared to be a special case.

The theory of the stellar nature of all but planetary nebulae was given a nasty shock when, in 1864, Huggins examined the spectra of eight nebulae, including the one in Orion, and found that they all had bright emission line spectra, indicating that they were gaseous in form. Six or seven lines were observed in each object, three being due to hydrogen, but the origin of the others was unclear. Huggins originally identified the brightest nebular line, that was green in colour, as being caused by nitrogen, but this was shown to be incorrect, as it was of a slightly different wavelength. Norman Lockyer suggested that it was due to magnesium then, in 1890, Keeler showed that the bright green line did not coincide with that of any known element. So it was attributed to a new element called nebulium, in the same way that the line of an unknown element in solar prominences had been attributed to a new element called helium (see Page 2).

In 1868, Huggins found with a much larger sample of nebulae that one third were gaseous, but two thirds (including the Andromeda nebula) gave a continuous spectrum, indicating that they consisted of stars, although some astronomers thought that the continuous spectrum was due to starlight reflected by the nebulae which were still gaseous. Simon Newcomb, for example, who was a professor at the US Naval Observatory, was still not convinced in 1882 that even the Andromeda nebula was a system of stars, and he was helped in this by the appearance of a very bright new star in that nebula in 1885. If the Andromeda nebula really was very distant galaxy of stars, this nova must have an almost impossible intensity.

The Structure of the Universe

Thomas Wright, an English clockmaker, suggested in 1750 that the Milky Way was a large disc of stars slowly rotating about its centre of mass and, in 1784, Herschel made the first attempt to estimate its size. He assumed, as a first approximation, that stars were equally dense in all parts of the sky, and the reason why there were more stars seen in some parts of the sky than others was because of the varying thicknesses of the star clouds. From his star survey, he concluded that the Sun was near the centre of the Milky Way, which had a diameter of about 800 times the mean distance of Sirius or Arcturus from the Sun, and a thickness of about 150 times that distance. These figures correspond to about 16,000 light years in diameter and about 3,000 light years in thickness, given the mean distance of Sirius and Arcturus which is now known to be about 20 light years.

A hundred years later, in 1890, Herschel's estimate of the size of the Milky Way was still the best available so, near the end of the nineteenth century, the Milky Way, or the Galaxy as it is sometimes called, was thought to be a disc of stars rotating about its centre of mass, with the Sun near the centre. Although the stars generally rotated around the centre of the Milky Way in this manner, measurements of the movement of stars near the Sun showed that there were deviations from this general trend, caused by the influence of local stars.

About 50 million stars were visible in the very largest telescopes of the day, leading astronomers to conclude that the Milky Way probably contained about 100 million stars, as a rough order of magnitude (modern value 200,000 million). Beyond the Milky Way, things were a complete unknown. Some astronomers thought that the universe consisted of innumerable galaxies like the Milky Way, but some thought that the Milky Way was the universe.

As far as the age of the universe was concerned, the age of the Sun appeared to be about 25 million years, although there was some doubt on this because the Earth seemed to be about 100 to 400 million years old. This led astronomers to conclude that the age of the universe was probably measured in hundreds of millions of years, rather than thousands of millions of years, although this was very much based on intuition, rather than on real data.

Dimensions and Structure

Initial Dimension Estimates

Between 1884 and 1909, Hugo von Seeliger, the director of the Munich Observatory, produced a new analysis of the structure of the Milky Way assuming, as a first approximation, that all stars have the same absolute magnitude, and that differences in apparent magnitude are due solely to distance. If the stars are distributed uniformly in space, the ratio of the number of stars with successive magnitudes should be 4* to 1. Seeliger found that the ratio varied between 2.8 to 3.4 per magnitude, indicating that the number of stars per unit volume continuously decreased with increasing distance from the Sun. He analysed the ratio of the number of stars with successive magnitudes for various galactic latitudes (i.e. distances from the plane of the Milky Way), and concluded that the Milky Way was 23,000 light years in diameter, and 6,000 light years in thickness.

Karl Schwarzschild extended Seeliger's analysis in 1910 by taking into account the fact that all stars are not of the same absolute magnitude, and deduced that the Milky Way is a flattened ellipsoid of rotation, with the Sun at its centre. The density is highest at the centre, and decreases faster towards the galactic poles than along the galactic equator. His size estimate of 30,000 light years diameter, and 6,000 light years thickness, was very similar to that of von Seeliger. Schwarzschild and von Seeliger both recognised that their picture of the Milky Way was oversimplified, and that it was very unlikely that the Sun was at the centre. They also realised that more extensive observational data and more sophisticated statistical analysis would be required to get much further in understanding its structure.

In parallel with this work, the Dutch astronomer J. C. Kapteyn at

---

*This figure is actually $2.512^{3/2} = 3.98$, or approximately 4, as stars of one magnitude are 2.512 times less intense than those of the preceding magnitude, intensities decrease as the square of distance, but volumes increase as the cube.

Groningen University had also tried to analyse the structure of the Milky Way, but he had quickly realised that there was a serious lack of statistical data on stellar parallaxes and proper motions. In 1906, therefore, he decided to ask observatories around the world to co-operate in a programme to determine the positions, magnitudes, spectral types, proper motions, parallaxes and radial velocities of as many stars as possible, in 206 selected areas of the sky. He later added 18 areas to give a more extensive coverage of the galactic plane and, although his Plan of Selected Areas was generally accepted, the project took several decades to complete.

In the meantime, Kapteyn undertook a thorough analysis of the data that were available in the early years of the century and concluded, in 1912, that the distribution of stars near the Sun was virtually uniform up to a distance of 300 light years, but at greater distances the distribution decreased slowly in the galactic plane and rapidly towards the galactic poles. His analysis showed that the Milky Way was 55,000 light years in diameter, and 11,000 light years thick, which was somewhat larger than previously estimated by von Seeliger and Schwarzschild.

The Apparent Movement of the Sun

The first reliable Doppler shifts showing the velocity of the Sun relative to the nearby stars, had been made by Vogel and Scheiner at Potsdam between 1887 and 1890 (and published in 1888, 1891 and 1892). These were quickly followed by similar measurements made independently by Belopol'skii in Russia, Keeler in the United States, Henri Deslandres in France and H. F. Newall in England, but the most far-reaching work was undertaken by William Campbell when he was appointed to replace Keeler at the Lick Observatory in 1896.

In 1893 Paul Kempf at Potsdam deduced a solar velocity of 13.0 km/s, relative to the nearby stars, from Vogel's and Scheiner's measurements, and in the same year A. D. Risteen in the United States produced an estimate of 17.5 km/s. Campbell then published the radial velocities of 280 stars in 1901 and deduced a solar velocity of 19.9 km/s, with a target area or apex of motion of RA 268°, declineation +25°, very similar to that estimated over a hundred years earlier by William Herschel from stellar proper motions.

Lewis Boss, the director of the Dudley Observatory, New York produced his Preliminary General Catalogue in 1910 which listed the proper motions of about 6,000 stars. On analysing these he estimated the apex of solar motion was somewhat north of that deduced 9 years earlier by Campbell, which had been based on the radial velocities of just 280 stars. Campbell increased his sample of stars to over 1,100 in 1911, and found an apex closer to that of Boss, but still to the south. It was generally assumed that Boss' results were more accurate than Campbell's, as Boss had analysed six times as many stars but, in 1925, H. Raymond and Ralf E. Wilson showed that Boss' proper motions had a small systematic error in them which, when eliminated, brought them into agreement with Campbell's.

Hidden small systematic errors have often produced erroneous results in astronomy, as such errors are often hard to recognise and locate. The fortuitous prediction of the position of Pluto and the erroneous prediction of its size being a case in point.

Star Streaming

In 1871, the Swedish astronomer H. Gyldén had analysed the proper motion of stars and found that, in one part of the sky they tended to have a maximum movement across the sky in one direction, while in the opposite part of the sky they moved in the opposite direction. At positions half way between these two places there was no general drift. He correctly interpreted this as showing that the Milky Way was rotating.

Thirty years later, Kapteyn undertook an extensive analysis of the proper motion and radial velocity of stars and found, in 1904, that they appeared to be streaming in two different directions, in a manner similar to that discovered earlier by Gyldén. Kapteyn had, unlike Gyldén, been able to use radial velocity measurements, giving him a three-dimensional picture of stellar movements, and when he corrected for the movement of the Sun, he found that the vertices of the two star streams were in opposite parts of the sky, with the line joining these two points being in the plane of the Milky Way. Kapteyn thought that these results showed that there were two groups of stars streaming in opposite directions, and that this was probably true of the Milky Way as a whole, and not just for that part near the Sun that he had been able to measure. Arthur Eddington and Frank Dyson extended Kapteyn's studies to the proper motions of more distant stars and showed, in 1908, that Kapteyn's results were still valid at these greater distances. Then, in 1915, Campbell, Walter Adams and Kapteyn confirmed the streaming for even more distant stars by measuring their radial velocities.

Schwarzschild pointed out, in 1908, that Kapteyn's conclusion on stellar streaming had been made on the basis of the *preferred* direction of movement of the stars. There was, in fact, a considerable amount of dispersion of stellar directions about these two preferred directions and so it was not necessary to postulate that the stars moved in two streams at all. He explained that there was just a *tendency* for about twice as many stars to move along the two preferred directions of motion, compared with along any other direction.

H. H. Turner of Oxford University explained this preferential streaming in 1912 by analogy with the movement of comets. He pointed out that a comet in a highly elliptical orbit around the Sun would appear to have a preferential motion directed towards or away from the Sun. He proposed, therefore, that the stellar motions discovered by Kapteyn could be due to stars being in highly elliptical orbits around the centre of the Milky Way. In the following year, Eddington suggested that the streaming could be due to stars moving into, or out of, the centre of the Milky Way, along two spiral arms on opposite sides of the nucleus. Kapteyn, on the other hand, thought that the observed effect could be due to stars moving at different speeds around the centre of the Milky Way.

If Kapteyn was correct, the preferred motion would be perpendicular to the direction of the centre of the Milky Way, whereas Turner's theory predicted preferred motions directed towards and away from the centre. Six years later, Harlow Shapley showed, from his analysis of globular clusters (see below) that the centre of the Milky Way was aligned with the direction of preferential star drift, and so Turner's theory seemed to be vindicated.

Finally, in the 1920s, Bertil Lindblad of Sweden showed that only relatively small deviations from circularity in stellar orbits were needed to produce the observed star streaming effect. He further explained the small proper motions of white stars, that had first been noticed by W. H. S. Monck in 1892 (see Page 129). These white stars and, in particular the B type white stars, had also been shown by Campbell in 1911 to show no preferential streaming. Lindblad pointed out that B type stars are concentrated more in the plane of the Milky Way than any other type of star and, because of this, they should have the most circular orbits, and hence show virtually no streaming.

In 1922 Erwin Freundlich and Emanuel von der Pahlen at Potsdam further studied distant B-type stars and found that those at about galactic longitude 0° and 180° appeared to show a maximum recession from the Sun, whereas those at 90° and 270° appeared to show a maximum approach velocity. Gustav Strömberg at Mount Wilson studied the motion of the Sun relative to the faintest globular clusters, and found in 1924 that the Sun appeared to be approaching galactic longitude 70° with a velocity of 300 km/s, compared with a velocity relative to local stars of only 20 km/s. A few years later, Edwin Hubble found that the Sun moved at a velocity of 280 km/s relative to the nearest spiral galaxies, which by then were known to be galaxies of stars outside of the Milky Way. So the explanation of star streaming appeared clear. The Sun was moving at 280 km/s around the centre of the Milky Way, and stars in the vicinity of the Sun were producing an apparent slow-speed streaming effect, relative to the Sun, because they were following orbits of slightly different eccentricities.

The Distance of Star Clusters and the Size of the Milky Way

Lewis Boss had spent many years analysing the proper motions of stars when, in 1908, he found that the stars in the Hyades open star cluster were travelling together in the sky and, apparently, converging on one point. He concluded that this convergence was due to perspective, and that the stars in the cluster were really travelling parallel to each other at the same velocity relative to the Sun. Boss measured their radial velocities and proper motions and deduced the distance of the cluster as 120 light years, and its diameter as 30 light years. The self-consistency of these data gave the first clear evidence that stars in open star clusters move as a group in the Milky Way.

A few years later, Shapley produced H–R diagrams for individual open clusters, and used these diagrams to estimate the absolute magnitudes of the stars, and hence the distance of the clusters. This method gave good agreement with the distances for nearby clusters estimated using Boss' technique, and so it was applied, with some confidence, to estimate the distance of clusters up to 10,000 light years away, which were far beyond the limit (of 500 light years) for which Boss' method could be used.

John Herschel had noticed, in the early nineteenth century, that globular clusters were far more prominent in the Sagittarius region of the sky than any other, but it was not until the twentieth century that globular clusters were examined in any detail. Karl Bohlin suggested, in 1909, that they formed a compact group at the nucleus of the Milky Way, with the Sun just outside of it. Shapley disagreed with this because, although there are many globular clusters in Sagittarius in the southern sky, there are clusters in the northern sky also, so he concluded that the clusters surrounded the Milk Way about a centre in Sagittarius. Shapley then measured the distance of the globular clusters, using Cepheid variables and other techniques, and found that the centre of the group of globular clusters was 50,000 light years away from us in the direction of galactic longitude 325° (in Sagittarius). They were distributed symmetrically on both sides of the plane of the Milky Way and had distances from Earth ranging from 20,000 light years to 220,000 light years.

Shapley also measured the distances and diameters of the Large and Small Magellanic Clouds of stars in the southern sky, which compared with the distances of the globular clusters as follows:

	Distance (light years)	**Diameter (light years)**	**Centre of group (light years)**
Large Magellanic Cloud	85,000	13,000	
Small Magellanic Cloud	95,000	7,000	
Globular Clusters	20,000 to 220,000		50,000

Shapley concluded, in 1920, that the centre of the group of globular clusters was the centre of the Milky Way, which meant that the Sun was 50,000 light years from the centre. The Magellanic Clouds appeared to be small galaxies, and the globular clusters were clusters of stars, all just outside the Milky Way but associated with it. He assumed that the globular clusters defined the limit of the Milky Way along its equator, calculating that it has a diameter of 260,000 light years, and a thickness of 30,000 light years.

Shapley's analysis had suddenly increased the estimated diameter of the Milky Way by a factor of 5, and had placed the Sun appreciably off-centre. These results were revolutionary when they were announced, and they were not generally accepted for some years. Heber D. Curtis, in particular, considered that Shapley's estimate of the size of the Milky Way was far too large, and participated in a debate with him on the subject at the National Academy of Sciences in April 1920 (see Section 11.1).

Both Shapley and Curtis agreed that there is no appreciable interstellar absorption, but Curtis had a basic distrust of any distance estimates produced using Cepheid variables, as he did not accept their period–

luminosity relationship. Curtis' main counter-argument to Shapley centred on the absolute magnitudes of the bright yellow and red stars in the globular clusters, which Curtis considered to be yellow and red main sequence stars, whereas Shapley maintained (correctly) that their spectra showed that they were highly luminous giants and supergiants. Curtis accepted that the relative distances of the globular clusters estimated by Shapley were correct, but thought that they were all exaggerated by a factor of 10, which he reduced to a factor of 4 during the debate.

We now know that Shapley was basically correct in his estimate of the distances of the globular clusters, although modern estimates put them slightly closer to us, when the effects of interstellar absorption are included.

The Rotation of the Milky Way

In 1927 Jan Oort, of the Leiden Observatory in the Netherlands, finally solved the star streaming problem when he proposed that the Milky Way is rotating about a point where the star clouds are brightest, at about 330° galactic longitude, with an angular rotation near the edge slower than near the centre. This results in the Sun approaching very distant objects (i.e. galaxies) at right angles to 330°, i.e. 60° galactic longitude, and in local stars showing a maximum recession at 15° and 195°, and maximum approach at 105° and 285° (see Figure 10.1), which is very close to the effect observed by Freundlich, von der Pahlen and Strömberg.

Oort calculated that the centre of the Milky Way, with a central attraction equivalent to 60,000 million solar masses, was at a distance of about 17,000 light years from the Sun (or a third of that estimated by Shapley), and that the Sun revolves about the centre once every 120 million years, with a velocity of 270 km/s. This was virtually the same velocity as determined independently by Strömberg and Hubble (see above). Lindblad using a similar analysis found the distance to the centre was 30,000 light years, the central attraction was of 150,000 million solar masses, and the Sun's period of revolution 200 million years.

J. S. Plaskett and J. A. Pearce of the Dominion Observatory in Canada studied the spectra of O and B stars in the plane of the Milky Way, and found in 1930 that the radial velocities shown by the ionised calcium lines, which were known to be due to interstellar gas (see Page 222) increased with distance, and were half of those deduced from the shift in the stars' own lines. The parts of the sky showing maximum recession and maximum approach velocities for the gas, were the same as those found by Oort for the stars. So Plaskett and Pearce not only confirmed the rotation of the Milky Way, but showed that the interstellar gas was approximately uniformly dispersed* in

---

*The fact that the calcium lines showed half of the Doppler shift of the stars' lines, and that the shifts of both sets of lines increased linearly with distance, indicated that the average distance of the intervening gas was half way between the Earth and the star. This is exactly what would be found if the gas were uniformly distributed between the Earth and the stars under investigation.

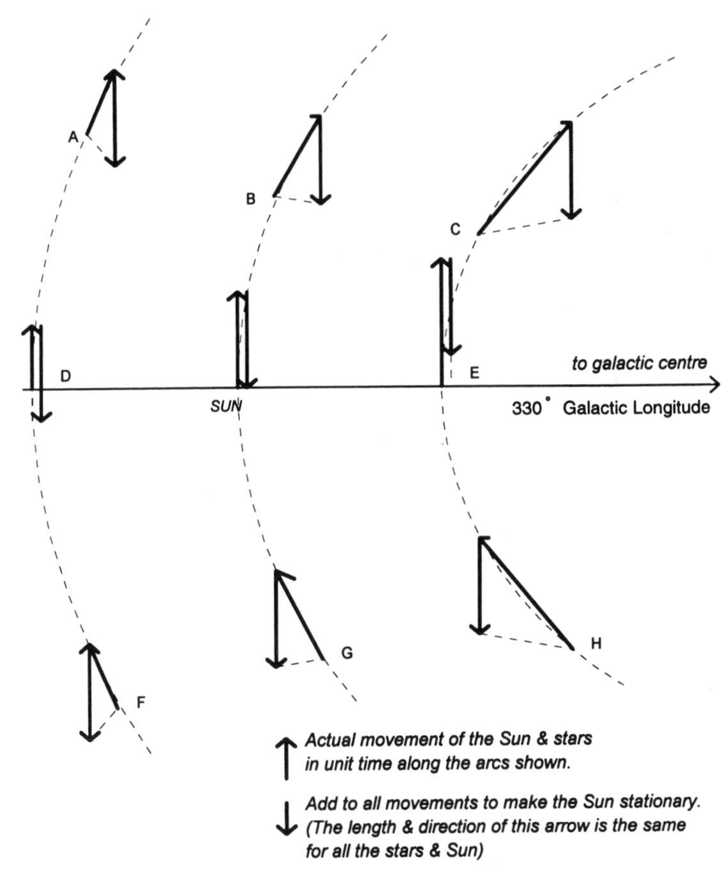

The net motion of the stars above, relative to the Sun, is shown by the thin lines above, to produce (approximately) the net effect shown by the arrows below:-

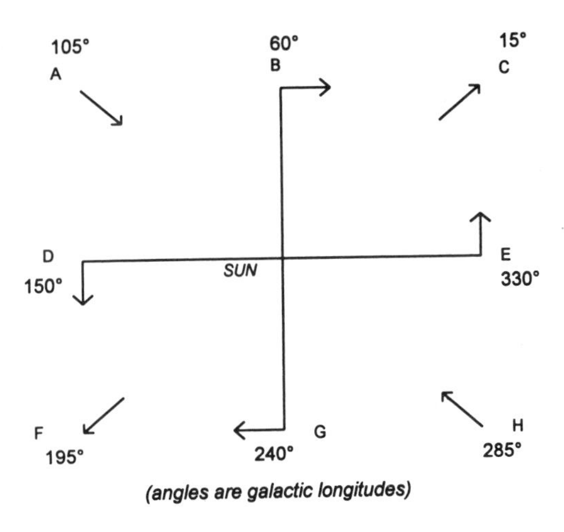

Figure 10.1 *Apparent movement of stars close to the Sun.*

the plane of the Milky Way, and was rotating with the same velocity as the stars.

Shapley had, in the meantime, continued with his analysis of the size and structure of the Milky Way and, in 1930, he reduced his estimate of the diameter to 180,000 light years. So, he pictured the Milky Way as being surrounded by globular clusters in a cloud which was shaped like a flattened sphere, with a maximum diameter, in the plane of the Milky Way, of 180,000 light years, and a minimum diameter, perpendicular to that plane, of 90,000 light years. In that same year, however, Robert Trumpler at the Lick Observatory showed that there was an absorbing layer of dust in the plane of the Milky Way (see Pages 224–225), and this had the effect of reducing Shapley's estimate of the distance of globular clusters in the plane of the Milky Way, while leaving the distance estimates perpendicular to this plane virtually unchanged. Joel Stebbins modified Shapley's estimates, taking this absorption into account, and concluded that the globular clusters were in a spherical halo, of diameter 90,000 light years, around the disc of the Milky Way which had the same diameter.

So by 1933 the estimates of the size of the Milky Way had evolved as follows:

		Distance of Sun from centre (light years)	Diameter of Milky Way (light years)
Kapteyn	1912	Assumed 0	55,000
Shapley	1920	50,000	260,000
Oort	1927	17,000	
Lindblad	1927	30,000	
Shapley	1930	50,000	180,000
Stebbins	1933	30,000	90,000

In the 1930s Shapley investigated the distribution of Cepheids in the Milky Way in the opposite direction to the centre, in an attempt to find out whether the star system extended up to the halo of globular clusters in the plane of the Milky Way. Of the 52 Cepheids measured, 75% were within the confines of the halo 16,000 light years away, but other Cepheids were up to 42,000 light years away, or 26,000 light years beyond the halo. So, although the globular clusters defined an approximate limit to the size of the Milky Way, there were still some stars that extended beyond these clusters in the plane of the Milky Way.

Shapley also used Cepheids to study the maximum thickness of the Milky Way and found, in 1934, that, of the 54 Cepheids that he studied near the galactic poles, 18 were over 30,000 light years from the galactic plane, and two were over 60,000 light years from that plane.

This analysis helped to paint the general picture of the Milky Way in the mid 1930's as being a disc of stars about 3,000 light years thick, except in the centre where there was a bulge of about 12,000 light years thickness, that was surrounded by a spherical halo of globular clusters some 90,000 light years in diameter. Some stars were in the space between the disc of stars and the halo of globular clusters, and a few stars were up to 26,000 light years outside this halo.

The Milky Way was thought to have a mass of 150,000 million suns, with the Sun about 30,000 light years from the centre, located at 3300 galactic longitude in Sagittarius. The Sun was estimated to be moving with a velocity of 270 km/s around the centre, taking about 200 million years to complete one circuit.

Frank Drake investigated the nucleus of the Milky Way in 1959 using the 85 ft (26 m) radio telescope of the National Radio Astronomy Observatory at Green Bank, West Virginia, concluding that it consists of a small, very dense, cluster of stars with a density of more than 30 stars/cubic light year (i.e. more than a thousand times the density of stars in the vicinity of the Sun). The French astronomers A. Lallemand and M. Duchesne, and Merle Walker of the Lick Observatory also found a higher density of about 150 stars/cubic light year in the central 25 light years (radius) of the Andromeda nebula. Then Burbidge suggested in 1961 that the density of stars in the nucleus of some galaxies could be as high as 300,000 stars/cubic light year, and here stars would be so close that one supernova could trigger a supernova explosion in an adjacent star, a few hundred astronomical units away.

In 1971, the Uhuru spacecraft detected X-rays coming from the general direction of the centre of the Milky Way in Sagittarius and, towards the end of that decade, the Einstein Observatory spacecraft located the position of the source to within an accuracy of 1 arcmin. These X-rays were found to vary over a period of about 3 years, thus indicating that the source was a maximum of only 3 light years in diameter. This may indicate the presence of a non-stellar black hole.

Douglas Lin of the University of California analysed the movement of about 250 stars in the Large Magellanic Cloud and concluded, in 1993, that it was orbiting the Milky Way at 236 km/s, and approaching the Milky Way at 55 km/s. Lin concluded that the mass of the Milky Way is 600,000 million solar masses, or about four times the previous estimate, and he attributed the extra mass to a halo of dark matter about 600,000 to 800,000 light years in diameter, surrounding the Milky Way. If this is the case, the LMC, which is only 169,000 light years from the centre of the Milky Way, would be within the halo. Clearly there is still some way to go before we can be sure of the size of the Milky Way.

The Spiral Arms

Easton had suggested, in 1900, that the Milky Way may have a spiral structure, like that of the Andromeda galaxy, although 20 years later, Shapley considered the Milky Way to be more of a collection of star clouds and star clusters without any clear shape. In the late 1920s, after the Andromeda nebula had been proved to be a spiral galaxy some distance from the Milky Way, astronomers also thought of the Milky Way as a spiral galaxy, although there was no clear evidence of our galaxy's spiral structure until the 1950s.

In 1951, Walter Baade and N. U. Mayall showed using Mount Wilson and Palomar photographs that the spiral arms of the Andromeda galaxy could best be delineated by blue supergiant O and B type stars, by hydrogen (Hα) emission nebulae, and by clouds of galactic dust. Later that year, W. W. Morgan of the Yerkes Observatory described measurements of O and B stars, and of Hα emission nebulae in the Milky Way, which clearly showed what he called the Orion and Perseus arms. There were suggestions of a third arm, the Sagittarius arm, and this was confirmed shortly afterwards by Bart Bok, M. J. Bester and C. M. Wade, using data from the southern Ha survey at the Arequipa station of the Harvard College Observatory.

Interstellar absorption limits visual surveys of the Milky Way to those parts within about 10,000 to 15,000 light years of the Sun. Such absorption

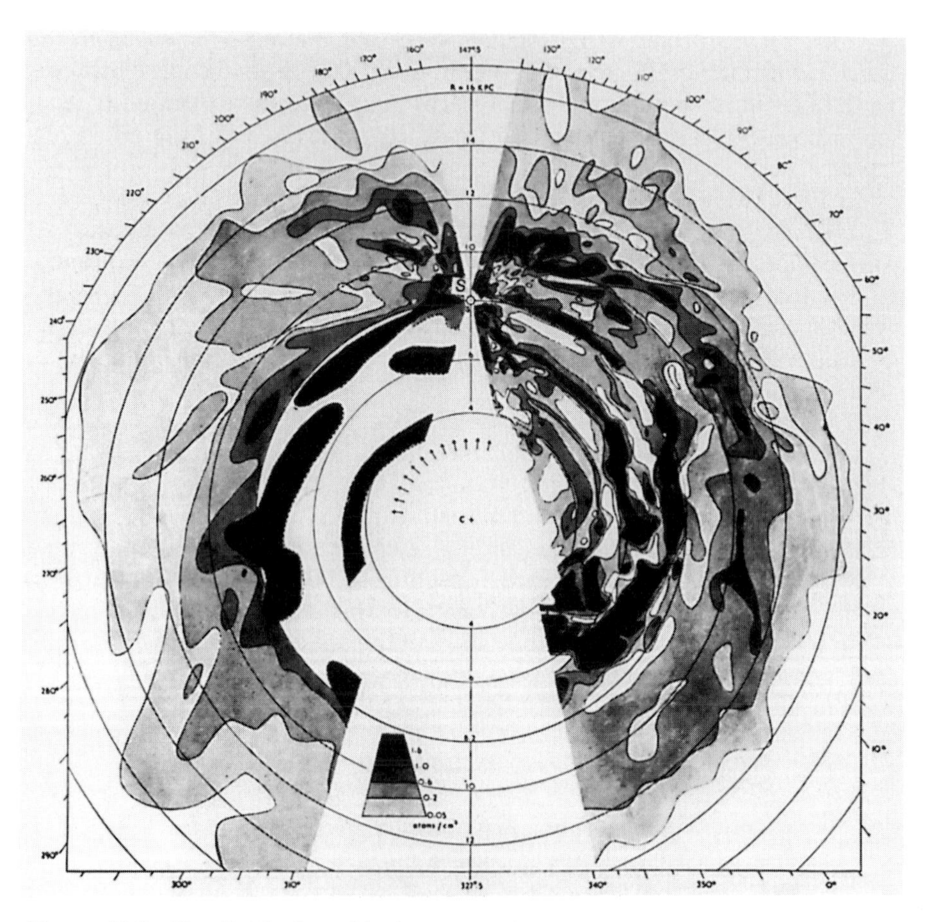

Figure 10.2 *The distribution of hydrogen gas in the Milky Way, as deduced by Jan Oort, Gart Westerhout and F. J. Kerr. (From* Monthly Notices of the Royal Astronomical Society, *Vol. 118 (1958), Plate 6, courtesy* The Royal Astronomical Society.*)*

has no significant effect on radio waves, however, and so the development of radio telescopes played a key rôle in unravelling the overall structure of our galaxy. Interstellar atomic hydrogen was found, in 1951, to emit radio waves at a wavelength of 21.1 cm (see Pages 313–314), and so surveys of the Milky Way at this wavelength showed the structure of the hydrogen clouds. The velocity of the hydrogen clouds along our line of sight could be determined directly by measuring the magnitude of the Doppler shift of the radio waves, and this enabled the structure of these clouds to be determined.

In the 1950's Jan Oort and Gart Westerhout of the Leiden Observatory, and F. J. Kerr of the Sydney Radiophysics Laboratory in Australia, used radio telescopes to produce our first view of the overall structure of the Milky Way, shown in Figure 10.2, assuming that the Sun (shown as S) is about 8,200 parsecs* or 26,700 light years away from the centre. Parts of the spiral arms of the Milky Way are clearly shown.

When Oort and his colleagues mapped the Milky Way using the 21 cm emission line of atomic hydrogen (H), they did not realise that half of the hydrogen in the Milky Way is molecular (H_2), which is invisible to radio telescopes. Fortunately, when conditions are favourable for the formation of H_2, they are also favourable for the formation of carbon monoxide (CO), and H_2 molecules excite CO molecules in these hydrogen clouds. The CO molecule emits energy at a wavelength of 2.6 mm, when it falls back to its lowest energy level, and so measuring emissions at this wavelength has enabled radio astronomers to trace the clouds of molecular hydrogen in our galaxy. Molecular hydrogen is thought to pinpoint areas of star formation, unlike atomic hydrogen which is more generally dispersed, and in 1988, Thomas Dame produced a survey map of the Milky Way showing these carbon monoxide–molecular hydrogen clouds. Work is still under way refining the results.

The Interstellar Medium

Gas

Why is the night sky dark? The answer to this seemingly simple question has far-reaching consequences.

The Englishman Thomas Digges had suggested in 1576 that, although he thought that the universe was infinite in extent, the night sky was dark was because the distant stars were too far away and faint to be seen. Kepler disagreed with this, and reasoned in 1610 that if the universe were infinite in extent and filled with stars then, no matter where one looked, one would see a star. As this was not the case,† Kepler had concluded that the stellar universe was finite in extent.

---

*1 parsec ≡ 3.26 light years.

†This is called Olbers' paradox, after the astronomer who popularised the question in the early nineteenth century.

Jean Phillipe de Chéseaux proposed an alternative theory in 1744. He realised that Thomas Digges' explanation was wrong because, although the intensity of a star seen from Earth is reduced by a factor of 4 when its distance is increased by a factor of 2, the star also appears to be four times smaller in area, and so its apparent surface brightness does not change. De Chéseaux calculated that, if the stars are spaced uniformly throughout the universe with the same density as those near the Sun, and if all stars are the same size as the Sun, then we would see a star wherever we looked provided the universe was at least 10^7 light years in size, in today's units. Unlike Kepler, de Chéseaux believed that the universe was infinite in extent, and concluded that the night sky was dark because there was gas and dust between the stars that absorbed their light. Wilhelm Olbers came to the same conclusion in the nineteenth century.

The concept of gas and dust pervading the universe received little attention until, in 1904, J. F. Hartmann, who was then at Potsdam, discovered the stationary K line of ionised calcium in the spectrum of the *binary* star δ Orionis, which Eddington attributed to absorption by interstellar gas. The same stationary line was discovered in other binary stars and, in 1919, the same effect was found for sodium in the binaries δ Orionis and β Scorpii by Mary Heger of the Lick Observatory.

Not all astronomers accepted Eddington's explanation, as the stationary calcium lines appeared to be present only in the spectra of hot stars (B3 and earlier). This led some people to suggest that the calcium gas may be in a small stationary cloud surrounding the binary system. In 1924 Plaskett, while accepting that the absorption lines were due to large clouds of interstellar gas, proposed that the cloud's calcium atoms were only ionised in the vicinity of the hotter stars, thus explaining why they were only seen near to these stars.

Eddington solved the problem 2 years later, when he showed that the ionisation could not be produced in the vicinity of hot stars, as near these stars the calcium gas would be doubly ionised. He showed, instead, that the ionised calcium lines produced by the interstellar gas are concealed for stars later than B3 by lines generated in the stars themselves, whereas, for earlier stars, the stars' lines are broad and diffuse and are easily distinguished from the sharp narrow lines produced by the intervening gas. Eddington also pointed out that only the very bright O and B stars could be seen at the very large distances where interstellar absorption is appreciable, also explaining the apparent preference for these stars to show interstellar absorption lines.

At the same time, Otto Struve showed that the ionised calcium lines were generally more evident in stars near to the plane of the Milky Way, although there were places outside of this plane where they were also quite strong. So, although the calcium gas clouds are generally in the plane of the Milky Way, there are also some clouds away from this plane. Four years later Struve and Plaskett showed, at low galactic latitudes, that the calcium lines are most evident for the most distant stars, thus proving that they are to due a galactic absorption effect, and Plaskett and Pearce showed that the gas rotated with the stars in the Milky Way.

It was also of interest to see over how small a distance absorption could be detected and, in 1941, Theodore Dunham, Jr. detected ionised calcium lines in a Virginis at a distance of 170 light years and, in 1961, Albrecht Unsöld and Guido Münch, working at the Palomar Observatory, found an interstellar line in α Ophiuchi at a distance of only 60 light years.

C. S. Beals found, in 1936, that interstellar lines are sometimes multiple, which suggested that more than one cloud is between the star and the Earth, and that these clouds are moving with different velocities. Walter Adams at the Mount Wilson Observatory then measured interstellar absorption lines in hundreds of stars and concluded, in 1949, that the strongest component in the multiplet nearly always corresponds to the lowest velocity, while some of the weaker components showed velocities of up to 100 km/s. Measurements of the absorption lines of neighbouring stars showed the same strongest component of the interstellar lines, but, sometimes, the highly displaced lines were different. This indicated that the high velocity clouds are generally smaller than the low velocity clouds and, although the average cloud size was estimated at about 25 light years, the range of sizes was found to be very large.

Guido Münch discovered, in 1957, that the absorption lines for the distant stars in the plane of the Milky Way, and located between galactic longitudes 60° and 160° (see Figure 10.2), usually showed two strong components, one due to clouds in our spiral arm (the Orion arm), and the other due to clouds in the Perseus arm, which is the next arm out from ours. Struve and others had shown that, although clouds are much scarcer in high galactic latitudes than in the plane of the Milky Way, they are still found there. As the general gas density outside of the clouds is very low at high galactic latitudes, these clouds ought to have dispersed by now, but in 1961 Münch and Harold Zirin suggested that this dispersal could have been restrained by a hot, very tenuous galactic corona.

Although interstellar gas had first been found to consist of calcium and sodium, Dunham and Struve suggested that it is probably mostly hydrogen. In 1944, the Dutch astronomer Hendrik van de Hulst, predicted that interstellar hydrogen should emit radio waves at a wavelength of 21.1 cm, but it was not until 1951 that this radiation was discovered by Harold Ewen and Edward Purcell at Harvard University. Jan Oort et al. then used the Doppler shifts of this 21 cm radiation to detect and measure the slow rotation of the thin sheet of neutral hydrogen gas in the Milky Way. This showed that the gas in the central 3,000 light years of the galaxy rotates as if it were a solid body, with an almost constant angular velocity, but outside of this region, it has an almost constant linear velocity.

In the 1930s, Dunham and Adams at Mount Wilson discovered that interstellar gas consisted, not only of atoms, but also of molecules when they found methylidyne (CH) and cyanogen (CN). Further discoveries had to await the development of infrared and ultraviolet detectors mounted on spacecraft, and of radio telescopes, but in 1963 Weinreb and Barrett discovered the hydroxyl (OH) molecule, and in 1968 water (H_2O) and ammonia (NH_3) molecules were found. These were followed by the discovery of formaldehyde (H_2CO) in 1969, and that of the hydrogen molecule H_2 and of carbon monoxide (CO) in 1970. Since then ever more complex molecules have been discovered including formic acid (HCOOH) in 1970 and methylamine (CH_3NH_2) in 1977, which can react together to form the amino acid, glycine (NH_2CH_2COOH). Over 50 different molecules have now been discovered in interstellar gas, a large number of which are organic (i.e. contain carbon).

Dust

William Herschel had noticed, in the late eighteenth century, that there were a number of areas in the sky in which he could see virtually no stars, even with his largest telescopes. He interpreted these as regions devoid of stars, suggesting that the material that would have condensed to form stars had been drained away by the gravitational attraction of adjacent nebulae. These views were described by the Reverend Thomas Dick in his book *The Sidereal Heavens*, which Edward Barnard had read in the 1870s, so when Barnard started to photograph the Milky Way in 1889, he was not surprised to discover a number of black voids.

Barnard published his first photographs of these dark regions of the Milky Way in *Knowledge*, a popular science magazine, in 1894, still assuming that they were regions devoid of stars. The editor, A. C. Ranyard, disagreed with this interpretation, however, and suggested that they were clouds of absorbing material blotting out the background stars. Unknown to Ranyard, a similar suggestion had been made by Angelo Secchi in 1853.

Until his death in 1923, Barnard oscillated between the two theories of the dark areas (i.e. areas empty of stars, or regions of absorbing matter). In 1905, Barnard had almost convinced himself that the sharp outlines to some of the areas indicated that they were clouds of obscuring material, but two years later he favoured the void theory. In his last paper on the subject in 1919 he concluded that maybe both theories were correct.

In 1908, Gavril Tikhov and Charles Nordmann in Europe found that the minimum light of Algol's eclipse occurs 16 minutes earlier in red light than in blue, and a similar effect, of magnitude 4 minutes, was found for RT Persei. They concluded that this was caused by the velocity of light in the interstellar medium varying with wavelength, but the Russian physicist Pyotr Lebedev pointed out that the star with the apparently larger effect was much closer to Earth, so the suggested cause seemed unlikely. Later observations showed that Tikhov's and Nordmann's results were in error, and the effect did not exist. In particular, Shapley showed in 1917 that the times of the maxima of eleven Cepheids in a globular cluster in Serpens were the same in both red and blue light.

In the late 1920s, Trumpler of the Lick Observatory measured the apparent diameters of 100 open star clusters, and the intensity of their stars, and found that the intensities reduced faster with distance than expected, by about 0.7 magnitudes/kiloparsec (abbreviated to mag/kpc), assuming that the clusters were all of the same size. He published his results in 1930, concluding that the Milky Way contained interstellar dust like that causing the dark bands in spiral galaxies.

Trumpler also showed that stars of the same spectral type were redder, the further away they were. He found an average selective* absorption of 0.3 mag/kpc for stars in open star clusters. This appeared to contradict Shapley's observations of 10 years earlier, that showed no such effect for stars in globular clusters, which are much further away than open star clusters. Open star clusters are generally in the plane of the Milky Way, however, whereas the globular clusters are generally far away from this plane, so Trumpler concluded that this selective absorption is confined to a relatively thin layer of material extending about 400 light years on either side of the plane of the Milky Way.

Three years later, van de Kamp of the Leander McCormick Observatory in the United States showed that this layer of dust explained the relative paucity of globular clusters in the plane of the Milky Way, and Stebbins showed that those clusters in the plane of the Milky Way are redder than those out of the plane; the reddening being due to dust. Stebbins found that the absorption made the globular clusters in the plane of the Milky Way appear four times further away than they really are, but for clusters just 5° above or below the plane this correction factor had reduced to 2, and it rapidly reduced to 1 (i.e. no correction) with a further small increase in angle.

These conclusions were broadly confirmed and extended in 1934 by Hubble's analysis of 80,000 spiral galaxies in different parts of the sky, which showed that the absorption varied from about 2.2 mag/kpc in the plane of the Milky Way to about 0.25 mag/kpc perpendicular to that plane. The dust seemed to be much more constrained to the plane of the Milky Way than was the gas.

Subrahmanyan Chandrasekhar predicted, in 1946, that early-type stars should exhibit a small amount of plane polarisation near their limbs, and the American astronomers Albert Hiltner and J. S. Hall set out to try to measure this effect in 1947, using the Wolf–Rayet eclipsing binary CQ Cephei. They found some evidence for the effect but, because their photometer was unreliable, they both agreed to try to produce an improved instrument. In the following year, Hall was transferred to another observatory, and so they then observed separately.

Hiltner found, much to his surprise, that the polarisation for CQ Cephei was quite large and that it was independent of the relative positions of the two stars in the binary. He measured other Wolf–Rayet stars, and found that some showed this effect, but others did not. He concluded that the polarisation that he was measuring had nothing to do with the stars, but was caused by the interstellar medium. Hall confirmed this, and showed that increased polarisation is accompanied by increased selective absorption

---

*Selective absorption is due to scattering by dust particles and depends on wavelength, general absorption (often referred to just as "absorption") is, over optical wavelengths, basically independent of wavelength.

(i.e. stellar reddening). Further work by Hiltner, Hall and others in the early 1950s showed that the degree of polarisation varied according to galactic longitude, in such a way that the alignment force on the dust grains* appears to have its axis parallel to the spiral arms of the Milky Way. Chandrasekhar and Fermi were able, in 1953, to calculate the strength of the galactic magnetic field, which they assumed to be the source of the alignment force, from the degree of polarisation.

Nebulae in the Milky Way

Emission and Reflection Nebulae

Huggins had found, in the nineteenth century (see Page 210), that some nebulae showed continuous spectra with absorption lines, whereas others showed emission line spectra. The former type were compact nebulae with a generally spiral pattern,† but the latter, which are called emission nebulae, were highly irregular in shape with a cloud-like appearance.

Vesto Slipher found another type of cloud-like nebula in 1912, when he observed the spectrum of the nebulosity in the Pleiades star cluster. Much to his surprise he found, not an emission line spectrum, but a continuous spectrum that was an exact copy of that of the brighter stars in the cluster. This reflection nebula, as it is called, consists of very small particles which scatter the starlight and, as the shorter wavelength light is generally scattered more than the longer wavelengths, the nebula tends to appear blue. (This is the same scattering process that makes the sky appear blue in daytime.) By 1919 Slipher had found six other reflection nebulae.

Hubble became interested in these reflection nebulae, and undertook a detailed survey of both emission and reflection nebulae, concluding in 1922 that:

(i) The vast majority of the 29 emission and 33 reflection nebulae that he examined surrounded relatively bright stars.

(ii) Stars of type B0 or earlier are associated with emission nebulae, whereas stars of type B1 or later are associated with reflection nebulae.

(iii) The size of both the emission and reflection nebulae correlate with the magnitude of the brightest star associated with them. This was to be expected for reflection nebulae, as the brighter the star, the more of the nebula it would illuminate, but it was unexpected for emission nebulae.

It had always been a mystery why emission nebulae should emit any light at all, as they seemed to be too diffuse to be self-luminous, but here, at last,

---

*The alignment force was assumed to be acting on those dust grains that were slightly elongated.

†These were shown in the 1920s to be galaxies of stars. They are discussed in detail in Chapter 11 below.

Hubble had found an explanation. The emission nebula was in some way excited by the very energetic type B0, or earlier, star embedded within it. Russell suggested that this excitation could be produced either by the energetic short-wavelength light, or by fast moving electrons, emitted by the star. Hubble tried to explain the brightness of planetary nebulae in the same way, but was surprised to find that the central star of the planetary appeared to be too dim to generate the amount of light observed in the nebula.

If the emission and reflection nebulae were of the same composition, then the former should be brighter, as the scattered starlight should be the same in both, whereas the emission nebulae also produce their own light. Hubble showed in the 1920s that these two types of nebulae are, in general, of the same intensity, thus showing that their composition must be different, adding to the evidence that reflection nebulae are made of dust, while the emission nebulae are gaseous.

In the nineteenth century, some of the spectral lines of emission nebulae were known to be due to hydrogen, but there were others that could not be identified. Ira S. Bowen finally identified these lines in 1927 as being due to oxygen, nitrogen and other atoms making forbidden transitions (see Page 10), which could only occur in a very tenuous, hot gas.

So, by 1930 the diffuse emission nebulae were known to be composed of very tenuous gas that was being illuminated by the hot O, B and, sometimes, A stars embedded within them, but did these stars come from the nebulae or vice versa? Were these nebulae the birthplace of the hot stars, or had the nebulae been ejected by them?

In the 1930's, it was clear that the hot O, B and A stars associated with the diffuse nebulae were on the main sequence of the H–R diagram. For some years Russell had maintained that this main sequence was an evolutionary track, with the O, B and A stars being the youngest but, in 1927, he accepted the prevailing view that stars on the main sequence are all mature stars where they stay for the majority of their lives (see Pages 138–139). So the diffuse nebulae could not be giving birth to stars as, otherwise, there should be stars of all types within them. It was not until the 1950s that Merle Walker and Allan Sandage showed (see Pages 150–152) that hotter stars evolve faster than the others, which could explain why the diffuse nebulae only contain these hot and young O, B and A stars. These nebulae could thus be the birthplace of stars after all, although the nebulae could have been emitted by the hot stars, like the corona emitted by the Sun.

Bengt Strömgren extended Zanstra's ionisation theory of planetary nebulae (see below) to estimate the effect of ultraviolet light emitted by hot stars in emission nebulae. He showed that all of the hydrogen in an emission nebula is ionised near to the star but that, at a certain distance, the ultraviolet light ceases to have enough energy to ionise the hydrogen. In particular, for a main sequence O5 star, the hydrogen would be ionised up to about 200 light years from the star, assuming a nebula density of 3 hydrogen atoms/cm^3, for a B0 star, the region would extend to about 40 light years, whereas for an A0 star it would extend to less than 1 light year. This explained why Hubble had found that only early-type stars are associated with emission nebulae.

Planetary Nebulae

A number of planetary nebulae were known in the nineteenth century. They were generally torroidal in appearance, and surrounded a central star which was thought illuminated them. Planetary nebulae were gaseous and had the same emission spectrum as emission nebulae.

In the early twentieth century, as the number of planetary nebulae grew, it was found that they tend to congregate in the plane of the Milky Way, with more in the direction of the centre of the Milky Way than elsewhere. In other words, the distribution is similar to that of the stars. The Russian astronomer Boris Vorontsov-Vel'iaminov undertook a survey of 119 planetary nebulae in the early 1930s and, in 1934, concluded that they have diameters ranging from 1/50 to 15 light years, and are at distances ranging from 600 to 30,000 light years.

William Campbell, Joseph Moore and Ralf Wilson at the Lick Observatory showed in the 1920s, by measuring Doppler shifts, that the planetary nebulae are rotating about their central star, and that the masses of the planetary nebulae, deduced from the speed of this rotation, are between four and 150 times the mass of the Sun. They also found that the spectral lines are often double at the centre of the nebula, but single at the edge, indicating that some planetary nebulae are still expanding.

As early as 1894, Campbell had found that the size of a planetary nebula was not the same at all wavelengths of light and, in the 1920s, W. H. Wright of the Lick Observatory showed that the relationship between wavelength and size is the same for all planetaries. Once Bowen had explained the origins of the lines in the spectra of emission and planetary nebulae, as being due to forbidden transitions in a very tenuous hot gas, the reason for these differences in size with wavelength became immediately apparent. For those images taken in the wavelengths of oxygen, for example, the largest image was due to singly ionised oxygen, the next largest was due to doubly ionised oxygen, and the smallest was due to triply ionised oxygen. Clearly, the higher degree of ionisation occurs nearer the central star, thus proving that the gas in planetary nebulae is being stimulated by the central star. Bowen's calculations showed that the hottest of the stars surrounded by a planetary nebula had a surface temperature of the order of 100,000 to 150,000 K. At these temperatures, the stars emit much of their energy in the ultraviolet, a large amount of which is absorbed by the Earth's atmosphere, and this explained why Hubble had found that the central stars of planetary nebulae appeared to be too dim to produce the amount of light observed coming from the nebulae.

H. Zanstra of the Dominion Observatory in British Colombia suggested, in the 1920s, that ultraviolet light can ionise interstellar hydrogen, and he developed Bowen's theory that predicted the temperature of stars in the centre of planetary nebulae. Zanstra estimated that the temperature of the coolest of these stars is 28,000 K, and that of the hottest is 140,000 K, and both Zanstra, and Louis Berman at the Lick independently found that the hotter the central star appears to be, the fainter it is in comparison with the nebula. They attributed this to the fact that, at these temperatures, the hotter the star, the more of its radiation is in the ultraviolet which is absorbed by the Earth's atmosphere. Berman even suggested that the reason why some planetary nebulae do not seem to have a central star is because the star is so hot that virtually none of its light gets through the Earth's atmosphere.

So the central stars of planetary nebulae are very hot O type stars, and the surrounding nebulae appeared to have masses of from four to 150 times that of the Sun, although this mass range was later revised substantially downwards. In 1934, Vorontsov-Vel'iaminov calculated that the density of these stars ranged from 3 to 1,000,000 g/cm^3, on the assumption that they weigh the same as the Sun. So they are very dense, and some of them are white dwarfs with densities of up to, at least, 10 times that of the dwarf star Sirius B (see Pages 139–140).

The similarities between planetary nebulae and the gas shells around novae were recognised in the 1920s and 1930s. They both have the typical nebula emission spectrum, and the central star, in both cases, is an early-type star. It was thus thought possible that gas shells around novae could eventually turn into planetary nebulae. There were three main problems with this idea, however:

(i) The gaseous envelope of a nova was found to expand much faster than a planetary nebula, although this could be explained by the nova envelope slowing down with time.

(ii) There were much fewer planetary nebulae than there should be if most novae produce planetary nebulae. Only 150 planetaries were known in 1935.

(iii) The amount of gas in a planetary nebula is much greater than that in the cloud ejected by a nova.

This led astronomers in the 1930s to suggest that planetary nebulae are the remnants of the much rarer and much more powerful supernova explosion, rather than that of a nova. Later work reopened the debate, however, when it was realised that red giants lose mass, some of it substantial, by a strong stellar wind en route to becoming white dwarfs. This is the favoured theory of today, although the origin of planetary nebulae is still not really clear.

11 Galaxies

The Nature and Distance of Spiral Nebulae

Immanuel Kant had suggested in the eighteenth century that nebulae were galaxies of stars, like the Milky Way, but at very great distances (see Page 209), and William Huggins had shown in the 1860s that two thirds of the nebulae that he had investigated exhibited continuous spectra like stars. These nebulae, which tended to have a compact spiral structure, were thought by some astronomers to be distant star systems like the Milky Way, but others thought they were gaseous objects in the Milky Way, with their continuous spectrum being produced by scattering starlight. In 1899, James Keeler discovered hundreds more small spiral nebulae on photographs taken with the Crossley reflector at the Lick Observatory.

One of the key problems that exponents of the distant star system or "galaxy" theory had to explain was why, if these galaxies were not associated with the Milky Way, were there many more spiral galaxies away from the plane of the Milky Way than in it. If this effect was real, spiral galaxies would seem to be intimately connected with the Milky Way even if they were outside of it. In 1904, J. F. Hartmann discovered the existence of interstellar matter in the Milky Way (see Page 222), and so the paucity of spiral galaxies close to the plane of the Milky Way could simply be due to their light was being absorbed by this interstellar matter. So the evidence for the stellar nature of spiral galaxies was beginning to build up, but the proponents of this theory got a nasty shock when, in 1912, Vesto Slipher of the Lowell Observatory discovered that the nebulosity in the Pleiades had a stellar spectrum, because it was scattering light from its associated stars.

A new star had appeared in 1885, apparently in the bright central regions of the Andromeda nebula, and although it was possible that it was a nova in front of the nebula, it seemed an unlikely coincidence of alignment, particularly as a similar new star had also been observed in the M 80 globular cluster in 1860. In 1911, the American F. W. Very estimated the distance of the Andromeda nebula as 8,000 light years, which would put it within the confines of the Milky Way, by comparing the brightness of Nova S Andromedae, as this new nova was called, with that of Nova Persei, which was a nova in the Milky Way. In fact Nova Persei was a nova, while Nova S Andromedae was later found to be a supernova, but the idea that supernovae could exist with absolute luminosities about ten magnitudes higher than those of novae, seemed ridiculous. So Very, Harlow Shapley and others concluded that the Andromeda and other spirals were part of the Milky Way.

Adriaan van Maanen at the Mount Wilson Observatory announced in 1916 that he had detected the circular rotation of the spiral nebula M101 in Ursa Major. The rotation period he measured, of 85,000 years, would imply rotation speeds of greater than the velocity of light, if M101 was very distant. As velocities greater than the velocity of light are impossible, the conclusion was clear, M101 must be part of the Milky Way.

In the following year, four novae were found in the Andromeda nebula that were about ten magnitudes fainter than Nova S Andromedae. Heber Curtis suggested that these four novae, rather than Nova S Andromedae, should be compared with Nova Persei in the Milky Way, on the assumption that Nova S Andromedae was a nova of exceptional intensity. Using this approach, Curtis first estimated the distance of the Andromeda nebula as 10 million light years, but he later reduced this figure to 500,000 light years, after he corrected his distance estimates for Milky Way novae. This distance clearly put the Andromeda nebula well outside of the Milky Way.*

Not everyone accepted Curtis' conclusion. If the four Andromeda novae were like the Milky Way novae, he was correct, but that implied that Nova S Andromedae was unusual and would have an incredibly high intensity. If, on the other hand, Nova S Andromedae was a nova of normal intensity, he was wrong, and the Andromeda nebula would be about 100 times nearer than he had estimated. In this case, however, the four novae seen in the Andromeda nebula must be very much dimmer than those seen in the Milky Way.

On 26th April 1920 there was a famous debate on the merits of these two alternative scenarios between Harlow Shapley and Heber Curtis at the National Academy of Sciences. Shapley maintained that spiral galaxies were masses of gas associated with the Milky Way, and probably just outside of it, and that their continuous spectrum was due to the scattering of light from nearby stars. Curtis, on the other hand, maintained that they were stellar systems like the Milky Way, but at very great distances.

Shapley's argument was based on:

(i) Van Maanen's estimate of the distance of M101, based on his measurement of its rotation. (The rotational velocities were later

*In 1919 Knut Lundmark independently estimated the distance of the Andromeda nebula to be 650,000 light years, using a similar approach.

found by the Swedish astronomer Knut Lundmark to be overstated by at least a factor of 10.)

(ii) Frederick Seares' estimate of the surface brightness of our galaxy compared with that of the spirals, that showed the Milky Way to be much less bright, assuming that the spirals were at large distances. (Seares' estimate of the surface brightness of the Milky Way was in error, however, as it was based on J. C. Kapteyn's estimate of the size of our galaxy, which was too small.)

(iii) Nova S Andromedae would have to have an incredible absolute magnitude if the galaxy was a great distance from the Milky Way. (This was found to be true when the existence of incredibly bright novae, or supernovae, like Nova S Andromedae, were recognised in both the Milky Way and other galaxies in the 1920s and 1930s.)

(iv) The lack of spiral nebulae in the plane of the Milky Way. (This was correctly explained, at the time, as being due to the obscuring effect of interstellar matter in the plane of the Milky Way.)

(v) Slipher's spectroscopic evidence of the fast recession velocity of spirals, which seemed difficult to accept if they are such large objects.

Curtis argued that:

(i) If Shapley's estimate of the distance of 20,000 light years for nearby spirals was correct, then the most distant spirals, which subtend only about 10^{-3} of the angle of the nearest ones, must be about 20 million light years away, assuming that they are all of approximately the same size. Shapley's conclusion was only self-consistent if the spirals had a range of diameters varying by a factor of about 1,000, which was thought to be unlikely for objects that looked so very similar in structure.

(ii) The majority of novae in the Andromeda nebula indicate that it is well outside the Milky Way, as explained above.

(iii) Spiral galaxies could not be seen in the plane of the Milky Way, because of the large amount of dust there.* Many spirals that are edge-on to us also show such dust in their central plane, indicating that they are similar in structure to the Milky Way.

(iv) Curtis saw no reason why spiral galaxies should not have large velocities with respect to the Milky Way.

The debate was useful in bringing the evidence on both sides clearly into the open, but the result was largely inconclusive. The nature of spiral nebulae was, however, finally resolved by Edwin Hubble (Figure 11.1), when between 1919 and 1924 he made a series of photographs at Mount

---

*At the time, only interstellar gas was known in the plane of the Milky Way, although Edward Barnard thought that there may also be clouds of dust there. It was not until 1930 that Robert Trumpler showed that the Milky Way had a layer of dust in its plane. Four years later, Edwin Hubble showed that this explained the observed distribution of spirals.

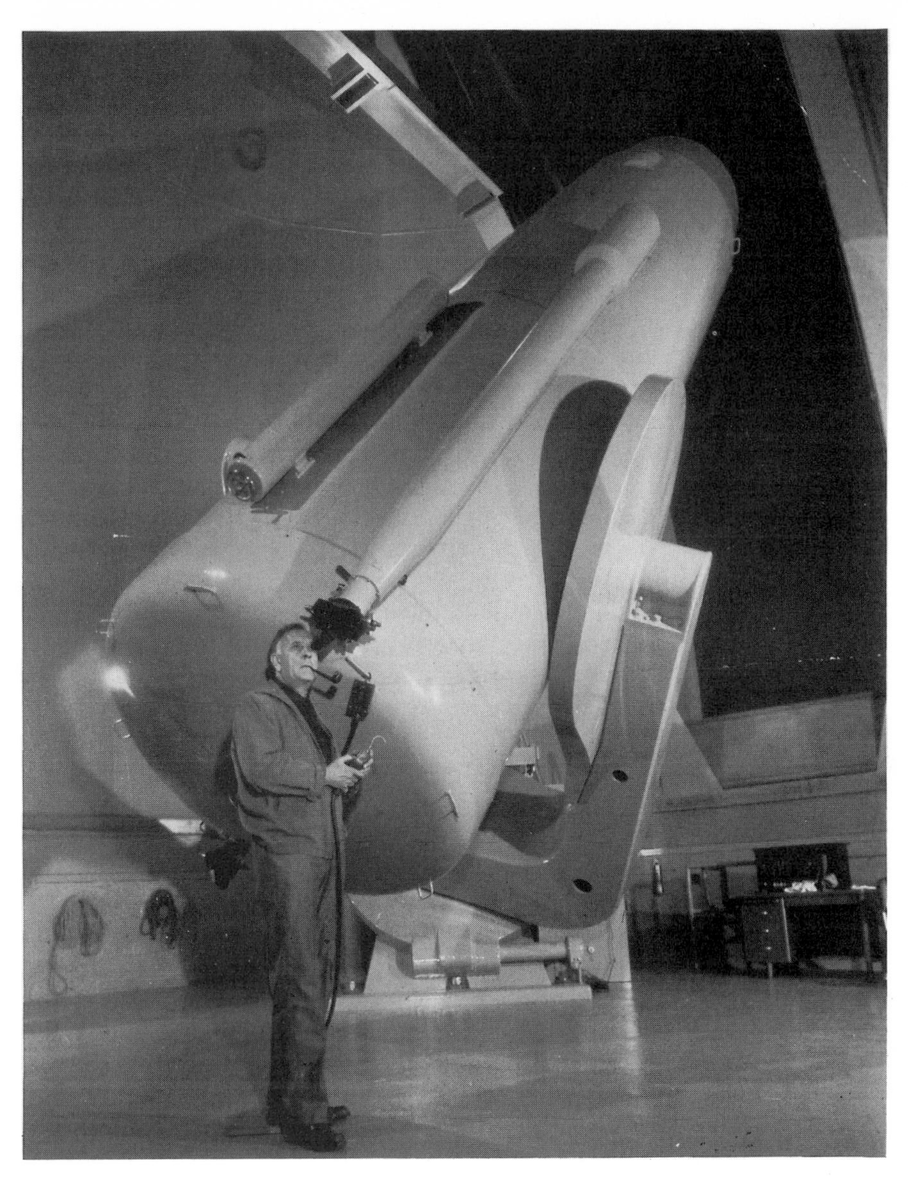

Figure 11.1 *Edwin Hubble using the 49 inch (1.2 m) Schmidt telescope on Palomar Mountain. (Courtesy* California Institute of Technology.*)*

Wilson of M31 (the Andromeda nebula) and M33 (the triangulum nebula). Hubble discovered his first Cepheid in a spiral nebula (M31) in February 1924, and he found a further 33 Cepheid variables in M31 and M33 during the year. This enabled him to determine the distances of the Andromeda and triangulum galaxies as 900,000 and 850,000 light years, respectively, and to estimate their diameters as 40,000 and 16,000 light years. So spiral nebulae were finally shown to be stellar systems like the Milky Way but at very great distances.

Hubble found Cepheids in a few more galaxies, which also enabled him to estimate their distances. He also found, in these galaxies, that the intensity of the brightest, non-variable stars, was about 20 times that of the average Cepheid. This enabled him to estimate the distances of those galaxies that were just too far away for their Cepheids to be seen, from the intensity of their brightest stars, increasing the number of spiral galaxies whose distances he had estimated to 40, with distances of up to 6 million light years.

Galaxy surveys in the early part of the twentieth century indicated that the spiral galaxies often appeared to occur in groups, which Slipher called super-galaxies, the nearest of which appeared to be in Virgo, numbering about 500 galaxies. Some very bright stars could be seen in some of these Virgo galaxies, enabling the distance of the Virgo super-galaxy, or galaxy cluster, to be measured at 7 million light years. Hubble found that the vast majority of the galaxies in this galactic cluster (to use a modern term) had intensities that varied between half and twice the average intensity. He also found a relatively narrow intensity distribution for galaxies in 20 other galactic clusters, giving him confidence that the distances of galaxies can be estimated, at least to an order of magnitude, by measuring their apparent brightness. Hubble and Milton Humason found, in 1931, the distance of the Pegasus galactic cluster to be 23 million light years, and by 1936 had measured the distance of some clusters at over 200 million light years.

These three steps in estimating the distances of galaxies, namely intensities of Cepheids, then of the brightest stars, and then of the whole galaxy, were exactly the same as used by Shapley in 1919 when he estimated the distances of globular clusters (see Page 215).

Hubble found that the average intensity of all the spiral galaxies that he had measured was about 100 million times that of the Sun. His estimate of 40,000 light years for the diameter of the Andromeda galaxy, was much less than the estimate of 260,000 light years for the diameter of the Milky Way that Shapley had produced a few years earlier. Hubble found that other galaxies seemed to be even smaller than Andromeda, so, although the spiral nebulae had been established as galaxies of stars like the Milky Way, the latter seemed to be larger than any other galaxy. It seemed unlikely that we should fortuitously be living in the largest galaxy in the universe, so what was wrong? Shapley, Lundmark and Trumpler independently suggested that maybe the Milky Way was not a galaxy at all, but a cluster of galaxies. The other alternative was that there was something wrong with the distance estimates.

Shapley had thought that interstellar absorption was insignificant, and so it was ignored in this early work. Including it in the 1930s brought the distance of the Andromeda galaxy down a little to 750,000 light years, and reduced its diameter to 30,000 light years, compared with the new diameter estimate of 90,000 light years for the Milky Way.

In 1910, George Ritchey had taken a series of photographs with the new 60 inch (1.5 m) Mount Wilson reflector of about a dozen spiral nebulae and noticed, on each one, a number of small spherical condensations which he called "nebulous stars", thinking that they were stars in the process of formation. Starting in 1919, Hubble photographed these spiral nebulae and spherical condensations, which he called "globular nebulae", although he thought that they were probably stars also.

Hubble realised in 1930 that the globular nebulae were really globular clusters, photographing 140 of them surrounding the Andromeda galaxy. He found that these clusters were twice as far away from the centre of the galaxy, compared with the edge of the stellar part of the galaxy. Two years later Shapley, using a much more sensitive recording and measuring system, found that the stellar part was much larger than had been shown on previous photographs, so that it reached to the globular clusters at least. This instantly doubled the diameter estimate of the Andromeda galaxy to 60,000 light years, making it much closer in size to the Milky Way.

In the 1920s, Hubble had estimated that the average diameter of spiral and elliptical galaxies as about 10,000 and 2,000 light years, respectively. Shapley showed in the early 1930s that the fainter parts of spiral galaxies can increase the diameters up to 20,000 light years, and of elliptical galaxies up to 15,000 light years. In the Virgo cluster he measured the diameter of the largest spiral as 24,000 light years and the largest elliptical as 15,000 light years, whereas in the Centaurus cluster he measured diameters for the largest spirals of up to 45,000 light years. So, although the Milky Way was still thought to be the largest galaxy, the differences were getting smaller.

Shapley's estimates were the best available until 1952, when Walter Baade showed (see Page 254) that the distances of the galaxies needed to be doubled, because the Cepheid variables seen in the spiral arms were of a different type to those seen in the globular clusters. This increased the diameter estimates for all the galaxies (except the Milky Way) by a factor of 2, finally eliminating the diameter problem. In particular the Andromeda galaxy was seen to have a diameter of 120,000 light years, compared with that of 90,000 light years for the Milky Way. Shapley re-estimated the distances and sizes of the two Magellanic Clouds in 1952, following Baade's discovery of the two Cepheid populations, and concluded that they were both 150,000 light years away, with diameters of 30,000 and 20,000 light years.

Hubble was able to estimate the mass of galaxies in the 1930s, by measuring their rate of rotation, and concluded that the Andromeda galaxy (M31) had a mass of 3,500 million suns, and the Sombrero galaxy some 2,000 million suns. By 1957, after successive modifications, mainly caused by the increase in its diameter by a factor of 4 (and a volume increase of $4^3 = 64$), the generally-accepted mass of M31 was 340,000 million suns, which is very close to today's estimate.

So in the early half of the twentieth century the distances to the nearest galaxies was becoming known, but how many galaxies were there in the universe? Keeler estimated in 1899 that about 120,000 nebulae were within photographic range of the 36 inch (91 cm) Crossley reflector that he was using at the Lick observatory. Later this estimate was increased to 500,000 by Charles Perrine, also of the Lick.

A programme of galaxy counts was carried out at the Lick from 1947 to 1964 by Shane and Wirtanen, using a double astrograph that enabled wide-field photographs to be taken simultaneously in blue and yellow light. The survey covered the whole sky down to −23° and registered 800,000 galaxies down to magnitude 18.8. Then in 1983, when J. Anthony Tyson and Patrick Sweitzer of the Bell Laboratories in New Jersey were experimenting with deep-sky imaging techniques, they discovered that what appeared to be almost blank areas of sky near the galactic poles were, in fact, teeming with faint distant galaxies. This led them to conclude that there were about 20 billion galaxies in the range of their equipment, which was an enormous increase on previous estimates. So we appear, even today, to be a long way from understanding the size and complexity of the universe.

Red Shifts

Vesto Slipher estimated the radial velocities of 13 spiral nebulae between 1912 and 1914, by measuring the Doppler shifts of the H and K lines of ionised calcium on exposures lasting up to 40 hours each at the Lowell Observatory. Harlow Shapley examined these radial velocities and noticed, in 1917, that the nebulae generally seemed to be moving away from us. By 1925, Slipher had produced a list of 41 spiral galaxies with radial velocities ranging from 300 km/s towards us, to 1,800 km/s away from us. In the same year Lundmark, who was then director of the Upsala Observatory in Sweden, found that the smaller the apparent size of the spiral galaxies, the faster they were moving away from us. Assuming that the galaxies with the smallest apparent size were the furthest away, Lundmark concluded that the velocity of recession increased with increase in distance from the Earth. This was consistent with the prediction made in 1916 by the Dutch astronomer Willem de Sitter, using the theory of relativity, that nebulae should show a receding motion which would increase with increasing distance from the Earth.

Slipher's work in measuring the radial velocities of galaxies, was continued by Milton Humason in 1928, when it was decided to use the 100 inch (254 cm) Hooker telescope on Mount Wilson to extend the observations to more distant galaxies. In the mid 1920s, Hubble was able to estimate the distances of galaxies for the first time, and in 1929 Hubble and Humason showed that the recession velocity, or red shift, for distant galaxies, increased linearly with increasing distance, at a rate of about 500 km/s per million parsecs (written km/s/Mpc) (see Figure 11.2). This is called the Hubble constant, written as H_o. By 1931 Humason had extended the distance measurements of galaxies to 150 million light years (45 million parsecs), and had found that the relationship of recessional velocity to distance was still valid. Five years later, velocity measurements of up to 42,000 km/s had been made, and distances of up to 230 million light years (70 million parsecs) had been measured.

The 200 inch (508 cm) telescope on Palomar Mountain was first used for

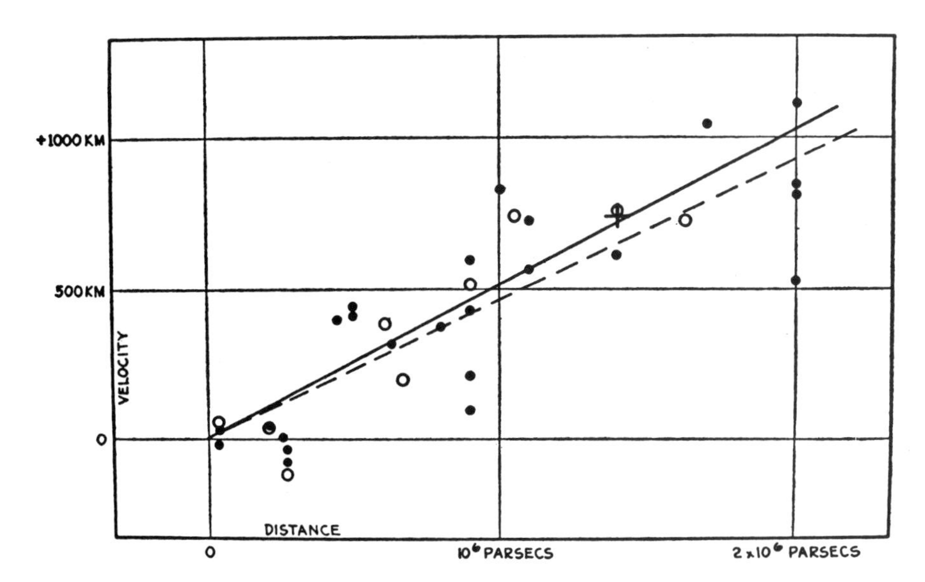

Figure 11.2 *The distance–velocity relation published by Edwin Hubble in 1936. (From* The Realm of the Nebulae, *by E. Hubble, New Haven Press, p. 114; courtesy Yale University Press.)*

this work in 1950, and 3 years later it confirmed that the linear expansion relationship was still valid up to the furthest galaxies measured which were at distances of up to 350 million light years (107 million parsecs). At about the same time, Walter Baade showed that the distance estimates of galaxies using Cepheids needed doubling, resulting in a halving of the rate of expansion of the universe, or Hubble constant H_o, to 250 km/s/Mpc. Then, in 1958, Allan Sandage produced an even smaller estimate for H_o of 75 km/s/Mpc.

Some 14 years later, Sandage at Mount Wilson and Gustav Tammann of Basle in Switzerland jointly produced a new estimate for H_o. They recalibrated the period–luminosity relationship for Cepheid variables, to redetermine the distances of nearby galaxies, and then measured the sizes of the ionised hydrogen regions in the nearby spiral galaxies of type Sc (see Page 249). This provided another criterion for measuring the distances of more distant galaxies, yielding a value for H_o of 53 km/s/Mpc.

Today (1994) the value of H_o is still unclear, with estimates ranging from 45 to 95 km/s/Mpc (see Page 255).

Quasars

Discovery

In 1943, Carl Seyfert at the Mount Wilson Observatory discovered a number of galaxies, now known by his name, to have small, intensely bright, very blue nuclei. In short photographic exposures only the nucleus could be seen, looking very much like a star, but in long exposures the spiral arms appeared. The galactic spectra were characterised by broad emission lines. Some years later, Merle Walker of the Lick Observatory made a detailed examination of M77 (NGC 1068), which is one of the brightest Seyfert galaxies, and found a number of clouds, each weighing about 10 million suns, being ejected from the central region at velocities of up to 600 km/s. The processes then-known were insufficient to produce so much disruption, and so there must be a new, unknown source of energy present in the nucleus.

Then, in 1960, Allan Sandage, John Bolton and Thomas Matthews correlated one radio source, called 3C 48, with what appeared to be a dim blue star. This seemed to be of no major significance but, in the following year, another radio source, 3C 273, was also identified with a faint blue star. A search was carried out of previous photographs of the area, and these showed that the blue star, identified with 3C 273, varied in intensity about a mean photographic magnitude of 12.5. Harlan Smith and Dorrit Hoffleit of Yale found two types of variation by measuring old Harvard plates, one of about half a magnitude, that occurred over a timescale of about 10 years or so, and another, much more rapid variation, which appeared to be quite random.

Fortunately the Moon sometimes occults (passes in front of) 3C 273, and Cyril Hazard in Australia used the occultations of 5th August and 26th October 1962 to measure the position of the radio source accurately, which was found to be double, with its two components separated by some 20 arcsec. A few weeks later, Maarten Schmidt of Caltech showed that the position of one radio source correlated with the faint blue star, but the position of the other source coincided with the end of a faint bluish jet coming from the star. The spectrum of the jet was too faint to be recorded, but the spectrum of the star was unlike anything that had been seen before. It was only when Schmidt was writing a scientific paper in the following year, summarising his results, that he realised that the lines in the spectrum were hydrogen lines with a very large red shift of 0.16.*

The red shift of 3C 273 showed that it is receding† from us at 15% of the speed of light, assuming the red shift is due to the Doppler effect, and that 3C 273 is about 2.0 billion light years away, assuming a value of the Hubble constant of 75 km/s/Mpc.

One of Schmidt's colleagues, Jesse Greenstein, immediately checked the spectrum of the other radio source, 3C 48, and found that it had an even larger red shift of 0.36, indicating a distance of 3.9 billion light years. In addition, the radio brightness of 3C 48 was found to vary substantially

---

*The red shift z is given by the Doppler shift $\Delta\lambda$, divided by the rest wavelength λ, i.e. $z = \Delta\lambda/\lambda$.

†The recession velocity v is given by $v = [(z + 1)^2 - 1] \times c/[(z + 1)^2 + 1]$, where c is the velocity of light.

during a day, indicating that it is only about one light day across. Over the next few years, many of these quasi-stellar objects, or "quasars" as they are now called, were discovered at ever greater distances, many of them showing similar short-period variations in their radio output, indicating their very small physical size.

Martin Ryle of Cambridge University presented an interesting analysis of quasars and radio galaxies to the IAU Congress in Prague in 1967, where he showed that the radio sources in both quasars and radio galaxies tended to be double, with the optical image in between the two radio images. This led Ryle to suggest that the optical objects are the sites of explosions which eject radio-emitting clouds of gas in opposite directions. He further explained that, assuming the gas is ejected at equal speeds in both directions, the cloud ejected towards us would have moved more than the further one when we see it. This is because the light from the one nearer to us has less distance to travel, and so the gas in that is seen at a later time. As a result, the optical image is not exactly in the centre of the two radio images.

Geoffrey and Margaret Burbidge of the University of California suggested, in the early 1960s, that quasars seemed to be the more violent relations of Seyfert galaxies, which were themselves known to have very energetic nuclei ejecting large clouds of gas. Some of these Seyfert galaxies were found to be radio sources, and some were found to have variable radio emissions. This led astronomers to see if the light emissions also varied and, in 1967, W. S. Fitch, A. G. Pacholczyk and Ray Weymann using the 36 inch (0.9 m) reflector on Kitt Peak found a 0.25 magnitude variation in the intensity of the central region of Seyfert galaxy NGC 4151. In the following year William Morgan, Harlan Smith and D. Weedman showed, using the 82 inch (2.1 m) Mc Donald reflector, that the spectrum of NGC 4151 had changed since it had been recorded in 1956; although the bright emission lines were still present, some absorption lines had disappeared. J. B. Oke and Wallace Sargent of Caltech suggested shortly afterwards that the emission lines are mostly produced by rapidly moving clouds of gas in the central region of the nucleus, and estimated, using the 200 inch (5 m) Palomar reflector, that these clouds have a temperature of 20,000 K and a weight of 200,000 suns. They also suggested that the remainder of the nucleus consists of low density gas at a temperature of about 1 million K producing forbidden lines of nine and 13 times ionised iron atoms, like those produced in the Sun's corona.

Since their first identification in 1963, more than 5,000 quasars have been discovered at ever greater distances, so that today the maximum wavelength shift is 4.90, implying a recession velocity of 94% of the speed of light, and giving a distance of 12.3 billion light years (assuming the Hubble constant to be 75 km/s/Mpc).

If quasars are at the distances outlined above, they must be very luminous objects, and currently they are thought to be the nuclei of active galaxies. 3C 273, for example, appears to be 400 times as bright in the optical waveband as the Milky Way, but the majority of its radio energy, like that of 3C 48, seems to be coming from a very small core about 1 light day across. In 1979, the Einstein Observatory found that the X-ray intensity of 3C 273 is about a million times greater than that of the Milky Way, but it is varying over periods as short as half a day, indicating that it is being emitted from a source no larger than the diameter of the Solar System. This source could well be a black hole. In the following year, 3C 273 was found by the Cos-B satellite to be emitting γ-rays, with the energy emitted in γ-rays being even more than in X-rays.

Not everyone accepts that quasars are at great distances, and suggest, instead, that the quasar spectral lines are being red shifted by a non-Doppler effect. In particular, James Terrell of the Los Alamos laboratory proposed, in the 1960s, that quasars could have been ejected by a gigantic explosion at the centre of the Milky Way about 5 to 10 million years ago but, if that was the case, some of them should show motion across our line of sight. No such motion has been found. In addition, if quasars had been formed by an explosion in the Milky Way, which is a pretty ordinary galaxy, then some quasars ought also to have been ejected by other galaxies. Some of these quasars should be approaching us, but no blue-shifted quasars were found.

Halton Arp who was then at Caltech also published evidence, starting in the mid 1960s, that indicated that quasars are relatively* near objects, and, although his theory is not generally accepted, it is more difficult to disprove than Terrell's. Arp showed that some high red shift quasars are apparently interacting with low red shift galaxies, and some galaxies with radically different red shifts also appear to be physically linked to each other (e.g. NGC 7603, which is apparently receding at 8,700 km/s, appears to be linked to a smaller companion galaxy which is receding at 17,000 km/s). Most astronomers reject such linkages, however, suggesting that they are nothing more than line-of-sight effects.

Superluminal Radio Sources

David Robertson in Australia, and Moffat in Canada, discovered, in 1967, that a quasar appeared to be expanding at faster than the speed of light. These so-called superluminal radio sources were also discovered independently by Irwin Shapiro et al., in 1970, when they linked a radio telescope in California with one in Massachusetts to make a very large radio interferometer, and found that Quasar 3C 279 appeared to be expanding at ten times the speed of light. Images of a jet in the core of 3C 273 (see Figure 11.3) show a similar effect. These jets, associated with what are now called "active galaxies", are not really moving faster than the speed of light, but the effect is due to a trick of geometry caused by the high velocity jet pointing almost straight at us.

For 10 years, 3C 273 was the only quasar known to emit γ-rays, and then,

---

*Terrell thought that quasars were just outside the Milky Way. Arp put them at the distances of ordinary galaxies.

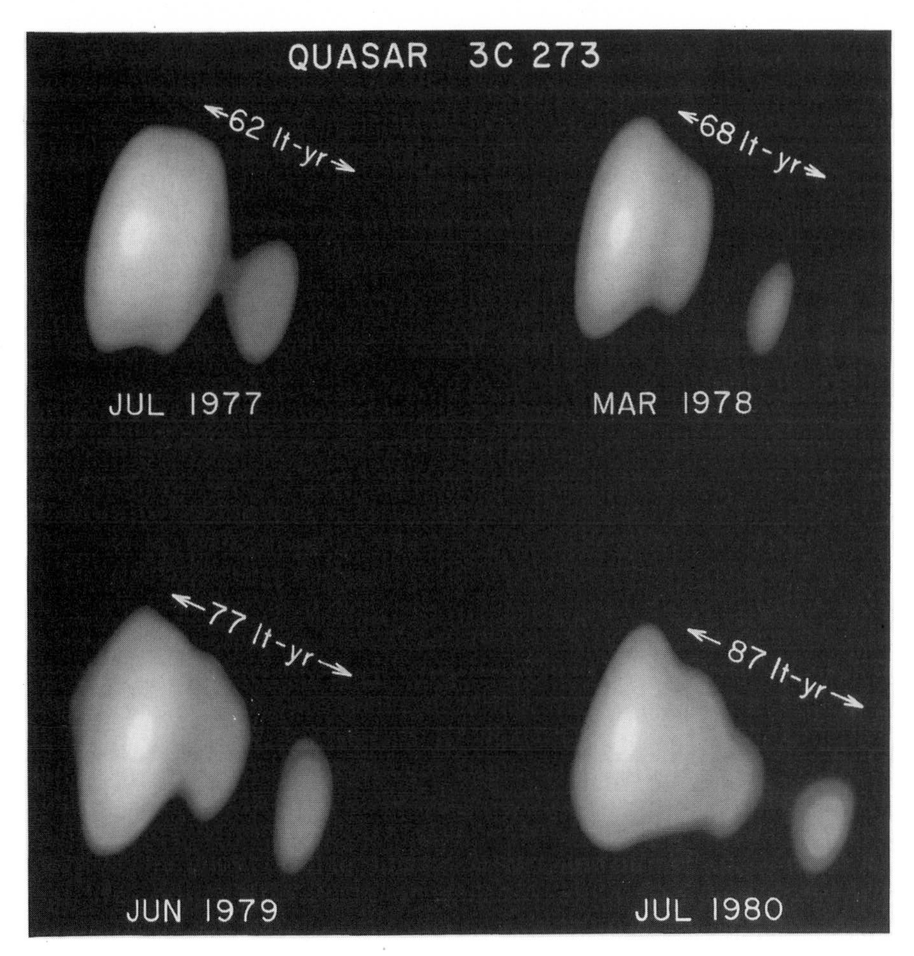

Figure 11.3 *Images of the jet emitted from the quasar 3C 273, indicating an apparent expansion at superluminal velocity. Image produced by the VLBI arrangement of radio telescopes at Effelsberg, Haystack, Green Bank, Fort Davis, Owens Valley and Hat Creek. (Courtesy* California Institute of Technology.*)*

in 1991, the GRO spacecraft found that the quasar 3C 279 was also a source of γ-rays, emitting 10 million times more energy than the Milky Way, which made it the most luminous source of γ-rays in the sky. Although 3C 279, at a distance of 7 billion light years, is over three times as far away as 3C 273, it is surprising that it had not been found to emit γ-rays before, and this was attributed to possible variations in its intensity.

Gravitational Lensing and Einstein Rings

An apparently ordinary double radio source was detected at Jodrell Bank in the 1970s, but, in 1979, David Walsh, Robert Carswell and Ray Weymann found, using the 2.1 m telescope on Kitt Peak, that the position of this source coincided with that of two 17th magnitude quasars. Surprisingly, the optical spectra of the two quasars were identical. It was thought that the two images, which are only 6 arcsec apart, were probably of the same quasar, one image coming directly from the quasar, and the other coming past an intervening galaxy which was bending the light because of the galaxy's large gravitational field. A few months later, Peter Young discovered the image of this intervening galaxy, and proved the existence of what is today called "gravitational lensing". Later data on this double quasar (called 0957 + 561 A and B) showed that the brightness of the B quasar image varies over a period of months, whereas that of A appeared to be constant. This threw doubt on the gravitational lensing theory for this object but, in 1991, David Roberts showed that the variation in the radio intensity of one of the images was the same as the other, but with a time delay of 1.4 years. He was able to use the magnitude of this time delay, together with the geometry of the sources, to deduce a value of the Hubble constant of 57 ± 12 km/s/Mpc.

A small number of other possible examples of gravitational lenses have since been found, the most interesting of which is two adjacent quasars 1146 + 111 B and C, that Cyril Hazard and Halton Arp found to have identical red shifts of 1.01. The problem is that the two images are separated by an unprecedented 157 arcsec, which implies an enormous intervening mass, no other evidence of which has been found. Bohdan Paczynski suggested in 1986 that the intervening mass could be in the form of a "cosmic string", which is thought to be an incredibly thin and incredibly massive structure left over from the Big Bang.

Edwin Turner of Princeton University showed, using the new 4 m telescope on Kitt Peak, that the red shifts of the two quasars 1146 + 111 B and C were identical to within one part in 1,000, but then John Huchra, of the Harvard–Smithsonian Center for Astrophysics, found clear differences in their ultraviolet spectra. In 1986 Peter Shaver and S. Christiani at the European Southern Observatory also found differences in their near infrared spectra. As the two path lengths to the quasar, if this is a case of gravitational lensing, could be thousands of light years different, the quasar could have changed its spectrum over that amount of time. On the other hand, they may be two different quasars. At present the answer is still open.

Probably one of the most convincing visual examples of a gravitational lens was found in 1992 by Richard Ellis and his colleagues of Durham University, using the Hubble Space Telescope. They were observing the galactic cluster AC 114, which is about 4 billion light years away, when they noticed two distant galaxies, separated by 10 arcsec, that are remarkably similar in their visual structure. Their spectra, which were measured by the Anglo-Australian telescope, were also virtually identical, so Ellis et al. concluded that the cluster AC 114 is bending light from a more distant galaxy, to produce two identical images. They deduced that the foreground cluster must be about 10 times as massive as its visible light indicates, leading to the conclusion that it must have this amount of dark matter (see Page 257).

In 1936 Einstein suggested that if a distant galaxy lies exactly behind a closer massive galaxy, then the image of the distant galaxy would be distorted into a ring by the nearer one. Fifty years later, in 1986, Roger Lynds, of the Kitt Peak National Observatory and Vahe Petrosian of Stanford University discovered giant luminous arcs in the galaxy clusters 2242–02 and Abell 370, which were thought could be incomplete Einstein rings caused by the imperfect alignment of two galaxies. Then, in early 1988, the first complete ring was found by Jacqueline Hewitt of the Haystack Observatory, using the Very Large Array radio telescope in New Mexico, in the object MG 1131 + 0456.

In November 1988 another possible Einstein ring was discovered, also using the VLA. Radio waves from the quasar MG 1654 + 1346, which has a red shift of 1.74, were found to be imaged on either side of the quasar. One of the two radio images was a ring shape with a giant elliptical galaxy of red shift 0.25 in the centre. This was the first ring candidate to have the red shift measured of both the distant quasar and the lensing galaxy, giving added confidence that higher red shifts are associated with larger distances.

In 1986, Georges Meylan et al. at the European Southern Observatory correlated the radio source PKS 1145–071 with two faint starlike objects at visible wavelengths, separated by 4.2 arcsec. Spectroscopic observations made with the Multiple Mirror Telescope showed that the two objects were both quasars about 12 billion light years away, with masses of 100 billion solar masses, which is the typical mass for a normal galaxy like the Milky Way. Observations with the Very Large Array clearly showed that only one of the quasars is producing the radio emissions, indicating that these are two different quasars, rather than an example of gravitational lensing. This was the first binary quasar found.

BL Lac Objects

BL Lacertae was identified as a variable star in 1929 and then, in 1968, Maarten Schmidt also identified it as a radio source. Its optical spectrum was strange as it showed no absorption bands. Other similar objects were found, and then, in 1976, Joe Wampler, at the Lick Observatory, found faint absorption lines in a few of the BL Lac objects, as they came to be called, enabling him to estimate their red shifts. He concluded that they are, like quasars, the nuclei of active galaxies, and it is now thought that BL Lac objects are quasars oriented so that we are looking straight down the emitted jet. In the 1980s BL Lac objects were generally found to emit energy at all wavelengths from X-rays to radio waves, inclusive, then in 1991 the GRO spacecraft also found γ-rays coming from three BL Lac objects, adding to the evidence of their similarity to quasars.

BL Lac objects were thought, for a number of years, to be only associated with the nuclei of elliptical galaxies but, in 1991, McHardy and colleagues found that the BL Lac object PKS 1413+135 was at the centre of a large flattened disc galaxy. This was the first indication that high velocity ("relativistic") jets, that are associated with active galaxies, may not be limited to the nuclei of ellipticals.

Centaurus A

Centaurus A was found to be a strong radio source in 1948 and, in the following year, John Bolton, G. J. Stanley and O. B. Slee in Australia identified it with the seventh magnitude galaxy NGC 5128. This galaxy appears in visible light (see Figure 11.4, top) to be an almost circular disc of stars, 6 arcmin in diameter, with a dark dust band across its centre. More recent radio data have shown that the radio emissions cover an astonishing 9° of the sky or 90 times the linear size of the normal optical image. The orientation of the radio lobes is perpendicular to the dust cloud, and highly enhanced optical images have also shown similarly-oriented faint optical lobes, although the latter are only about 10% of the size of the radio image. Centaurus A, which is 16 million light years from the Milky Way, is the nearest so-called active galaxy.

The lobes of the radio galaxy cover 2½ million light years, so the electrons which cause them have probably been flowing from the central galaxy for about 100 million years. A high resolution radio image of the central galaxy, made with the Very Large Array, is shown in Figure 11.4 (bottom) at about the same scale as Figure 11.4 (top). It shows two jets extending to about 15,000 light years from the galaxy centre, indicating that activity is still continuing.

The core of the galaxy is both a powerful X-ray and gamma-ray emitter and, in 1973, the OSO-7 spacecraft found that the nucleus brightened in X-rays over just a few days, indicating that the object at the centre is only a few light days in diameter. Many astronomers believe that the source of so much energy is gas being sucked into a black hole at the centre of the galaxy.

Cygnus A

The Swedish astronomer Erik Holmberg suggested in 1940 that, as the distances between galaxies were only about 10 to 100 times the diameters of the galaxies, there should be a number of cases where two or more galaxies are gravitationally attached to each other. In these cases, the galaxies would orbit their common centre of mass, but tidal forces would eventually cause their orbits to shrink, and eventually the galaxies would merge.

Grote Reber discovered, in the early 1940s, that a region of the sky in Cygnus was a source of radio waves (see Page 311), and the position of this source, called Cygnus A, the second brightest radio source in the sky, was measured by the English radio astronomers James Hey, S. J. Parsons and J. W. Phillips in 1946. Five years later, Baade and Minkowski identified an inconspicuous 16th magnitude galaxy as the originator of these radio signals. At optical wavelengths the galaxy appeared, at first, to be a double galaxy and Baade was convinced that it was two galaxies in collision. Minkowski disagreed, and these two colleagues bet a bottle of whisky on who would be proved to be correct. A few months later, Minkowski paid up, convinced that the discovery of high energy emission lines had proved Baade to be right, but further investigations showed Cygnus A to be a single

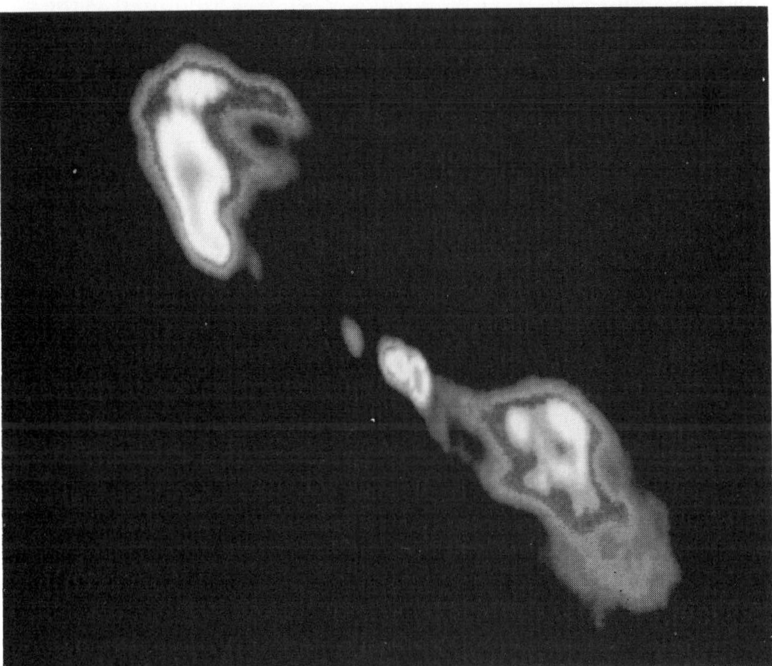

Figure 11.4 *Centaurus A in visible light (top), imaged by the Anglo-Australian Telescope, and at radio wavelengths (bottom), imaged by the Very Large Array. The scale and orientation of the two images is almost identical. (Top: Courtesy* Anglo-Australian Observatory, *photography by David Malin. Bottom: Courtesy* National Radio Astronomy Observatory/AUI, *observers J. O. Burns, E. J. Schreier and E. D. Feigelson.)*

galaxy crossed by a central dust band, like Centaurus A. As Cygnus A is 750 million light years away from us, however, its structure is much more difficult to see than that of Centaurus A.

Roger Jennison and Mrinal Das Gupta discovered at Jodrell Bank in 1953 that the radio emissions of Cygnus A came, not from the galaxy, but from two diffuse patches on either side of it. More recent radio images produced with the Very Large Array shows two jets extending up to 200,000 light years from the centre of the galaxy, the ends of these jets being the source of radio emissions.

In 1979, the Einstein Observatory showed that Cygnus A is an intense source of X-rays, indicating a central gas temperature of 100 million K. In fact, Cygnus A is an active galaxy with radio emissions 1,000 times more powerful than Centaurus A, which is, in turn, 1,000 times more powerful than a normal galaxy like the Milky Way. Like Centaurus A, the source of energy in Cygnus A is believed to be gas being sucked into a black hole at the centre of the galaxy.

Virgo A

Another early radio source, Virgo A (or 3C 274), was identified as the galaxy M87 by Bolton, Stanley and Slee in 1949. A small jet had been found extending from M87, a galaxy in the Virgo cluster, by Heber Curtis as long ago as 1916. Baade and Minkowski rediscovered this jet in 1954, and Arp discovered a counterjet in 1966.

In optical images, M87 appears to be a moderate size elliptical galaxy, 40,000 light years in diameter, but sophisticated image-processing techniques have shown that this galaxy has a diameter of 1 million light years at optical wavelengths. In X-rays, the Einstein Observatory showed that the gas of the galaxy has a temperature of 30 million K, and that it extends to a diameter of half a million light years. The core of the galaxy is extremely bright at both optical and radio wavelengths, and the jet, which is also visible at optical and radio wavelengths, extends just 5,000 light years from the nucleus. So Virgo A has clear signs of recent activity.

In 1978, astronomers using the 200 inch (508 cm) Palomar telescope concluded, from measurements of stellar velocities in M87, that there are 5 billion solar masses of material in a volume just 10 light years in diameter at the core of this galaxy. This high density of stars in the centre of M87 was confirmed by the Hubble Space Telescope in 1991, indicating the presence of a black hole.

Dwarf Galaxies

In 1937, Mrs Lindsay of the Harvard Observatory was examining a photographic plate taken two years earlier at the Boyden station in South Africa, when she noticed something that she had not seen before. There appeared to be a faint extended object in the constellation of Sculptor, that looked like a group of unusually faint stars arranged nearly uniformly in a circular region. A search though the Harvard archives showed that the object had been recorded on a plate taken at the Harvard station in Peru 30 years earlier.

Photographs were then taken with the large, 60 inch (152 cm) Harvard telescope in 1938, showing that the object was a cluster or small galaxy of stars, 10,000 stars being counted in two square degrees of sky. In the following year, Hubble and Baade used the 100 inch telescope on Mount Wilson to photograph the object, and discovered 40 Cepheid variables, enabling its distance to be measured as 270,000 light years. It was clearly a small (dwarf) galaxy, further away than the Magellanic Clouds but nearer than the Andromeda galaxy.

In the meantime, astronomers at Harvard had found a similar, but fainter, object in the constellation of Fornax. Hubble and Baade could not detect any Cepheids in this new object, as it was too far away, but there seemed to be a few globular clusters associated with it, enabling its distance to be determined as about 610,000 light years. The diameters of the Sculptor and Fornax objects were estimated as 6,000 and 10,000 light years, respectively. Their masses of 2 million and 20 million suns were much less than those of the Magellanic Clouds.

Baade and Hubble suggested in 1939 that the Sculptor and Fornax dwarf galaxies were similar to extended globular clusters, and this was confirmed in the 1960s for the Sculptor system whose H–R diagram was found to resemble that of the globular cluster M3. In many ways it appeared as though the Sculptor object, which had no globular clusters around it, was like a very large globular cluster, but the Fornax object, which had its own globulars, was like a genuine (dwarf) galaxy.

Galactic Evolution

Jeans' Condensation Theory

Isaac Newton had speculated in 1692 that if the infinite universe had been originally filled with evenly distributed, very low density material, that material would have condensed into a large number of clouds (or "condensations") which would then have contracted further to become the Sun and stars.

The English astronomer James Jeans began in 1901 to analyse what would happen if the universe had originally consisted, as Newton had speculated, of such a very low density of material. He refined his analysis over the next three decades, as observational results became available about the distances, sizes and densities of galaxies, and published his conclusions in his *Astronomy and Cosmogony* which was published in 1928.

Hubble estimated that if the whole of the matter in the visible universe were spread evenly throughout the universe, it would have a density of 1.5

$\times 10^{-31}$ g/cm^3. Jeans thought that the original density would have been a little higher than this, because a fair amount of energy would have been dissipated in the meantime by stars, and he considered an original density of about 10^{-30} g/cm^3 more reasonable. He also thought that it was highly unlikely that the material would have been spread absolutely uniformly thoughout the early universe, and suggested that such a lack of uniformity would lead to condensations of material. Jeans showed that, if the original material had had a velocity of 2 km/s, for example, the minimum condensation that could have formed would have had a mass of about 4 billion suns, or that* of a typical galaxy. Gravity would have been insufficient with smaller mass condensations to overcome the gas velocity, so smaller condensations could not have formed at those low original densities. Jeans then went on to explain that a protogalaxy would have been unlikely to condense absolutely symmetrically, and any lack of symmetry would have resulted in a slow spinning. This rotational rate would have increased as the galaxy condensed, because of the principle of conservation of angular momentum.

At first the protogalaxy would have been irregular in shape as it started to condense, but it would have tended to become spherical as condensation got under way. This tendency would have been compromised, however, as the rotation rate increased, causing the galaxy to become lenticular† in shape. Any further increase in rotational rate would have resulted in material being ejected from the edge of the galaxy. Jeans pointed out that galaxies with irregular, spherical and lenticular shapes were known, some of them also having spiral arms.

Jeans then went on to consider what size of smaller condensations there could be within the galaxies themselves, using densities of 10^{-21} and 10^{-22} g/cm^3, which Hubble had estimated to be the densities of typical galaxies. Jeans found that the minimum condensations ranged from 2 to 600 solar masses, assuming internal gas velocities of 100 to 500 m/sec, which led him to conclude that stars could be formed in this way either directly, or through an intermediate stage of condensation.

Jeans had to continually modify his theory as new observational evidence became available. For example, in the 1930s Hubble increased his estimate for the mass of a typical galaxy to 100 billion suns, and the density of the universe to 10^{-28} g/cm^3, while Arthur Eddington estimated the density of the universe as 10^{-27} g/cm^3. Jeans then showed that the original gas velocity could not be greater than 40 km/s in order to produce minimum condensations of the mass of typical galaxies (i.e. 100 billion suns). This maximum velocity of 40 km/s seemed much more reasonable than his original limit of 2 km/s.

Jeans' theory was thus able to explain, in broad terms, how galaxies could

---

*At that time Hubble had estimated the Andromeda galaxy to have a mass of 3.5 billion suns, for example.

†Like a convex lens.

form, and how stars could be formed within them. It did, however, have two weaknesses. Although it showed that rapidly rotating galaxies would lose material from their edges, the older spirals should have longer and more tightly wound arms than the younger ones, and this is not the case. The other problem was that all the galaxies, even the irregular ones, appeared to be made up of stars, whereas the irregular galaxies, at least, should still be gaseous.

Density Waves

Once Hubble had shown that spiral galaxies were distant accumulations of stars, he devised a classification system for galaxies in 1925, based on their shape. Although Hubble was emphatic that his scheme was not an evolutionary sequence, most astronomers considered it as such, believing that galaxies started life as spheroidal galaxies (type E0, see Figure 11.5), once the initial gas condensation had taken place. These E0 galaxies were then thought to become more and more flattened (to E7), before either following the normal spiral or the barred spiral path. Hubble described his scheme in detail at a series of lectures that he gave at Yale University in 1935, and which he consequently used as the basis for his book *The Realm of the Nebulae*, that was published in the following year.

It became clear, in the 1950s, that the outer part of the Milky Way and of the Andromeda galaxy were rotating about their centre more slowly (in angular velocity) than were their inner parts. Given the velocities measured, it was concluded that the spiral arms would be wound up in a period of less than 100 million years, and so would disappear. This led astronomers to conclude that galaxies start off life as spiral galaxies (types Sc or SBc), which have tighter and tighter arms as they evolve, before the spiral arms virtually disappear completely in galaxy type S0. So it was concluded that the evolutionary sequence was the reverse of that outlined in the 1930s.

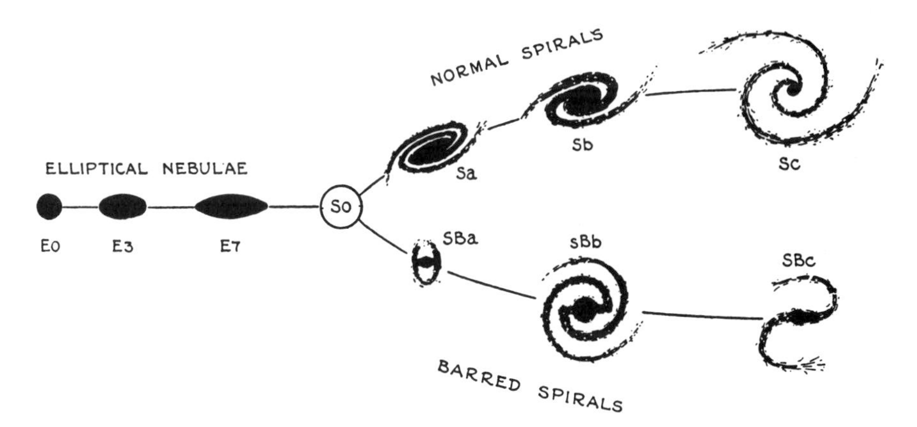

Figure 11.5 *Edwin Hubble's galaxy classification scheme. (From* The Realm of the Nebulae, *by E. Hubble, New Haven Press, p. 45: courtesy Yale University Press.)*

Then, in 1964, Chia Chiao Lin of the Massachusetts Institute of Technology (MIT) and Frank Shu of the University of California developed an idea, first suggested in the 1940s by the Swedish astronomer Bertil Lindblad, that the spiral arms are produced by density waves moving around the nucleus of the galaxy. Stars and gas, which move about the centre of the galaxy faster than the density wave, pass through the density wave. When the gas enters the density wave the gas is compressed, leading to the formation of stars. The heaviest of these stars are the brightest, with lifetimes of only a few million years (see Pages 150–151) and, during this time, they do not move very far around the galactic nucleus, being born and dying in the spiral density wave. It is these very bright stars that make the density wave visible. Normal stars have much longer lifetimes, however, so they move out of the density wave, passing through successive density waves as they orbit the centre of the galaxy a number of times in their lifetime. So the density waves, or spiral arms, do not wind themselves up with time, as the velocities of the stars and of the density waves are completely different,

There were two major problems with the density wave theory. First, Alar Toomre of MIT showed in 1971 that the density waves should only be stable for about 1 billion years, and second, stars should form preferentially on the trailing edge of the density wave where gas entering the wave becomes compressed. Although some galaxies do show the latter effect, many do not, and both problems are still under active investigation. Lin, Bertin, Lowe and Thurstans suggested in 1989, for example, that the spiral pattern could be maintained by two sets of density waves moving in opposite directions.

In 1991, doubt was also cast on Hubble's separation of galaxies into normal and barred spirals, when David Block and Richard Wainscoat found that the galaxy NGC 309 was both. In blue light it was known to be a normal Sc spiral, but, at a wavelength of 2.1 μm (in the infrared), Block and Wainscoat found it to be a barred spiral. This showed that the older red K and M stars are formed into a barred spiral, whereas the younger bluer stars are in a normal spiral pattern.

Galactic Collisions

About 50 years ago, Erik Holmberg had suggested that some galaxies should be in the process of merging, and Walter Baade had proposed that Cygnus A was caused by the collision of two galaxies, although Cygnus A was later shown to be a single galaxy crossed by a central dust band. In the 1960s Halton Arp identified hundreds of colliding and interacting galaxies and, in 1972, the brothers Alar and Juri Toomre of MIT modelled galactic collisions using very large computers, and were able to generate galactic configurations similar to those observed in space. In the following decade the IRAS satellite showed that many starburst galaxies (which show unusual amounts of star formation) appeared to be colliding galaxies.

The case of the colliding galaxy 4C 12.50 was unravelled in the late 1970s and early 1980s, for example. This bright compact radio source, known both as PKS 1345 + 125 and 4C 12.50, was known to vary in intensity by about 30% over a period of 4.5 years, when it was identified in 1977 as an unusual 17th magnitude galaxy at the centre of a cluster of galaxies in the constellation of Boötes. Gerard Gilmore of Cambridge University and Martin Shaw of the University of Edinburgh examined the object optically, using the 3.9 m Anglo-Australian telescope, and found that the nucleus of the galaxy has two components, separated by 1.8 arcsec. The eastern nucleus coincides with the radio source, whereas the western nucleus coincides with an infrared source discovered by the IRAS satellite. Gilmore and Shaw obtained the spectra of both nuclei, using the 2.6 m Isaac Newton telescope on La Palma, and found that the radio source is a giant elliptical galaxy, which is in collision with an active spiral (Seyfert) galaxy associated with the infrared source. So the 17th magnitude galaxy is really two galaxies in collision.

12 Cosmology

Early Cosmological Theories

James Jeans examined the dynamics of open star clusters like the Pleiades in 1913, and concluded that the gravitational attraction of other stars would cause any cluster to disperse in the order of 1,000 billion years. As the average mass of stars in these clusters appeared to be much above average, he concluded that the clusters had already been substantially disrupted, and were near to the end of their lifetimes. An analysis of the expected progressive disruption of binary star systems yielded a similar timescale, leading Jeans to conclude that the universe was of the order of 1,000 billion years old. This estimate was consistent with his estimate of the age of stars, assuming their energy is produced by the annihilation of matter, giving added credibility to his estimates.

In 1916, Albert Einstein published his general theory of relativity, and, in the following year, he used it to examine the structure of the universe. Much to his surprise, his equations predicted that the universe was either expanding or contracting, but as he was convinced that the universe was static, he introduced a term, the cosmological term, into his equations to make them predict a static universe.

At about the same time, Willem de Sitter developed his theory of the expanding universe based on Einstein's equations. Then, in 1922, the Russian mathematician Aleksandr Friedmann wrote a paper called *On the Curvature of Space* in which he developed Einstein's theory without the cosmological term, and showed that, if the average density of the universe is greater than 5×10^{-30} g/cm^3, gravity will force the universe to collapse on itself, but if the density is less than this value, the universe will expand for ever. Einstein dismissed the paper as mathematical speculation, as he was still convinced that the universe was static.

Georges Lemaïtre was ordained as a priest in 1923 and, in the same year, he went to the University of Cambridge's solar physics laboratory to study.

Two years later, at Eddington's suggestion, he joined the Massachusetts Institute of Technology where he became acquainted with the work of Hubble, Slipher and Shapley on galactic red shifts. In 1927 Lemaître linked his knowledge of Einstein's theory of relativity with these galactic red shifts to propose what is now called the Big Bang theory of the universe, 2 years before Edwin Hubble and Milton Humason showed that the recession velocity of galaxies increased with increasing distance. In 1931, Einstein finally dropped his cosmological term, later referring to the original introduction of the cosmological term as his biggest blunder.

Hubble and Humason had showed that the recession velocity of galaxies increased with increasing distance, at the rate of 500 km/s/Mpc (see Page 236). If the galaxies had been moving away from us (and each other) at this rate for all time, they must have all been at the same point in space about 2 billion years ago*, when the Big Bang started the expansion of the universe. This timescale for the age of the universe, which was about 1,000 times less than Jeans' estimate, indicated that his calculations were in error. Unfortunately, however, there was a serious problem with the Big Bang theory as, in the 1940s, geologists estimated, by examining radioactive elements in rocks, that the Earth was at least 3.5 billion years old, or older than the apparent age of the universe!

In 1948, Hermann Bondi, Thomas Gold and Fred Hoyle working at Cambridge University solved this inconsistency between the age of the universe and the age of the Earth, when they introduced their Steady State theory of the universe, in competition with the Big Bang theory. In the Steady State theory, hydrogen atoms are created out of nothing throughout the universe, at a rate that keeps the density of the universe constant as the galaxies recede from each other. In the Steady State theory, the universe has no age, as it has never fundamentally changed, assuming the universe has no boundary.

The Steady State theory was modified in the 1950s to explain the isotropic X-ray background radiation that early sounding rockets were beginning to detect. In the modified theory, neutrons were continuously created, rather than hydrogen atoms, and these neutrons then decayed to produce protons and electrons of a high-enough temperature to emit X-rays. Calculations showed, however, that the flux of X-rays that could be produced in such a reaction was much less than that observed.

Revisions to the Hubble Constant

Walter Baade took advantage of the black-out of Los Angeles during the Second World War to resolve the nucleus of the Andromeda galaxy, and of

---

*1 Mpc $\equiv 3.09 \times 10^{19}$ km. A galaxy would move 1.58×10^{10} km/year travelling at 500 km/s. So the time to travel 1 Mpc at this rate = $(3.09/1.58) \times 10^{9}$ years = 1.96×10^{9} years.

its companion galaxies M 32 and NGC 205, into individual stars, using the 100 inch (2.5 m) Hooker telescope on Mount Wilson. He was surprised to find that the brightest stars in these nucleus regions were cool red stars (like those in globular clusters), instead of the hot blue O and B stars which had previously been found to be the brightest stars in the spiral arms of the Andromeda galaxy. The hot blue O and B stars and the bright red stars he called Population I and II stars, respectively. Baade also studied the Cepheid variables in these galaxies, and in 1952 announced that the Cepheid variables associated with the hot blue Population I stars in the spiral arms are 1.5 magnitudes brighter than the Cepheids associated with the red, Population II stars* in the nucleus region.

The distance scale of the universe had previously been estimated by measuring the distance–period relationship for Cepheids in globular clusters, and applying this to Cepheids seen in the spiral arms of the Andromeda and other spiral galaxies. As these latter Cepheids were, in reality, 1.5 magnitudes brighter than those in globular clusters, it was clear that the distances to the spiral galaxies were about twice† that previously estimated. So the Hubble constant was cut in half, to 250 km/s/Mpc, and the age of the universe doubled to 4 billion years at a stroke. Although the latter still appeared too short, when compared with the 3.5 billion year age of the Earth, at least the new figure was a great deal more plausible than the previous one.

Baade presented his results to the 1952 meeting of the IAU in Rome. At the same meeting, A. D. Thackeray, who had been using the 74 inch (1.9 m) Radcliffe Observatory reflector in South Africa, announced that the first cluster-type Cepheid variables seen in the Small Magellanic Cloud were 1.6 magnitudes fainter than expected, using the distance previously established by measuring what Baade called "classical Cepheids". As, according to Baade, these classical Cepheids were four times as bright in absolute terms as previously thought, the Magellanic Clouds were twice as far away as previously estimated. (The distance estimate of the globular clusters around the Milky Way was not affected, however, because that had been previously estimated using the intensity of the correct type of Cepheids, namely the cluster variables.)

Hubble's estimate of the distances of the galaxies was based on the apparent magnitudes of the brightest stars, when the galaxies were too far away for Cepheids to be detected. In the 1950s it became clear that some of the objects that Hubble thought were bright supergiants were, in fact, small gaseous nebulae or star clusters. Eliminating these tended, in 1960, to increase the estimated distances of the galaxies, and reduce the value of the Hubble constant to about 100 km/s/Mpc. This implied a far more plausible

---

*The Population I, or Type I, Cepheids are called Classical Cepheids, and the Population II, or Type IIs, are called Cluster-type Cepheids, or RR-Lyrae variables.
†Actually $\sqrt{(2.5^{1.5})} = 1.99$, as a one magnitude difference is equivalent to a difference in intensity of about 2.5.

age for the universe of 10 billion years, which is over double the age of the Earth.

In the 1970s it was found that there was a good correlation between the absolute infra-red intensities of galaxies and the width of their 21 cm hydrogen line measured by radio telescopes. This is because hydrogen gas in large massive galaxies rotates faster about the centre of the galaxy than in smaller galaxies, producing a larger Doppler effect. This correlation provided a good way of estimating the distance of galaxies that were too far away to see either Cepheid variables or nova in them, and these galaxy distances produced a value for the Hubble constant of 95 ± 4 km/s/Mpc, compared with 53 km/s/Mpc estimated by Allan Sandage and Gustav Tammann in 1972 by another method (see Page 237).

A great deal of work has been undertaken in recent years to try to improve on the accuracy of the distance scale of the galaxies. Edward Ajhar, G. A. Luppino and John Tonry at MIT developed a technique that measured the roughness of a galaxy's image, on the basis that it is easier to resolve individual stars the nearer the galaxy is to us, so the images of the nearer galaxies should appear more mottled than the more distant ones. Using M32 as a comparator they deduced, in 1989, an average distance to the Virgo cluster of 52 million light years (16 megaparsecs).

George Jacoby and Robin Ciardullo of the Kitt Peak National Observatory and Holland Ford of the University of Michigan used a completely different technique, based on the fact that the brightest extragalactic planetary nebulae have an absolute magnitude of −4.4. Using this approach they deduced, in 1989, a distance to the Virgo cluster of 48 million light years (15 megaparsecs) which agrees, within error, with the results of Ajhar's group. The average of the Jacoby and Ajhar groups' results gave a value of H_0 of 77 km/s/Mpc.

In 1992, Sandage, Tammann, Macchetto and colleagues used the Hubble Space Telescope to examine 27 Cepheid variables in the galaxy IC 4182, which lies about 16 million light years from the Earth, in an attempt to improve the estimate of H_0. A type Ia supernova had exploded in IC 4182 in 1937, and by accurately measuring the distance of this galaxy using Cepheids, Sandage et al. could determine the *absolute* magnitude of the 1937 supernova. This, in turn, allowed the distances of further galaxies to be determined more accurately, by measuring the *apparent* magnitudes of their type Ia supernovae. The value of H_0 announced in 1992 using this technique was 45 ± 15 km/s/Mpc.

These three recent determinations of H_0 still show a substantial difference in results and they, together with other recent determinations, have still not improved on the estimates ranging from about 50 to 100 km/s/Mpc produced in the 1970's.

If $H_0 = 100$ km/s/Mpc, the age of the universe would be about 10 billion years, assuming that H_0 has not changed with time, but if $H_0 = 50$ km/s/Mpc then the age of the universe would be about 20 billion years. As Allan Sandage has found that the oldest globular clusters (M 92 and M 15) are about 18 billion years old, the 20 billion year figure seems the most likely. If the universe is expanding faster now than billions of years ago, then the above estimates of the age of the universe would be too short, but if, as most astronomers think, the attraction of matter in the universe has slowed its expansion down, the above estimates are too long. So the 50 km/s/Mpc rate would imply an age of the universe of less than 20 billion years.

The Microwave Background Radiation

George Gamow, Ralph Alpher and Robert Herman of George Washington University had predicted, in 1948, that the Big Bang would have produced radiation that should still be observable throughout the universe at a temperature of about 5 K, having a radiation peak in the microwave region. In the early 1950s, Tanaka in Japan and Medd and Covington in Canada thought that they had found evidence for this microwave background radiation, but their results were not accurate enough to be convincing. Robert Dicke at Princeton was looking for evidence of this radiation when, in 1965, Arno Penzias and Robert Wilson of the Bell Laboratories in New Jersey accidentally found that the whole sky was radiating energy uniformly at a wavelength of 7.3 cm. Penzias and Wilson analysed the frequency dependency of the radiation, and showed that it was the same as from a black body at a temperature of 3.5 ± 1.0 K, which was very close to that predicted. This was a major discovery, putting a final nail into the coffin of the Steady State theory, and giving the Big Bang theory substantially increased credibility.

Over the last 30 years, astronomers have examined this microwave background radiation to find evidence for its anisotropy, or variation from place to place across the sky. The microwave background radiation should not be perfectly uniform, for two reasons:

(i) There should be an anisotropy caused by the Earth moving relative to it. An anisotropy of four parts in 10^5 was discovered in the late 1970s by experiments carried aloft by balloons and U-2 aircraft, indicating a velocity of the Sun of about 600 km/s relative to the radiation. Once the movement of the Sun in the Milky Way was taken into account, the data indicated that the Milky Way itself is moving towards a point, as seen from Earth, in the constellation of Serpens Caput.

(ii) If the microwave background radiation was perfectly uniform, it would be difficult to explain how galaxies and stars had been produced. Very small deviations from uniformity were discovered in April 1992 by astronomers using data from the COBE spacecraft, lending even more support to the Big Bang theory. The deviations were minute (one part in 10^5), but were enough to support the theory. The COBE spacecraft also showed that the background radiation perfectly followed a black-body curve with a temperature of 2.726 ± 0.01 K.

While on the subject of general radiation from the sky, the question of why the night sky is dark needs to be readdressed. We left this topic (see

Pages 221–222) with the discovery of interstellar gas and dust, but absorption by these is not sufficient to produce a dark night sky. The discovery, in the early twentieth century, of the red shift of light with increasing distance seemed to provide another answer, as such red shifts would take much stellar energy out of the visible waveband. This effect, like interstellar dust, undoubtedly has some effect, but calculations showed that it was not enough, and the real reason why the night sky is dark is that the visible universe is neither infinite in space nor in time. In other words, the density of stars in the finite visible universe is not sufficient to produce a bright night sky.

The Missing Mass

Aleksandr Friedmann had shown theoretically in 1922 that if the density of the universe is greater than 5×10^{-30} g/cm^3, then gravity will cause the universe to collapse in on itself, but if the density is less than this critical value, the universe will continue to expand for ever. Hubble and Jeans had originally produced density estimates (see Pages 247–248) of the universe close to this critical value, but in the 1930s Hubble and Eddington independently produced density estimates about some 10 to 100 times greater than the critical value, implying that the expansion of the universe would eventually stop and go into reverse.

Fritz Zwicky of Caltech measured the velocities of galaxies in the Coma galactic cluster in 1933, and concluded that the mass of the cluster is too small by about a factor of 10 to retain the cluster as the gravitationally-bound structure it appears to be. He concluded that there was a considerable amount of invisible material or dark matter (as it is called today) binding the cluster together.

Over the next few decades, estimates of the size of the galaxies increased, but so did estimates of the size of the distances between them. Then in 1974, Richard Gott, James Gunn, David Schramm and Beatrice Tinsley added up all the mass visible in the near universe at all wavelengths (from γ-rays to radio waves), and concluded that the density was only between 1% and 10% of that required to eventually stop the universe from expanding. That is, the ratio, Ω, of the actual density to the Friedmann critical density was between 0.01 and 0.1, so the universe would continue expanding for ever, if these results were correct.

According to the theory of the Big Bang, however, if Ω is between these values of 0.01 to 0.1 today, it must have been within 10^{-15} of 1, one second after the Big Bang. This led many theorists to conclude that such a tiny deviation from 1 was unlikely, and that Ω is currently 1, and has always been 1. In that case we can "see", at all wavelengths, only 1% to 10% of all matter, the remainder being the dark matter and missing mass.

Vera Rubin of the Carnegie Institution in Washington started in the early 1970s to study the rotation of the galaxies and, after analysing over 60 galaxies, concluded that there must be more mass in the galaxies than in their visible discs in order to explain their rotation rates. But dark matter is not in every galaxy. In 1993 Robin Ciardullo and George Jacoby of Kitt Peak measured the velocities of planetary nebulae in the elliptical galaxy M105, and found that they were consistent with the gravitational attraction of the stars in the galaxy. No dark matter is present.

The anisotropy of the microwave background radiation had indicated, in the late 1970s, that the Milky Way is moving towards the constellation of Serpens Caput. Michael Rowan-Robinson of London University, David Walker and Amos Yahill then calculated the net gravitational pull on the Milky Way from all galaxies within about 600 million light years, using 60 μm data from the IRAS satellite survey. Their calculation of the direction of pull on the Milky Way was consistent with the microwave background observations and, using the Milky Way velocity determined from the latter, they were able to estimate the density of the universe within 600 million light years of the Milky Way. If this is representative of the universe as a whole, $\Omega = 1 \pm 0.2$, and about 90% of the mass is in invisible form.

So evidence for the existence of dark matter or the missing mass looks quite promising but, if it exists, what does it consist of? There are various candidates including black holes, cosmic strings (see Page 242), brown dwarfs, black dwarfs and various elementary particles that are very difficult to detect namely neutrinos, photinos, gravitinos and axions.

Brown dwarfs are objects that would be intermediate in size between Jupiter and the Sun. They would be large enough to have retained some of the heat from their gravitational contraction phase, but would be too small to produce energy by hydrogen fusion like stars. Black dwarfs, on the other hand, are stars that have completely run out of fuel. So far (1994) no clear cases of brown or black dwarfs have been found, although there are some brown dwarf candidates.

Neutrinos have been discovered in small numbers (see Page 194) in underground detectors, and in high energy physics experiments. Neutrinos were originally thought to have no mass, but there are some reasons to suggest now that they may have an extremely small mass. In the early universe neutrinos would have had velocities that were close to that of light, and they are therefore called *hot* dark matter. Photinos and gravitinos should be heavy particles, heavier than a proton, and axions should be very light particles, lighter than an electron but much heavier than a neutrino. Photinos, gravitinos and axions, which are known collectively as WIMPs (Weakly Interacting Massive Particles) are the supposed possible constituents of *cold* dark matter, but so far no such particles have been found.

13 Optical Telescopes and Observatories

Early Telescopes

Traditional nautical-type telescopes, or refractors, use an objective lens to focus the image and an eyepiece to view it. Their main drawback is that the image suffers from chromatic aberration, because light of different colours is brought to a focus at slightly different distances from the objective lens, producing coloured fringes around the image. Isaac Newton was the first person to explain chromatic aberration in the seventeenth century but, because he thought that it could not be corrected, he produced a completely different design of telescope in which the image is formed be a parabolic mirror. Newtonian reflectors, as they are called, do not suffer from chromatic aberration.

In the eighteenth century Chester Moor Hall, an English barrister, solved the problem of chromatic aberration in refractors by producing an achromatic lens made with two elements of glass of different refractive indices; one element being made of crown glass, and the other of flint glass. It initially proved very difficult, however, to make flint glass lenses with diameters of more than a few inches, although at the beginning of the nineteenth century Pierre Guinand, a Swiss cabinet maker, solved the problem, enabling the German optician, Joseph Fraunhofer, to produce an achromat of 10 inches (25 cm) diameter. It is only theoretically possible to correct chromatic aberration at two wavelengths if two elements are used, and between and outside of these wavelengths images still showed coloured fringes. During the nineteenth century many attempts were made to produce glass of various properties to improve lens design, but progress was very slow, with the best results being produced by the Schott company in Germany in the 1880s.

It may be thought that the obvious solution to the problem of chromatic aberration was to follow Newton's example and use reflectors, but these had problems of their own. Probably the biggest of these problems, in the first half of the nineteenth century, was that the speculum metal mirrors tarnished quickly, and they had to be frequently refigured and repolished. While still at Cambridge, George Airy had suggested in 1827 the use of silver-coated glass mirrors, but the technology was not available until the 1850s. Then Karl Steinheil in Germany and Léon Foucault in France independently produced small silver-coated parabolic mirrors, using a process invented by the German chemist, Justus von Liebig. This process produced a surface with a reflectivity of 90%, compared with about 50% for a polished metal mirror, and it also tarnished at a slower rate. Foucault managed to produce a 31 inch (80 cm) specimen in the mid 1860s, but the first large telescopes to use this technology in the 1870s were Henry Draper's 28 inch (70 cm)* in America, with which he took the first photograph of a stellar spectrum, the Paris Observatory's 47 inch (120 cm), and A. A. Common's 36 inch (90 cm) in England, with which he took the excellent photograph of the Orion nebula shown in Figure 14.1. In 1885, Common sold his instrument to an English amateur astronomer, Edward Crossley, who donated it to the Lick Observatory in California 10 years later.

Both reflector and refractor telescopes of varying designs were in use in the nineteenth century. The largest *reflector* was that built by the Third Earl of Rosse in Ireland in 1845, that had an aperture of 72 inches (1.8 m), and there were six more with an aperture of 36 inches (90 cm) or greater, although two of these were dismantled around the middle of the century. The largest *refractor* was the 36 inch (90 cm) *f*/19† at the Lick Observatory on Mount Hamilton that was completed in 1888, and there were seven more with objectives of greater than 24 inches (60 cm).

As anyone who uses reflectors and refractors knows, one cannot compare their usefulness by just comparing their diameters. Generally speaking, the image seen through a well-corrected refractor is superior to that seen through a reflector of the same aperture. Which type of telescope is the better, however, depends on further details of their design (i.e. their focal lengths, the accuracy of the mirror, the colour correction of the lens, etc.) and on the objects that are to be observed or photographed. For instance, are these objects point sources (i.e. stars) or extended objects (i.e. planets or nebulae), and is high resolution, high contrast or large light-gathering power required? Is the telescope to be used for visual, photographic or spectroscopic work? There is also no point in having a beautifully made telescope on an inadequate mount.

The telescopes of the nineteenth century were of widely varying types and constructions. Lord Rosse's 72 inch reflector could be moved vertically

---

*Figures inside brackets are rounded equivalents.

†The *f* ratio of a telescope, or camera, is the ratio of the focal length (*f*) to the diameter (*d*) of its objective lens or mirror. So for the Lick refractor, where $f = 684$ inches and $d = 36$ inches, the *f* number is *f*/19.

but with a limited horizontal movement of about ±10°, whereas his 36 inch was initially of altazimuth* construction, with the movement in the horizontal plane being effected by rotation on a circular track. The Melbourne instrument was much simpler in this respect because it was of a Cassegrain† construction with the focus near the bottom of the tube, enabling the observer to stay close to the ground. The more modern telescopes had electric drives to follow the stars, but some of the older large instruments had to be moved manually by ropes and pulleys.

None of the largest reflectors built before 1890 had been built in the 1880s, whereas all six of the largest refractors had been completed in that decade. This was mainly because the reflectors had become too large and unmanageable, a fate that was shortly to befall the refractors. The 36 inch Lick refractor (see Figure 13.1), for example, had a tube length of 57 ft (17 m), which required a very tall pier for mounting and a very large dome for the observatory.

Early Observatories

The main observatories at the start of the nineteenth century had been in Europe, at Greenwich (England), Paris and Berlin but, in 1820, the British Board of Longitude decided to establish the first permanent observatory in South Africa and, 2 years later, Thomas Brisbane, the Governor of New South Wales and an amateur astronomer, set up the first observatory in Australia. About the middle of the nineteenth century, new observatories were founded on both sides of the Atlantic, in Pulkovo (Russia, 1835), Washington (1842) and Harvard (USA, 1843), and later in the century the two great astrophysical observatories of Potsdam (Germany, 1874) and Meudon (France, 1875) were established. Towards the end of the nineteenth century, the United States observatories took the lead from Europe both numerically, and in the quality of their equipment and observing sites.

---

*With an altazimuth mount the telescope can rotate independently about vertical and horizontal axes. With an equatorial mount the telescope can rotate about one axis, which points towards the north celestial pole (near the pole star), and a second axis, which allows a variation in angle above the celestial equator. Rotations about both axes have to be effected with an altazimuth mount to track a star as it moves across the sky at night, whereas with an equatorial mount tracking only requires a rotation about the polar axis.

†In a Newtonian reflector, the parabolic mirror at the bottom of the telescope tube produces an image at the top of the tube. In a Cassegrain reflector, a small convex mirror is placed at the top of the tube to intercept the light from the parabolic mirror and send it downwards once more. The image is viewed through a hole in the centre of the parabolic mirror. So a Cassegrain can have a longer focal length, and higher *f* number, for the same length of tube as a Newtonian.

Figure 13.1 *The 36 inch (90 cm) Lick refractor, completed in 1888.* (Lick Observatory Photograph.)

The questions of climate and altitude were beginning to be seriously discussed by the major observatories towards the end of the nineteenth century. William Lassell left the UK for Malta in 1852 because of its better atmosphere, for example, and James Lick chose Mount Hamilton (altitude 1,280 m or 4,200 ft) for his observatory in 1876, because of its clear skies.

The main observatories in Europe in the nineteenth century were state-run institutions, although wealthy amateur astronomers like William Herschel and Lord Rosse often set up their own observatories. In the United States, however, the US Naval Observatory was the only nationally-funded enterprise, with the others being funded by wealthy citizens and/or operating as university departments. Harvard, for example, received over a million dollars, in total, from a series of benefactors including Anna Draper,

Henry Draper's widow, and Edward Pickering (its Director), and in 1876 James Lick, a wealthy piano maker and landowner, left $700,000 to the University of California to build the largest telescope in the world. This was the 36 inch (90 cm) Lick refractor which saw first light at the new observatory on Mount Hamilton in 1888.

The Transition to Reflectors

The largest refractor available in 1890 was the 36 inch (90 cm) Lick on Mount Hamilton that had seen first light in 1888. When this telescope was completed, the University of Southern California at Los Angeles ordered a 40 inch (1.0 m) refractor from Messrs Warner and Swasey who had built the Lick. Unfortunately, when the glass blanks arrived from Paris, the University were unable to pay for them, and they were bought, instead, by George Ellery Hale at the University of Chicago. Hale had persuaded Charles T. Yerkes, the industrialist, to donate the money for the telescope,* which was installed in 1897 in a new observatory at Williams Bay, Wisconsin, 90 miles (150 km) from Chicago. It is still the largest refractor in the world.

The American Lewis Rutherfurd had, in 1863, been the first person to make a telescope that was optimised for photography. Then, in 1886, the brothers Paul and Prosper Henry of Paris had shown the merits of astrophotography, when they had photographed an astonishing number of stars in the Pleiades open star cluster with a small 13 inch (35 cm) refractor (see Page 286). Three years later, a 33 inch (85 cm) visual refractor, built by Gautier and the Henry brothers, had been installed in the Meudon observatory near Paris, and in 1893 a 24 inch (60 cm) photographic refractor was attached to make what is still the largest double refractor in the world. A few years later Steinheil and Repsold built a similar double refractor for the Potsdam Observatory although, in this case, the larger telescope was corrected for photography.

All of the large refractors had long focal lengths, and hence large *f* numbers, as these were easier to make optically, but they tended to be long and unwieldy and required long exposures to photograph nebulae. Alvan Clark and Sons were the first to make a large (24 inch, 60 cm) refractor with a low *f* ratio (*f*/5.5), when they built an instrument to this specification for the Harvard College Observatory. It was paid for by Miss Catherine Bruce, a wealthy New York spinster, and installed at the new high altitude observatory at Arequipa, Peru in 1896 to photograph the southern sky.

Refractors were the favourite telescopes of professional astronomers in the 1880s, because they produced brighter images, of higher contrast, than reflectors of the same aperture and focal length, and refractors did not need the frequent internal adjustments of reflectors. Unfortunately, as the lenses

---

*Yerkes, a wealthy streetcar and property tycoon, originally agreed to pay only for the lens, but he finally paid for the whole observatory that is now named after him.

of refractors got larger they became extremely expensive, mainly because of the low success rate in producing optical quality glass in such large sizes. Very large lenses also tended to distort under their own weight and the light loss through them was appreciable, particularly at the shorter photographic wavelengths (see Page 288), as they were so thick. It was also physically impossible to make a perfect achromat, and the competing wavelength requirements of photography and visual observing made matters worse. Achromats were also a nuisance for spectroscopists, as spectral lines were not all brought to a focus in the same plane.

Reflectors, on the other hand, could be made with glass of lower quality than refractors, and the mirrors could be supported across their whole diameter to make distortion less of a problem. Mirrors were naturally achromatic, and could be made of much shorter focal length, and hence smaller *f* ratios, than refractors, which was important for photographing faint nebulae (see Page 284). With the weight at the bottom of the tube, the reflector was also easier to balance, and with shorter focal lengths, the observatories could be smaller.

In spite of all the advantages of large reflectors, however, astronomers did not embrace this technology overnight because most of the large reflectors available in 1890 were relatively old and awkward to use. The mirrors needed frequent re-silvering, and the alignments of the various components needed frequent checking and adjusting. In the last decade of the nineteenth century, however, the requirements of photography and spectroscopy became more and more dominant, and so the preference for refractors began to change. In 1898 James Keeler used the 36 inch (90 cm) Crossley* reflector at the Lick Observatory to photograph nebulae, and in 1904 George Ritchey produced better photographs with his relatively small 24 inch (60 cm) reflector than with the new 40 inch (1.0 m) refractor at the Yerkes Observatory. Max Wolf used a 28 inch (70 cm) reflector to photograph nebulae at Heidelberg in 1906, and when Ritchey completed the great 60 inch (1.5 m) reflector on Mount Wilson in 1908, the reflector was well and truly back in fashion.

The Harvard College Observatory

Mention has been made many times in this book of the work of the Harvard College Observatory situated at Cambridge, Massachusetts, which was

---

*The Crossley reflector was, in 1933, the first reflector to have its silver coating replaced by an aluminium one, in a process mainly developed by John Strong of the California Institute of Technology. The aluminium surface has a much longer lifetime than silver, and has a much higher reflectivity in the ultraviolet. The secondary mirrors of the 60 inch and the 100 inch Mount Wilson reflectors were also aluminised in 1933, and their primary mirrors were aluminised during the next 2 years.

under the direction of Edward C. Pickering from 1877 to 1919, and Harlow Shapley from 1921 to 1952.

When Pickering took over as director, the total income of the observatory was insufficient to pay for any expansion in either research projects or equipment. Within a year, however, he had managed to secure addition public donations to fund a measurement programme of the magnitudes of all the naked-eye stars visible from the observatory, the results of which were published in 1884 (see Page 298). The work was extended to the southern hemisphere in 1889 when Pickering sent Solon Bailey to Peru, leading 2 years later to the establishment of the southern station of the Harvard Observatory at Arequipa, at an altitude of 2,450 m (8,000 ft).

The extension of photometry to the southern hemisphere, and Pickering's planned work on stellar spectroscopy (see Page 126), required a massive injection of new funds, which he secured over the period from 1886 to 1889. The Robert Treat Paine Fund of $300,000 covered the running costs of the observatory, the Henry Draper Fund provided almost $400,000 for the spectroscopy programme, the Uriah Boyden bequest of $238,000 paid for the Arequipa station, and Miss Catherine Bruce provided $50,000 for the 24 inch (60 cm) Arequipa photographic refractor. With this funding secure, the Harvard College Observatory became, at the turn of the century, the leading American centre for astrophysics, using both photography and spectroscopy to chart and analyse the stars in both hemispheres. In the process, it produced a wealth of photographic and spectroscopic data that were used regularly throughout the early decades of the twentieth century.

Harlow Shapley took over as director at Harvard after Pickering's death in 1919. Shapley's first visit to Harvard had been in 1913, when he was on his way to take up a research post at the Mount Wilson Observatory. He had discussed his planned research project with Solon Bailey, who had suggested that he study variable stars in globular clusters, following Bailey's earlier work on the subject. This suggestion was to have far-reaching consequences, as far as our understanding of the size and basic configuration of the Milky Way is concerned (see Pages 161 & 215).

Edward Pickering's programme of spectroscopic research required the analysis of thousands of stellar spectra, which was carried out by a team of women "computers", as they were called (see Figure 13.2), of whom Williamina Fleming, Antonia Maury and Annie Jump Cannon are the best known. Another woman, Henrietta Leavitt, undertook research into Cepheid variables in the Magellanic Clouds, leading to her discovery of the period–luminosity law for Cepheids, and it was yet another woman, Cecilia Payne, who was the first to get a PhD at the observatory, for her brilliant work on stellar atmospheres.

The Harvard College Observatory was forced to shut down the Arequipa station in 1918, because of the First World War, and in 1927 the equipment was moved to Bloemfontein in South Africa. Four years later, the Oak Ridge Station (now called the Agassiz Station) was also set up near Harvard, Massachusetts, where a 60 inch (1.5 m) reflector was installed in 1933,

Figure 13.2 *The women computers of the Harvard College Observatory under the direction of Williamina Fleming (standing), in about 1890. (Courtesy* Harvard College Observatory.*)*

◀ ***Figure 13.3*** *George Ellery Hale with his spectroheliograph (see also Section 14.1). (Courtesy* California Institute of Technology.*)*

using one of Common's mirrors that he had produced in the nineteenth century.

Mount Wilson

In 1902, George Ellery Hale (Figure 13.3), who was director of the Yerkes Observatory, read in a newspaper that Andrew Carnegie had decided to donate $10 million to fund the Carnegie Institution, dedicated to scientific research and development. Hale was, at that time, interested in building a new solar observatory, so he contacted Daniel Gilman, who had been appointed director of the Institution, with his ideas. Gilman was impressed with Hale's approach, and invited him to serve on the Advisory Committee on Astronomy as an astronomical advisor.

In the following year, the Institution asked W. J. Hussey, of the Lick Observatory, to look for a site for a solar observatory. Hussey was particularly impressed by Mount Wilson and Palomar Mountain in Southern California and, although Palomar had the better viewing, he recommended Mount Wilson, because of its relative ease of access. Without waiting for a decision on the location, Hale decided to set up his own solar telescope on Mount Wilson and, in the following year, the Carnegie Institution decided to follow suit, establishing the Mount Wilson Solar Observatory with Hale as its first director.

Hale first moved the horizontally-mounted Snow solar telescope from Yerkes to Mount Wilson in 1905, where the seeing was so much better, and then built a solar telescope in a 60 ft (20 m) high tower, using the experience gained with the horizontal Snow telescope. The 60 ft opened in 1907, quickly followed 5 years later by a 150 ft (45 m) tower telescope that produced a 17 inch (45 cm) solar image which, although similar in size to that produced by the Snow telescope, was of much higher quality. Since then the performance of both tower telescopes has been significantly improved by using quartz mirrors, which have a very low thermal expansion coefficient, and reflective paint on the tower structures.

In 1894 Hale had persuaded his father to pay for a 60 inch (1.5 m) glass disc to make a large reflector at the Yerkes Observatory for stellar and nebula research, but in 1898 his father had died and, as there was no money to finish the telescope, the unfinished mirror was put into storage. In 1904 Hale obtained funding from the Carnegie Institution for a 60 inch reflector at Mount Wilson, and so he moved the unfinished mirror from Yerkes. Ritchey, who had started work on the mirror at Yerkes, also moved to Mount Wilson to finish the task, and the new telescope (see Figure 13.4) saw first light in December 1908, as the most powerful telescope in the world.

The 60 inch was to set the standard for most large reflectors of the future allowing, as it did, operation in four different configurations. As a Newtonian it operated at *f*/5 for photography and low dispersion spectrography, as a modified Cassegrain it operated at *f*/16 for spectrography and

Figure 13.4 *The 60 inch (1.5 m) Mount Wilson reflector, completed in 1908. (Courtesy* The Observatories of the Carnegie Institution of Washington.*)*

f/20 for photography, and as a Coudé* it was used at *f*/30 for high dispersion spectrography.

With manufacture of the 60 inch well in hand, Hale had started planning the construction of an even larger instrument and, in 1906, persuaded J. D. Hooker, a Los Angeles business man, to pay for the main mirror. Initially Hooker agreed to fund an 84 inch, but thinking that this may well be exceeded soon elsewhere, increased his funding to pay for a 100 inch (2.5 m) mirror. The Carnegie Institution eventually agreed to pay for the mounting.

The 5 ton, 100 inch diameter mirror disc for the Hooker telescope was ordered from the St Gobain glass works in France, the same people who had supplied the 60 inch disc, but it was found to have a large number of small subsurface air bubbles when it arrived at Pasadena, California in 1908, and to have partially lost its rigidity. St Gobain tried to make a better disc, but

*In the Coudé configuration, the spectrograph can be kept stationary, allowing it to be rigidly mounted, thus avoiding flexure problems.

found it impossible to improve on their previous attempt. In the meantime, however, tests had shown that the mirror could probably be figured from the original disc without cutting into the bubbles, although Hale had to persuade Ritchey to start grinding the mirror, because Ritchey was convinced that the risk of cutting into the bubbles was too great. Ritchey started the long grinding process in 1910, and when, 4 years later, he had finished grinding the mirror to a spherical shape, W. L. Kinney took over the job of parabolising it, which was successfully completed in 1916. In the meantime, the St Gobain glass works had been destroyed by fire in 1914, and so a replacement mirror was out of the question in the event of problems. So it was, with more than the usual nervousness, that the telescope was first checked out on 2nd November 1917.

Shortly after dusk, George Ellery Hale and his deputy, Walter Adams, looked at Jupiter through the new telescope and their hearts sank as, instead of a crisp single image, Jupiter was seen as a number of partially overlapping images. They made enquiries and found that the dome had been left open during the day, while the workmen had been putting the finishing touches to the telescope, and thought that the problem may be due to thermal inertia of the mirror. They decided to let the telescope cool down and agreed to return at three o'clock in the morning, when the temperature should have stabilised. Hale and Adams both returned early, full of apprehension, but their first glance through the telescope showed them that everything was all right. The 100 inch Hooker telescope (see Figure 13.5) was a resounding success.

The 60 inch and 100 inch telescopes at Mount Wilson made this the premier observatory in the world from 1908, when the 60 inch was completed, to 1948, when the 200 inch was inaugurated on Palomar Mountain. The discoveries made were right at the cutting edge of astronomy, particularly where large light-gathering power was important, namely in the study of distant stars and nebulae and the structure of the universe. With the 60 inch, Walter Adams and Arnold Kohlschütter developed the technique of spectroscopic parallax to measure the distance of distant stars, Harlow Shapley photographed Cepheid variables in globular clusters and estimated the size of the Milky Way, and Eric Becklin and Gerry Neugebauer mapped the centre of the Milky Way in infra-red and discovered a superluminous protostar in the Orion nebula. With the 100 inch Hooker reflector, on the other hand, Edwin Hubble proved that spiral nebulae were not part of the Milky Way and estimated the rate of expansion of the universe, Walter Baade discovered Population I and Population II stars in the Andromeda galaxy, F. G. Pease and John Anderson made the first measurement of a stellar diameter, and Horace Babcock discovered the first evidence for magnetic fields associated with stars.

Light pollution and photochemical smog from Los Angeles and the surrounding area resulted in the Carnegie Institution deciding to moth-ball the 100 inch in 1985, and to redirect their financial resources to Mount Wilson's sister observatory at Las Campanas in Chile. The solar telescope and the 60 inch telescope are still in use, however, financed by the National

Figure 13.5 *The 100 inch (2.5 m) Hooker reflector on Mount Wilson, completed in 1917. (Courtesy* The Observatories of the Carnegie Institution of Washington.*)*

Science Foundation, NASA and the Office of Naval Research. Plans are now also in hand to reopen the 100 inch, funded by a charitable trust.

Palomar Mountain and the 200 Inch

Hale resigned as director of Mount Wilson in 1923 for health reasons, but this did not stop him dreaming of making an even larger reflector. Experiments on the 100 inch with a 6 m Michelson stellar interferometer had shown that atmospheric turbulence would not be insurmountable, and research on new low-expansion glass was also encouraging. Hale initially wanted to build a 300 inch (7.6 m), but he changed this to a 200 inch (5 m) when he and Pease realised that the mirror would be too large to be moved from the manufacturer. They drew up more detailed plans for a 200 inch, and Hale eventually persuaded the International Education Board, which was associated with the Rockefeller Foundation, to pay the $6 million estimated construction costs of the telescope. The Foundation gave the telescope to the California Institute of Technology in Pasadena to operate; the running costs being covered by an endowment provided by Henry Robinson, a Los Angeles banker.

Even as early as 1930, light pollution was becoming a problem at Mount Wilson, so a number of alternative sites in the south-western United States were tested for weather conditions and seeing, as possible sites for the new 200 inch. In 1934, Palomar Mountain was chosen as the site for the new observatory, and construction of the road to the observatory was begun the following year.

The contract for making the mirror disc was given to the General Electric Company in 1928 at an estimated cost of $250,000. The GEC design used fused quartz (pure silica), with a very low expansion coefficient, but fused quartz has a very high melting point, and its high viscosity meant that it was very difficult to avoid air bubbles. By 1931 only two questionable-quality 60 inch discs had been made and over $600,000 had been spent, so Hale, seeing little chance of success, cancelled the contract.

The new contract was placed with Corning to make a mirror blank out of Pyrex (borosilicate) glass which, although it had a somewhat larger expansion coefficient than fused quartz, had a lower expansion coefficient than traditional materials and it was considered to be probably quite adequate. Pyrex glass was also much easier to work with in the liquid state than fused quartz, so there should be no major problems with air bubbles. A number of exploratory trials were undertaken before a perfect 200 inch disc was cast in December 1934 on the second attempt, after polarisation tests had indicated strains in the glass of the first disc. The new disc took a year to cool, during which the glass works was first flooded and then subjected to an earthquake. Fortunately, neither adversely affected the mirror blank, and it was eventually transported to Pasadena in April 1936 by train, taking 2 weeks to cross the country.

Hale and Pease had decided to make the mirror with the low focal ratio of *f*/3.3, because this allowed relatively short exposures for nebula photography, and it also enabled the tube to be made shorter allowing a smaller dome to be used. Such a low *f* ratio was known to produce significant off-axis coma distortion at the Newtonian focus, however, and so a special Ross corrector lens was fitted.

Grinding of the mirror was started in the Pasadena optical shops in 1936, under the supervision of Marcus Brown. Two years later George Ellery Hale died, and then the Second World War interrupted work, and so it was not until November 1947 that the completed mirror was transported to the new observatory on Palomar Mountain. The telescope was finally dedicated to Hale on 3rd June 1948 by Lee Du Bridge, President of the Carnegie Institution, and it saw "first light" as an operational instrument on 26th January 1949.

The statistics of the 200 inch (see Figure 13.6) make impressive reading and show why no attempt was made to build an even larger telescope until comparatively recently. Sixty-five tons of molten Pyrex heated to 1,575°C was used to produce the disc, and 30 tons of abrasives were used to polish it. The glass of the primary mirror weighs 15 tons, the weight supported by the polar axis bearings of the telescope is 500 tons, and the observatory dome weighs 1,000 tons.

Figure 13.6 The 200 inch (5 m) Hale reflector on Palomar Mountain, completed in 1948. The telescope is so large that the observer can sit in the cage at the focus of the primary mirror, as shown. (Courtesy California Institute of Technology.*)*

Since 1970, the two observatories on Mount Wilson and Palomar Mountain have been known as the Hale Observatories. Other Hale Observatories are the Big Bear Solar Observatory near Riverside, California, and the Las Campanas Observatory in Chile, covering the southern hemisphere.

Schmidt Telescopes

Photography over large fields was a problem with traditional telescopes, particularly fast telescopes with short focal ratios as they exhibited strong off-axis coma. Hale and Pease solved the problem for the Palomar 200 inch by fitting a corrector lens to the standard Newtonian design, but Bernhard Schmidt of the Hamburg Observatory had adopted a completely different solution in 1930. Instead of using a parabolic primary mirror, as in a Newtonian, to eliminate spherical aberration, and a corrector lens to eliminate off-axis coma, Schmidt used a spherical primary mirror, and an aspherical glass plate to correct for the spherical aberration. He also made the corrector plate smaller than the primary mirror, and placed it at the centre of curvature of the primary, thus eliminating coma. The main difficulty was that the focal plane was curved, which meant that the photographic plate or film placed in that plane had to be bent to a fixed shape to keep the image in focus.

Schmidt's first telescope of this design had a 17 inch (45 cm) diameter *f*/1.7 mirror, and a 14 inch (35 cm) diameter corrector to produce a field of 16° diameter. Many small Schmidt telescopes were made by various people in the 1930s to various designs, but the first large Schmidt was installed at the Palomar Observatory in 1948 (see Figure 11.1 earlier). It has a 72 inch (1.8 m) *f*/2.4 primary mirror, with a 49.5 inch (1.25 m) corrector plate, both produced in the Mount Wilson workshops, and uses photographic plates covering an area of 6° × 6°, which are bent to the required shape. It was used for the National Geographic–Palomar sky survey carried out between 1949 and 1958 to photograph stars and nebulae down to magnitude 21.

In 1940, the Dutchman A. Bouwers experimented with a design similar to Schmidt's, but Bouwers' corrector had spherical surfaces and so was easier to construct. Dmitrii Maksutov of the State Optical Institute in Moscow independently considered a similar system in 1941, and a number of these so-called Maksutov telescopes have now been made. The Schmidt design gives better images at fast *f* ratios, however.

South Africa

The British Admiralty had founded the observatory at the Cape of Good Hope in 1820, to extend the work of the Greenwich Observatory to the southern hemisphere, but it was underfunded, and periodically forgotten by its patrons. When the Scottish astronomer David Gill took over as director of the Cape Observatory in 1879, it was disorganised and run down, but by the time of his retirement in 1907, the Cape was an internationally respected observatory. In particular, his work on stellar photography in the 1880s had been instrumental in persuading others of the enormous advantages of this new technique (see Pages 284–285), and his Cape Photographic Durchmusterung (CPD) star catalogue was second to none when it was completed in 1900.

The Cape had probably the best heliometer in the world in the nineteenth century, which Gill used in the 1890s to measure the solar parallax, and in 1901 the observatory was provided with a 24 inch (60 cm) photographic refractor which was coupled to an 18 inch (45 cm) visual refractor on the same mounting.

The Radcliffe Observatory at Oxford had been founded as long ago as 1771, but in the late 1920s William Morris offered to purchase the site of the observatory to expand the Radcliffe Infirmary, which was grossly overcrowded. The Observatory were keen to move to a more suitable climate, and readily agreed to the sale. So in 1929, the British astronomer W. H. Steavenson went to Pretoria to find a suitable site but, unfortunately, it transpired that the High Court in London would have to authorise the transfer of funds, and this delayed the establishment of the Radcliffe Observatory in Pretoria until 1938. A 74 inch (1.9 m) reflector was ordered from Grubb Parsons Ltd in the UK for the new observatory but, before the telescope could be completed, the Second World War had begun, and the mirror spent the war under sandbags in Newcastle. It was not until 1948 that the 74 inch became operational in Pretoria, as the largest telescope in the southern hemisphere.

David Gill, J. C. Kapteyn the Dutch astronomer, and Jöns Backlund, who was director of the Pulkovo Observatory in Russia, persuaded the Transvaal Government to establish an observatory just outside Johannesburg in 1905. It became known as the Union Observatory when the Union of South Africa was set up in 1910. A 26 inch (65 cm) visual refractor was installed in 1925 for double star work, and Leiden Observatory installed a twin 16 inch (40 cm) photographic refractor in 1938. From 1925 Yale Observatory also operated a 24 inch (60 cm) photographic refractor, for measuring stellar parallaxes, at Witwatersrand University, Johannesburg.

In 1927, the Boyden station of the Harvard College Observatory was moved from Arequipa, Peru to Bloemfontein, South Africa, where the 60 inch (1.5 m) Common reflector was installed on a completely different mounting. In the same year, the University of Michigan installed a 27 inch (70 cm) visual refractor for double star work at Bloemfontein.

So, in 1950 South Africa had four main observatories, at the Cape, Pretoria, Johannesburg and Bloemfontein. These observatories had equipment belonging to the Greenwich and Radcliffe observatories in the UK, Yale, Harvard and Michigan observatories in the USA, and Leiden observatory in the Netherlands. In 1953, the Yale station was closed and moved to Canberra, Australia, and the University of Michigan station was also closed at the same time. In 1972, the Royal Observatory at the Cape, the Radcliffe Observatory in Pretoria and the Union Observatory in Johannesburg all became part of the South African Astronomical Observatory. At this time South Africa's optical observatories were the best in the southern hemisphere but, in the 1970s, major observatories were set up in Chile and Australia.

Kitt Peak

Walter Baade from Palomar Observatory and Leo Goldberg of the University of Michigan suggested, in 1953, that an American National Observatory be set up whose bad observing season was out of phase with that of the large observatories in California, so that research could be carried on all the year round. Goldberg suggested that the observatory should cover all types of observing including photometric, spectroscopic, photographic and solar work. An advisory panel was set up in 1954 to review possible locations and, after analysing the viewing conditions on 150 mountains, selected Kitt Peak (near Tucson, Arizona) and Hualapai Mountain (near Kingman, Arizona) for further testing. The cost of the new observatory was estimated at $12 million.

The National Science Foundation had agreed, in principle, in 1956 to the setting up of a national observatory, but the launch of Sputnik I by the Soviet Union in October 1957 unlocked the funds, as America decided on a big push in scientific and technological research. In 1958, Kitt Peak was chosen by the AURA (Association of Universities for Research in Astronomy) as the site for the national observatory, and observations were started there with the two 16 inch (40 cm) reflectors that had been used for the site surveys. Three years later the first major telescope, a 36 inch (90 cm) reflector, was installed, and in 1964 an 84 inch (2.1 m) *f*/2.6 was also completed.

On 23rd September 1961, a committee chaired by the American astronomer N. U. Mayall met for the first time to consider the construction of a 4 m (160 inch) telescope on Kitt Peak. The telescope, now called the Mayall reflector, was completed in 1973 at a cost of $10 million. It is the most powerful telescope at the observatory with a 150 inch (3.8 m) *f*/2.8 Ritchey–Chrétien primary mirror, made by General Electric, of fused quartz (using the same type of glass which they had failed with before the war in the Palomar contract). The telescope also has an *f*/8 Cassegrain focus, and an *f*/190 Coudé focus of an astonishing 720 m focal length.

Kitt Peak also boasts the largest solar telescope in the world, the McMath Pierce, which became operational in 1962. The main and secondary mirrors were made from some of the rejected fused quartz blanks that had been produced by General Electric before the war as test pieces for the Palomar 200 inch. The McMath Pierce has a 63 inch (1.6 m) diameter main mirror at the bottom of a 600 ft (180 m) shaft, partly above and partly below ground, which points permanently at the north celestial pole. The solar image is 34 inches (85 cm) in diameter, allowing the study of individual solar granules which are about 0.4 mm in diameter on the image. The enormous 70 ft (20 m) solar spectrum has been used, among other things, to confirm the existence of fluorine and chlorine in the Sun.

The Multi-Mirror Telescope (MMT)

The MMT on Mount Hopkins, Arizona, which saw first light in 1979, was the first major telescope to use the principle of active optics, where motors are used to correct mirror alignments a few times per minute. The MMT design was based on the principle that it is easier to make six 1.8 m (70 inch) mirrors than one 4.4 m (170 inch) mirror, which would have the same light-gathering power; the theoretical advantage in resolution of the one large mirror being completely wiped out by atmospheric turbulence. Until the advent of lasers, however, it was impossible to align and position the six separate mirrors accurately enough, at all telescope orientations, to combine their images.

In the MMT the flexure of the structures of the six adjacent telescopes were continuously monitored and corrected as they occurred. Laser beams measured the flexures, and these data were fed to servo systems which could vary the tilt of each secondary mirror up to a few times per minute, to ensure that the six images coincided. The MMT also used an altazimuth design, and a computer to drive the telescope array in both the horizontal and vertical axes, and rotate the image plane to compensate for the Earth's movement. This computer-controlled altazimuth system enabled a much simpler telescope mounting system to be used, compared with the traditional equatorial mounts.

Mauna Kea

A powerful tidal wave devastated Hilo on Hawaii in 1960, causing serious problems for the local economy. As a result, ideas were put forward to bring in much needed money to the island, one of which involved setting up an astronomical observatory, taking advantage of the unpolluted skies and high mountains. Mitsuo Akiyama of the Hawaii Chamber of Commerce sent letters to many American and Japanese Universities outlining the possibility of setting up such a facility on either of the two large extinct volcanoes, Mauna Kea and Mauna Loa. Gerard Kuiper, of the University of Arizona and former director of the Yerkes Observatory, had already undertaken a testing programme at Haleakala on the island of Maui in Hawaii, funded by NASA and the Department of Defense, which indicated that its altitude of 3,000 m (10,000 ft) was not high enough above the normal cloud-tops in the area. Mauna Kea, on the other hand, is 4,200 m (14,000 ft) high, so Kuiper drew up a proposal in January 1964 to investigate Mauna Kea as a possible site for an observatory.

Progress then accelerated. An agreement was signed between the University of Hawaii and the University of Arizona for the establishment of a 12 inch (30 cm) test telescope and dome on Mauna Kea. A road was built up Mauna Kea in May 1964, testing of the viewing conditions commenced on 13th June, and the dedication of the Mauna Kea Observatory took place on 20th July. At this stage only the 12 inch reflector was on the mountain for site testing, but this showed excellent observing conditions. Kuiper then submitted a proposal to NASA to build a 60 inch (1.5 m) reflector on Mauna Kea, while the University of Hawaii submitted a parallel proposal to build an 84 inch (2.1 m). The University of Hawaii's proposal was accepted because the observatory was to be managed locally and, in addition, it would provide a larger telescope. In the end, the University of Hawaii built a slightly larger telescope of 88 inches (2.2 m) that was completed in 1970.

In 1977 a group of astronomers at the University of California (UC) started to examine the possible designs for the next generation of ground-based optical telescopes after the MMT. The MMT used six separate telescopes on one mounting, on the basis that it is easier to make six small mirrors than one large one, but Jerry Nelson, a member of the UC group, suggested that one large 10 m mirror (twice the diameter of the Palomar 200 inch) could be built from small interlocking hexagonal segments, so that only one telescope structure needed to be built. Nelson started to work on this so-called segmented mirror design in 1977, to try to solve the numerous technical problems such a mirror would present. For example, a segmented mirror would not be as rigid as one large thick mirror, and so the individual mirror segments would have to have their alignments measured continuously and corrected. Terry Mast and George Gabor from the Lawrence Livermore Laboratory proposed a solution using sensors and actuators to continuously align the individual half ton mirror segments to an accuracy of one millionth of an inch (25 nm), or about 1/20th of the wavelength of light.

The next question was how to polish the mirror segments as simply as possible. In the early 1930s Schmidt had developed a stress-polishing technique to produce corrector plates for his telescopes, in which a thin optical flat was placed over the end of a tube, which was then partially evacuated to distort the flat by a specified amount. The flat was then polished to make it flat again, and the air let back into the tube, causing the glass to take up its pre-designated shape. Schmidt's technique could only be applied to thin pieces of glass, but this was precisely what would constitute the segmented mirror of the Keck, so Jacob Lubliner of UC and Jerry Nelson proposed to use a similar stress-polishing technique to produce the mirror segments of the Keck. Each of the 36 hexagonal segments were to be 6 ft (1.8 m) across and just 3 inches (75 mm) thick, allowing the glass in the primary mirror to be no heavier than that in the Palomar 200 inch, even though the new telescope's mirror was four times the area.

Nelson presented his outline design solutions in 1979 which persuaded Caltech to join the team. In 1984 the Keck foundation agreed to provide the $70 million of the estimated $94 million to construct the 10 m telescope and observatory, and in the following year UC and Caltec founded the California Association for Research in Astronomy (CARA) to design, build and operate the telescope. Mauna Kea was selected as the site for the instrument, because it was to be used at both infrared and visible wavelengths, and the high altitude was essential for infrared observations.

Over the next 5 years many problems were faced and solved by the design and construction team managed by Jerry Smith of CARA, the project manager, and Jerry Nelson, the chief project scientist, and many technical innovations were introduced to optimise the design. Finally, on 24th November 1990 the Keck telescope saw first light, with nine of the mirror segments in place, and in April 1992 the last of the 36 mirror segments was installed. In the meantime, in 1991 the Keck Foundation announced that it had approved the construction of a second 10 m telescope on Mauna Kea, whose images will be combined optically with those of Keck I in an interferometric arrangement (see Pages 306–307).

Mauna Kea now has an unparalleled family of telescopes in operation (see Figure 13.7), including the 3.8 m (150 inch) UK infrared telescope, and the 15 m (50 ft) diameter James Clerk Maxwell telescope operating in the sub-millimetre band, making it the premier observatory in the world for infrared and sub-millimetre astronomy.

Figure 13.7 *Observatories on the summit of Mauna Kea, Hawaii. The 10 m Keck is in the large dome in the foreground, just to the right of centre. The strange-looking observatory near the right-hand edge of the photograph houses the James Clerk Maxwell telescope operating at sub-millimetre wavelengths. (Courtesy B. W. Hadley, Copyright* Royal Observatory.*)*

La Palma

The Astronomer Royal for Scotland, Charles Piazzi Smyth, went to Tenerife in the Canary Islands in 1856 to make astronomical observations from Guajara at an altitude of 2,700 m (8,900 ft), and showed the great advantage of observing at altitude. Jean Mascart of the Paris Observatory observed Halley's comet from the islands in 1910, but it was not until the total solar eclipse of 1959 that astronomers began to realise the potential of these islands just off the north coast of Africa.

The first observatory in the Canaries was set up on the island of Tenerife, but since 1982 all the large telescopes have been installed on the adjacent island of La Palma. During 1974–75 the UK had surveyed six sites in the Mediterranean, North Atlantic and Hawaii, but decided on La Palma as the place for its northern hemisphere observatory, to complement that in Australia. Although the site at the Roque de los Muchachos, at an altitude of 2,400 m (7,900 ft), is not as high as that on Mauna Kea, for example, it is much closer geographically to the UK, while being generally above the cloud ceiling of about 1,500 m.

The Greenwich Royal Observatory had been gradually moved to Herstmonceux Castle in Sussex during the 1950s to escape from the light pollution of London. The move had been completed in 1958, and the new 96 inch (2.4 m) Isaac Newton Telescope (INT), often incorrectly called the 98 inch, had been completed there in 1967. Unfortunately the seeing and weather conditions in England are not good enough for a large telescope, and the INT was moved to the Roque de los Muchachos Observatory at La Palma, where it saw first light in February 1984 with a new 101 inch (2.6 m) mirror. The Royal Observatory was itself moved from Herstmonceux Castle to Cambridge in 1990.

The largest telescope on La Palma is the 165 inch (4.2 m), William Herschel Telescope (WHT), that was built by Grubb Parsons of Newcastle, UK as their last telescope, using a mirror blank supplied by Owens–Illinois. This blank was made out of Cervit, a glass–ceramic material with a virtually zero expansion coefficient. The telescope, which saw first light in June 1987, was one of the first large modern instruments to use an altazimuth mount, like the MMT, which is computer controlled on both axes. The computer also rotates the image field as it tracks an object.

The WHT has one Cassegrain and two Nasmyth foci, the latter being stationary on the altitude axis, thus allowing the mounting of heavy detectors, like at the Coudé focus of an equatorially-mounted telescope. The focal ratio of the primary mirror is *f*/2.5, which is normally modified to *f*/2.8 when the coma corrector is added. In the Cassegrain and Nasmyth configurations, the telescope is *f*/11.

The Nordic Optical Telescope that came into operation at La Palma in 1989 has an active optics system that frequently corrects the shape of the primary mirror when it deviates from its required shape as the telescope turns. This allowed the designers to produce the thinnest single 2.5 m primary mirror up to that time, it being only 7 inches (18 cm) thick, and weighing only 2 tons.

The Anglo-Australian Observatory

The Australian Astronomer W. G. Duffield suggested, at a meeting of the International Solar Union at Oxford in 1905, that a solar observatory be set up in Australia to complement the one in England. The idea received widespread support from both British and Australian astronomers, but the First World War intervened before the solar observatory could be established; the Mount Stromlo Observatory being finally established near Canberra on 1st January 1924.

The Cambridge astronomer Richard v. d. R. Woolley, who was later to become the UK Astronomer Royal, was appointed director of the Mount Stromlo Observatory in 1939. Woolley wanted to change the work of the observatory from solar to stellar research, and in 1945 he refurbished the 48 inch (1.2 m) Great Melbourne Reflector which had been the last large telescope to be made with a metal mirror in the nineteenth century. He changed the mirror to Pyrex, reducing the focal length in the process, and added an electric drive, enabling the first infrared observations to be made in the southern hemisphere. Woolley also hoped to acquire a 74 inch (1.9 m) reflector, similar to that planned for the Radcliffe Observatory in Pretoria, but delays due to the Second World War meant that the 74 inch was not installed until 1953.

Light pollution near Canberra was becoming a serious problem after the war, and this led to the search for a new observing sight, which eventually resulted in the establishment of the new observatory at Siding Spring in New South Wales in 1962. Two years later, the Royal Society in the UK and the Australian Academy of Sciences began discussions on the joint construction of a major optical telescope in Australia. By then the observing conditions at Siding Spring, at an altitude of 1,150 m (3,800 ft), were known to be good, and the great Parkes radio telescope had been installed only 280 km (170 miles) away to the south-west, so Siding Spring was agreed to be the ideal location for the new 153 inch (3.9 m) Anglo-Australian Telescope (AAT).

With the agreement to construct the AAT, attention was diverted to the problem of finding suitable objects for it to study. The ideal solution was to build a wide-field Schmidt telescope that could also extend the Palomar sky survey into the southern hemisphere at the same time. In the event, the 49 inch (1.2 m) UK Schmidt was completed in record time in 1973 and installed at Siding Spring to do precisely that, one year before the AAT was brought on line.

The Anglo-Australian Telescope, with its 3.9 m very low expansion Cervit mirror, was one of the first telescopes to take full advantage of the development of CCDs and computer control technology. The observer no longer had to sit in a cold dome to make manual adjustments to the telescope tracking system, being able, instead, to supervise a computer in the luxury of a warm room to do the job for him. The computer also automatically slewed to the objects to be studied, taking account of refraction and other corrections, changed filters on the telescope, and so on. Today such systems are becoming available for amateur astronomers using PCs, but in the 1970s, when they were introduced on the AAT, they were very much state-of-the-art developments.

The European Southern Observatory

The idea of the countries of Western Europe building a major observatory in the southern hemisphere was proposed by Walter Baade in 1953. Such an observatory did not exist in the southern hemisphere at the time, the two largest telescopes being the 74 inch (1.9 m) reflectors in South Africa and Australia. The idea of European collaboration was becoming very fashionable in the 1950s, but it took until 1962 for the European Southern Observatory (ESO) convention to be signed by Belgium, West Germany, France, the Netherlands and Sweden. Denmark, Italy and Switzerland were to join the ESO later, but noticeable absentees are the UK, who decided to contribute to the Anglo-Australian Observatory instead, and Spain, who have their own high-altitude observatory at La Palma in the Canary Islands. The administrative centre of the ESO was initially at Hamburg, but the centre was moved to Garsching, near Munich, in 1977.

Initially the search for a suitable observing site concentrated on South Africa, but eventually the final choice was made of la Silla in the Chilean Andes, at an altitude of 2,400 m (7,900 ft). The atmospheric conditions at La Silla are spectacularly good, because of the proximity of the Atacama desert and a cool coastal current, and the ESO sought to control the local environment by purchasing 625 km^2 of land surrounding the observatory site. The construction of the observatory was begun in 1965 and, over the next ten years, numerous medium-sized instruments were installed. A 1.0 m (40 inch) Schmidt telescope was commissioned in 1972 and, a few years later, the first 3.6 m (140 inch) reflector was available. It was not long, however, before the ESO were planning the construction of a new 3.6 m reflector, called the New Technology Telescope (NTT), which needed the addition of Italy and Switzerland to the ESO group in 1983, to complete the $15 million financing. The Zerodur blank for the primary mirror was completed in the following year.

The NTT, which has a very thin primary mirror (24 cm, or 9.5 inches) weighing only 6 tons, uses an active optics system to measure the distortion of that mirror, and correct its surface to the ideal shape several times per minute. This enabled the NTT to achieve a resolution that was unequalled at the time of its first use in 1989.

The design of the building housing the NTT was also unique. It is an hexagonal structure that rotates with the telescope, having large openings and louvres in front of and behind the telescope to allow the outside air to pass directly over the instrument. This design enabled the telescope to cool down to the ambient air temperature relatively quickly, so minimising thermal problems.

On 1st July 1987, while the NTT was still being constructed, the European Southern Observatory inaugurated the world's first intercontinental system for the remote control of telescopes, when they linked the ESO in La Silla to the ESO Headquarters in Garching, Germany, via an Intelsat communications satellite. Astronomers were then able to control the 2.2 m (85 inch) telescope in Chile from Germany, and receive images directly from the telescope, taking 7 minutes to send a full CCD image of 320 $\times$ 512 pixels (picture elements).

A system of adaptive optics* was used at the ESO in 1990 to improve the resolution of their telescopes even more. These so-called diffraction-limited images were produced by introducing a small deformable mirror in front of the image detector. The image distortions caused by the Earth's atmosphere

---

*A similar system was developed at the Massachusetts Institute of Technology in 1988 for military use, but that system was only declassified by the Pentagon in May 1991.

are measured many times per second, and the deformable mirror's shape changed to cancel the distortions. Preliminary experiments were carried out at the Pic du Midi Observatory in the Pyranees in the late 1980s, before the system was installed at La Silla, working perfectly at both sites.

14 Tools and Techniques

Photography

Early Developments

Early astronomical photographs made in the 1840s were taken using the daguerreotype process, but it was messy (using iodine and bromine vapour, and mercury heated to 75°C) and produced images on metal that could not be copied. It was also relatively insensitive to light, and so was limited to very bright objects like the Sun or Moon.

A new photographic process, called the wet collodion process, was introduced by the English sculptor Frederick Scott-Archer in 1851. It was much more sensitive to light and produced images on glass, although the exposure was limited to about 15 minutes, as the plates had to be exposed while still wet. At Harvard College Observatory, G. P. Bond took photographs of the double star Mizar using this new process in 1857, and found that the photographs could be used to measure stellar intensities, as the brighter stars were found to produce larger images.

Warren De La Rue, an English amateur scientist and inventor, decided to take up astronomical photography after seeing Bond's daguerrotype of the Moon at the Great Exhibition in London in 1851. He began photographing the Moon in 1853 using the new wet collodion plates and, in the following year, the British Association asked him to design an instrument to provide daily photographs of the Sun, to help in the study of sunspots at Kew Observatory. The problem with taking photographs of the Sun was its great brightness, and De La Rue had to design a very rapid shutter to enable satisfactory photographs to be taken. His photoheliograph, as it was called, was used to provide a daily record of sunspots at Kew from 1858 to 1872, when it was transferred to the Royal Observatory at Greenwich.

In 1860, photography was used to settle the long-standing dispute as to whether the prominences seen during a total solar eclipse are connected with the Sun or Moon. Both De La Rue and the Jesuit astronomer Angelo

Secchi photographed the total solar eclipse of that year and showed that the prominences moved with the Sun, not the Moon.

Bond had realised in 1849 that the visual and photographic foci of a refractor differed, because photographic emulsions had a different spectral sensitivity to that of the eye. It was not until 1863, however, that Lewis Rutherfurd made the first refractor specifically designed for photographic work, enabling him to photograph ninth magnitude stars with an exposure of only 3 minutes. Then, in 1871, the wet collodion process was replaced by the far more sensitive and convenient dry photographic plates, which allowed photographs to be taken of nebulae and the dimmer stars for the first time, and allowed photographs of the Moon to be taken with an exposure of only a few seconds.

So by the early 1870s photography was beginning to make itself felt as a potentially powerful astronomical tool, but it had its first serious setback in 1874, when it was used to record the transit of the Sun by Venus, in an attempt to produce a more accurate value for the mean distance of the Sun from the Earth (the Astronomical Unit). Unfortunately, the edge of the Sun was not defined sharply enough on the plates to enable accurate measurements to be taken, so it was decided not to use photography for serious research during the next transit of Venus due in 1882.

The first photographs of a comet were those of Donati's comet in 1858 taken, not with a telescope, but with an ordinary portrait camera. Donati's comet was exceptionally bright, however, and good photographs of ordinary comets were not produced until the new faster dry plates became available. Jules Janssen in France and Henry Draper in the United States photographed Tebbutt's comet in 1881 using these new faster plates, but their exposures still ranged from 30 minutes to over 2 hours. So they had to have their telescope drives specially modified to allow tracking at a non-sidereal rate, as the comet moved appreciably relative to the stars during such long exposures.

When taking photographs of ordinary terrestrial subjects, the exposure is dependent on the *f* ratio of the lens. When taking photographs of point sources like stars, however, the exposure is determined by the aperture of the lens or telescope, and not by its *f* ratio.

It was not until 1882, however, that astronomers realised that, because comets are not point sources like stars, the exposure required depends on the *f* ratio of the lens, and not on its aperture. This explained why photographs of Donati's comet with an ordinary portrait camera with a fast lens (i.e. a small *f* ratio) had been so successful, and why the telescopic images of 1881 taken with telescopes of large apertures, but large *f* ratios, had needed such long exposures, even with the much faster photographic plates available then. This discovery led David Gill in South Africa to use a standard Dallmeyer portrait lens, which he borrowed from a local photographer, to photograph comet 1882 II. His photographs, taken with exposures ranging up to 100 minutes, not only produced excellent images of the comet, but also recorded so many stars that it encouraged many astronomers to use photography to record star fields.

It may be wondered why an ordinary portrait lens of small diameter recorded dimmer stars, compared with telescopes of much larger diameter. The answer is, it didn't, but because the lens was of such a short focal length, its magnification was low, so there were more stars crammed into a given area on the photographic plate, which made the photograph appear more impressive.

Draper took the first usable photograph of a nebula (the Orion nebula) in 1880, but the best early photograph was taken by the English engineer A. A. Common 3 years later, which won him the Gold Medal of the Royal Astronomical Society. Common's photographs (see Figure 14.1, for example), which required exposures of about an hour, were a big advance on the best drawings produced in the nineteenth century as, not only could photographs be produced more quickly, they eliminated the subjective element of drawings.

Encouraged by his photographs of comet 1882 II, Gill started work in 1885 at the Cape Observatory to produce a photographic star map of the southern sky. He was assisted in the photographic part of the project by two assistants, and in the analysis of the photographs by J. C. Kapteyn who volunteered to set up a laboratory in the Netherlands to measure the plates. Kapteyn, who was Professor of Astronomy at the University of Groningen,

Figure 14.1 *The Orion nebula photographed by A. A. Common with his 36 inch (90 cm) reflector in 1883, using an exposure of 1 hour. (From* A Popular History of Astronomy during the Nineteenth Century, *by Agnes Clerke, 1908, Frontispiece.)*

carried out the work almost single-handed, in two small rooms borrowed from the physiological department, although at one stage he was given some local convicts to help with the work! This mammoth photographic and analytical undertaking was not completed until 1900, when the Cape Photographic Durchmusterung or CPD catalogue was issued giving the positions and intensities of 454,875 southern stars. The magnitudes of the stars were determined by measuring the diameters of their images, and comparing them with those of standard stars whose magnitudes had been determined visually by Eduard Schönfeld in Bonn or Benjamin Gould in Argentina.

As the 1880s progressed, photographs improved in quality, until near the end of the decade they were recording detail that was invisible visually with even the best telescopes. For example, in 1886 the brothers Paul and Prosper Henry photographed 1,400 stars in the Pleiades open star cluster using a 3 hour exposure with a 13 inch (34 cm) *f*/10 refractor at the Paris observatory, and in the process proved the existence of the Merope* Nebula.

These excellent photographs produced by the Henrys, and those being made by Gill, led Admiral Mouchez, the Director of the Paris Observatory, to invite astronomers from many different countries to an Astrographic Congress in Paris in April 1887, to discuss the production of the first international collaborative star catalogue of the whole sky, the Carte du Ciel, based on photographs. The agreed plan was to photograph stars down to the fourteenth magnitude, and to produce a catalogue of their positions and intensities down to the eleventh magnitude. Nineteen nations attended the Congress, and the sky was split into 18 separate sections, with one section to be photographed and catalogued per observatory. The exposures were all to be made with a 13 inch (35 cm) *f*/10 refractor of identical design to that used by the Henrys, that would produce a scale on the 16 × 16 cm plate of 1 mm to 1 arcmin.

So by 1890 photography had finally come of age.

The Bunsen–Roscoe Law and the Spectral Sensitivity of Emulsions

The problem of determining stellar magnitudes from photographic plates was the subject of much research and confusion about the end of the nineteenth century. Although dry photographic plates had been first produced in 1871, there was considerable uncertainty, even in 1900, about the effect of exposure time and luminosity on the density of the images. The Bunsen–Roscoe reciprocity law, which states that image density depends simply on the total amount of light striking the emulsion was accepted by

---

*Sometimes called the Maia Nebula. It had first been observed by E. W. L. Tempel in 1859, and subsequently confirmed by others, but S. W. Burnham could find no sign of it with a much larger telescope, and its existence was questionable until the Henrys proved its existence with their photograph.

many. That is, the density of the image is proportional to $I \times t$, until the density reaches saturation, where I is the intensity of the subject, and t is the exposure time.

Adjacent stellar magnitudes have a brightness difference of 2.512 by definition (see Page 298). So if the Bunsen–Roscoe law is correct, an eighth magnitude star would, for example, have the same image density as a seventh magnitude star if the former were to be given 2.512 times the exposure of the latter, under identical conditions. This was the basis of Prosper Henry's proposal for determining stellar magnitudes photographically, that had been accepted by the Permanent International Committee (PIC) of the Carte du Ciel project in 1889.

In 1891, however, the PIC reopened the subject of the determination of stellar magnitudes, because there were doubts that the Bunsen–Roscoe law was correct. Because of these doubts, it was thought that it was safer to measure the size of the stellar images, rather than their density, to determine the magnitude of the stars, as it was considered that the size of the image was proportional to intensity, even if the density was not. Bond had suggested this technique as early as 1858, but it was, unfortunately, difficult to decide on the image size as there was no clear edge.*

After further work, the PIC decided in 1896 that there was no really satisfactory method for determining stellar magnitudes, and so they left it to the individual observatories participating in the Carte du Ciel project to devise their own systems. Karl Schwarzschild, in particular, working in Vienna, continued his search for a solution to the problem, and found in 1899 that if the photographic plate is placed a little in front or behind the focal plane of the telescope, the stellar images produced are all the same size but of different densities. He measured these densities at first against a scale of representative densities by a simple visual side-by-side comparison, but in the same year J. F. Hartmann of Göttingen produced a visual microphotometer that enabled Schwarzschild to produce much more accurate results.

In 1900 Schwarzschild showed that the Bunsen–Roscoe law was not valid for low intensity subjects owing to, what we now call, the reciprocity failure of the emulsion. He proposed, instead, that equal blackening of the emulsion is achieved for long exposures with low intensity subjects, not when $I \times t$ are equal, but when $I \times t^p$ are equal. For the emulsions in use at the time, Schwarzschild estimated a value of p of 0.86, but it is now known that the value of p can vary from 0.6 to virtually 1.0, depending on the emulsion and the conditions under which it has been prepared and used.

In the last decade of the nineteenth century many astronomers still assumed that the density of photographic images was independent of the wavelength of the incident light, although this was not so. The first emulsions made were most sensitive to blue light with a wavelength of

*The first major photometric survey to use this technique was the CPD of Gill and Kapteyn published in 1900.

480 nm.* Green sensitive plates were first produced in 1884, panchromatic plates (with relative densities per colour that looked right to the eye) in 1904, and infrared sensitive plates in 1919, with a peak sensitivity at 820 nm. By 1934 the peak sensitivity had been pushed well into the infrared at 1,350 nm (1.35 μm).

Originally, because photographic plates were only sensitive to short (blue) wavelengths, refractors had to be optimised for these wavelengths if they were to be used for photography. This necessity largely disappeared in 1904 with the advent of the panchromatic plate having a spectral sensitivity similar to that of the eye, although it meant that those refractors optimised for the old blue-sensitive plates would not perform very well with the new plates. Eventually the requirement to produce large telescopes that could be used visually largely disappeared, and so telescopes were then optimised for the spectral sensitivity range of the detectors to be used, whether they be photographic plates, photocells or CCD arrays.

Photographic Magnitude Standards

Research in the astronomical community continued in parallel with the above developments, and in 1895 William Pickering of the Harvard Observatory reported that he had standardised the photographic magnitudes of stars in his analysis by making them equal to their visual magnitudes in the case of stars in Secchi Category I (white or blue stars). Fifteen years later, the International Solar Union set the first international standard, based on a similar proposal put forward by Schwarzschild, that the visual and photographic magnitudes of class A stars (which are white) should be made the same between magnitudes 5.5 and 6.5.

Astronomers do not always want to have to go back to the basic (primary) standard during their research, preferring a convenient but less accurate secondary standard instead. Edward Pickering, William's brother, obliged in 1912 when he issued a catalogue that described the accurate photographic magnitudes of 96 stars near the north celestial pole, ranging from the fourth to the 21st magnitudes. This "North Polar Sequence", as it was called, was designed to be an absolute reference for measuring stellar magnitudes. Sydney Chapman and P. J. Melotte at Greenwich extended it to 262 stars in 1913, and F. H. Seares at Mount Wilson produced a new north polar sequence that was accepted as an international standard by the IAU in 1922.

Sky Surveys

James Keeler started a programme in 1898 at the Lick Observatory of photographing nebulae with the 36 inch (90 cm) Crossley reflector. After

---

*The wavelengths corresponding to the colours as seen by the eye are about 400 to 490 nm, blue; 490 to 560 nm, green; 560 to 590 nm, yellow; 590 to 630 nm, orange; and 630 to 700 nm, red.

2 years Keeler had recorded thousands of nebulae, ranging in size from the great Andromeda nebula down to objects that were scarcely distinguishable from a stellar disc. He also noted that most of the nebulae had a spiral structure, and estimated that there were about 120,000 nebulae within the photographic range of his telescope.

The first international photographic sky survey, the Carte du Ciel, had started in a blaze of publicity in 1887, but the plan to photograph all stars down to magnitude 14, and to catalogue them down to magnitude 11, quickly ran into trouble. Unfortunately the production of the plates and the cataloguing were much more time-consuming and costly than originally foreseen, and a number of observatories dropped out before they had finished their task, which then had to be allocated elsewhere. As a result, the publication of the catalogue was not completed until 1964.

The National Geographic–Palomar Sky Survey of the 1950s was started before the Carte du Ciel was completed, taking photographs in blue and red light, down to a stellar magnitude of 21.1 (in blue light), using 14 inch (35 cm) square plates on the Palomar 49 inch (1.2 m) Schmidt. The emulsions used were quite fast, requiring only a 10 to 15 minutes exposures for the blue-sensitive Kodak plates, and 40 to 60 minutes for the red-sensitive plates. In 1986, a new Palomar survey was started, reaching down to magnitude 22.5 in blue light, using modern photographic emulsions of the Kodak 111 series on the improved Palomar Schmidt.

Although, today, CCDs (see Pages 308–309) have taken over from photography for much astronomical work, photography is still used in surveys, enabling vast amounts of data to be acquired and stored cheaply. This is especially true where fine image detail needs to be recorded over the wide field-of-view of Schmidt telescopes.

Solar System Photography

Solar photography took a great step forward when George Ellery Hale started experimenting with his spectroscope and a special photographic system in March 1891. He found that he could photograph the surface of the Sun in the light of a single wavelength, if the solar image was made to scan across the first slit of his spectroscope, and the photographic plate was moved simultaneously across the second slit. This spectroheliograph, as it was called, was first used to record the solar disc in the light of a line of hydrogen, but Hale then decided to use the H and K lines of calcium, as photographic plates were more sensitive to the ultraviolet end of the spectrum, at that time. These photographs enabled prominences and sunspots to be examined at will across the solar disk when the system was in full operation in 1892.

Independently, and virtually simultaneously, Henri Deslandres invented a similar spectroheliograph in Paris, but it was in operation a few months later than Hale's. Although Deslandres was not the first to build such an instrument, he played a major rôle in developing its use, and suggested that, as different parts of the calcium H and K bands originate in different layers of the Sun, photographs taken at these slightly different wavelengths would show the Sun at different depths of its photosphere.

Photographic observation of the planets was found to be disappointing for most of the twentieth century, owing to turbulence in the Earth's atmosphere and the small size of the images. The eye, on the other hand, can see a good amount of detail during a few tenths of a second of "good seeing", making it superior to photography in observing planetary detail from sea-level observing sites. In recent years, however, the introduction of high-altitude observatories, and the use of adaptive optics, which reduce the effect of atmospheric turbulence, have revolutionised imaging of the planets from the Earth. The majority of planetary observations are now made from spacecraft, however, where electronic image recording using CCDs has replaced photographic emulsions.

Spectroscopy

Early Developments

In the early nineteenth century, Joseph Fraunhofer tried to develop an accurate method of measuring the refractive index of various types of glass at his laboratory in Munich, to enable the design of achromatic lenses to be improved. In one experiment, where he produced a spectrum, he used sunlight that first passed through a slit, and then through a prism that was mounted just in front of the 1 inch (2.5 cm) diameter objective of a theodolite. Previous experiments with flames in the laboratory had produced a bright line spectrum, and he expected to find that the continuous solar spectrum would contain many more bright lines. He was astonished to find, instead, that the coloured solar spectrum was crossed by a great many dark,* rather than bright lines. These dark lines helped him to measure refractive indices to an unprecedented accuracy, but from 1814 he decided to concentrate his efforts on studying these dark solar lines, using what is now called objective prism spectroscopy. A few years later, he changed his theodolite for a telescope with a 4.5 inch (11 cm) diameter objective, and replaced the prism, as the dispersing medium, by a simple diffraction grating made of very fine parallel wires.

The main dark lines seen by Fraunhofer were labelled A to H, and in 1849 Léon Foucault, working from his home in Paris, showed that the D absorption line in the solar spectrum, which was double, coincided with a double emission line in a laboratory arc. He also observed that the darkness of the solar absorption line was intensified when sunlight was passed

*Unknown to Fraunhofer, these lines had first been seen in 1802 by William Wollaston, an English doctor and scientist, but Wollaston carried his discovery no further.

through the gas of the laboratory arc. In 1859, Gustav Kirchhoff and Robert Bunsen at Heidelberg repeated and extended Foucault's work, although they were in ignorance of it, and concluded that a luminous solid or liquid (or, as discovered later, a highly compressed hot gas) give a continuous spectrum, while a tenuous hot gas gives a bright line spectrum. This bright line spectrum was found to be unique to each element in the gas. When light with a continuous spectrum passes through a cool gaseous layer, this layer absorbs its characteristic lines from the continuous spectrum, but if light with a continuous spectrum passes through a very hot gas, the hot gas produces it own bright lines superimposed on the continuous spectrum.

Kirchhoff measured the positions of thousands of the dark Fraunhofer lines in the solar spectrum from 1859 to 1862, and established that these coincided with bright lines produced by various chemical elements when in a hot gaseous state in the laboratory. This enabled him to deduce which elements were producing the dark Fraunhofer lines above the surface of the Sun; the first elements to be recognised being sodium and iron. Kirchhoff used a prism spectroscope in his research, in which a slit is placed in the focal plane of (not in front of) the telescope objective, followed by a prism, but this produced a non-linear wavelength scale. In 1868, the Swedish physicist A. J. Ångström used a diffraction grating instead of a prism, however, which produced a linear wavelength scale. Ångström's scale, based on a unit of 10^{-10} m, was soon accepted as the standard, until it was overtaken by the SI unit of 10^{-9} m (or nanometre) in the twentieth century.

Angelo Secchi published his spectral classification of stars in 1863 (see Page 122) and, in the same year, William Huggins showed, at his observatory in London, that some of the dark lines in the spectra of stars correlated with those of hydrogen, iron, sodium, calcium and magnesium. This indicated, for the first time, that stars were made of the same elements as those found on Earth and in the Sun.

Although Fraunhofer had used both his prism and diffraction grating in front of the telescope objective, the majority of spectroscopes used in the mid nineteenth century used arrangements similar to those of Kirchhoff and Ångström in which the prism or diffraction grating are placed after the objective. These spectroscopes were improved in a number of ways as the nineteenth century progressed. For instance, J. W. Draper, a professor in Chemistry in New York, found that his glass diffraction gratings gave him brighter spectra when coated with a mercury–tin amalgam and then used as reflection, rather than transmission, gratings. Lewis Rutherfurd produced both silver-coated glass and speculum metal reflection gratings in the 1840s.

In 1842 Edmond Becquerel who, like his father and his son, was professor of Physics at the Musée de l'Histoire Naturelle in Paris, linked the two technologies of spectroscopy and photography, shortly after the latter was invented, when he photographed the solar spectrum in what we now call a spectrograph. Then, 30 years later, Henry Draper, J. W. Draper's son, replaced the standard glass prism in his spectrograph with a quartz prism, which transmits more ultraviolet light than does ordinary glass, and used this new device to photograph *stellar* spectra with the wet collodion process.

Four years later, Huggins used the new faster dry plates to photograph stellar spectra.

In 1885 Edward Pickering reverted to the objective prism arrangement, originally employed visually by Fraunhofer, and placed a large shallow-angle prism in front of his telescopic objective. This enabled stellar spectra to be photographed in quantity, as each of the images on the plate was that of a stellar spectrum rather than of a star. No slits or cylindrical lenses* were required as the movement of the stars, with the discuss the production of the first international collaborative star catalogue of the telescopic drive disconnected, produced widths to the spectra, and this new arrangement enabled Pickering to publish the spectra of over 10,000 stars in 1890 (see Page 124). The spectra produced in this way were bright but of low dispersion. Using a 13° prism in front of the 8 inch *f*/5.6 Bache refractor, for example, produced dispersions of the order of 30 nm/mm at a wavelength of 450 nm, which is an order of magnitude worse than with a standard spectrograph, but the exposure times required were an order of magnitude better. These low dispersion spectra were thus ideal for stellar classification, but not for studying the detailed composition of stars. That was facilitated in 1882 by Henry Rowland at Johns Hopkins University in Baltimore when he replaced the standard planar diffraction grating with a concave grating, greatly increasing the dispersion and resolution of his spectrograph. Because the ruled surface was concave, it brought the spectrum to a focus without the need for a further lens, which meant that the spectrum could be studied further into the ultraviolet, which was normally absorbed by the lens. With this new device Rowland was able to photograph 20,000 lines in the solar spectrum in 1888.

The interpretation of planetary spectra was more difficult than for stars, as the light coming from the planets originated in the Sun, rather than being produced by the planets themselves. So the planetary light started out with solar absorption lines, to which were added absorption lines produced by the planetary atmospheres, before the light finally passed through the Earth's atmosphere which added its own lines. Planetary spectra were analysed by comparing them with the spectrum of the light reflected by the Moon, with any differences being attributed to the planetary atmospheres. William Huggins, Walter Maunder, Jules Janssen and Hermann Vogel all used this technique and independently concluded that both oxygen and water vapour were present in the atmospheres of both Venus and Mars. Secchi and Huggins also found absorption lines in the spectra of Jupiter, Saturn and Uranus but they did not appear to correlate with those of any element known on Earth.

Spectroscopy was beginning to yield real dividends in the second half of the nineteenth century, particularly for stars, but in solving some problems it raised new ones, the most fundamental being, how were the spectral lines

---

*Previously, cylindrical lenses had had to be used when producing stellar spectra, to widen the spectrum (perpendicular to the wavelength scale).

produced and why were they different for each element? In order to understand this, attempts were made to find a mathematical relationship between the wavelengths of the lines produced by each element.

The first breakthrough came in 1885 when Johann Balmer, a Swiss schoolmaster with a love of numerology, discovered that the wavelengths λ of the four visible lines in the spectrum of hydrogen were given by the formula:

$$\lambda = 364.56 \times \frac{n^2}{n^2 - 4} \text{ nanometres}$$

where $n = 3, 4, 5$ or 6. Before publication of his paper he heard that William Huggins and Henry Draper had both found more hydrogen lines in the ultraviolet, and Balmer found that these had wavelengths given by his formula when n had values from 7 upwards.

In his paper Balmer speculated that similar formulae may apply to different elements, but he took this no further. Johannes Rydberg, a Swedish physicist, then showed in 1890 that the wavelengths of the lines in the spectra of a number of other elements could be given by a similar formula to that of Balmer, although it was slightly more complicated.

Technical Developments

An achromatic doublet can only be truly achromatic at two wavelengths, and these have to be chosen with care. Early photographic plates were more sensitive at the blue end of the spectrum, and most of the large refractors near the end of the nineteenth century were corrected for visual observations, so spectra produced by these refractors were never really satisfactory. Photographic corrector lenses were fitted, but these caused a further problem as the transmission of glass at the short photographic wavelengths is not very good. A better solution was to design refractors specifically for photography and spectroscopy, along the lines pioneered by Rutherfurd in the 1860s (see Page 284). A 24 inch (62 cm) photographic refractor was installed at the Meudon Observatory near Paris in 1893, and a 31 inch (80 cm) photographic refractor saw first light at the Potsdam Observatory in 1901.

Light transmission through a typical large refractor–spectrograph system at the turn of the century was only about 1%, at photographic wavelengths, because of:

(i) reflection, absorption and scattering losses in the telescope objective and the spectrograph prisms;
(ii) the lack of full achromatism of the telescope objective;
(iii) the very narrow slits that had to be used to produce sharp lines not allowing all the starlight to pass through.

Stellar scintillation (or twinkling), caused by non-stable atmospheric conditions, for example, produced stellar images appreciably larger than the 0.25 arcsec width that slits often had when projected onto the sky. Clearly the problem was even worse for faint nebulae which had images, even in perfect atmospheric conditions, larger than this projected slit width.

The gradual transition to reflecting telescopes in the early twentieth century helped to solve some of these problems, although refractors were still used for some applications. Typical exposures for spectrographic systems on *refractors* in the early twentieth century were 90 minutes, for a fourth magnitude star and a spectroscopic dispersion of 2.0 nm/mm. The spectrograph attached to the Coudé focus of the 60 inch (1.5 m) Mount Wilson *reflector* in 1911 gave a dispersion of 0.14 nm/mm, however. These high dispersion Mount Wilson spectra were of very high quality, but they were so dim that they could only be obtained for the very brightest stars; the spectrum of a fifth magnitude star requiring an exposure lasting all night. With very dim stars, clearly only modest dispersions could be usefully employed, although in 1952 Ira S. Bowen managed to photograph the spectrum of 15th magnitude stars, using the newly-completed Palomar 200 inch (5 m) telescope, with a dispersion of 3.8 nm/mm.

The best gratings produced in the nineteenth century were made by Henry Rowland, being up to 6 inches (15 cm) in size with 14,000 lines/inch (5,600 lines/cm). In 1912 George Ellery Hale established a laboratory at Pasadena to produce his own high-quality gratings for the new reflecting telescopes on Mount Wilson, and he recruited John A. Anderson, who had rebuilt Rowland's grating ruling machine, to the laboratory. Various other experimental work was also undertaken in this laboratory, and in 1931 John Strong developed a vacuum-evaporation process for aluminising telescope mirrors. Aluminium was found to be much better than silver for mirrors, as it reflects 50% more light at photographic wavelengths, and is more durable, requiring recoating at much less frequent intervals. Strong also found that thick aluminium coatings deposited on glass allowed the production of far better reflection gratings than did speculum metal, enabling so-called blazed gratings to be ruled with a sloping groove profile to concentrate an unprecedented 75% of light into the first order spectrum. After the Second World War, high precision blazed gratings became available that, when used with the faster camera objectives then available, produced spectra over ten times as bright as the old prism spectrographs.

Theory

Balmer and Rydberg had been able to produce mathematical formulae that gave the wavelengths of the spectral lines of a number of elements, and these formulae were developed further in 1908 for another series of lines that had just been found in hydrogen. Meanwhile, in 1900 the German physicist Max Planck had been able to explain the energy–temperature–wavelength relationship for an idealised black body by proposing that energy is emitted by atoms in discrete amounts, or quanta. In 1911 Ernest Rutherford at Manchester University in England proposed that an atom consisted of a positively charged nucleus surrounded by a small number of negatively charged electrons, and then in 1913 the Danish physicist Niels Bohr produced his master stroke.

Instead of the electrons orbiting the nucleus of an atom in orbits of any radius or energy level, as envisaged by Rutherford, Bohr proposed that they could only be in certain discrete states with discrete energies, and that when an electron moves from one state to another it emits a quantum of energy whose frequency is directly proportional to the energy difference between the original and new state. Thus when electrons go from a higher to a lower state they emit a quantum of energy of a discrete frequency or wavelength, and when they absorb energy and go from that lower state back to that same higher state they absorb a quantum of energy of exactly the same frequency or wavelength. This explained how the dark and bright lines in the solar spectrum are produced, and produced a much fuller understanding of the radiative processes in the visible parts of the Sun, stars and nebulae. Bohr's theory also enabled him to derive the formulae of Balmer and Rydberg from first principles.

Comparative Planetary Observations

In 1890 it was thought that the spectra of Venus and Mars indicated that both planets had both oxygen and water vapour in their atmospheres, but in 1894 William Campbell at the Lick Observatory could find no such evidence in their spectra. In 1908 Vesto Slipher took spectra for Venus and looked for the Doppler shift in the lines of these gases that should have been caused by the difference in orbital velocities of the Earth and Venus, particularly when the latter was at greatest elongation. He found no such effect, and concluded that oxygen and water vapour were not present, at least in the upper part of the Venusian atmosphere, the lines that he had observed being due to the Earth's atmosphere. Later, in 1932 Walter Adams and Theodore Dunham at Mount Wilson discovered evidence of carbon dioxide on Venus, but not oxygen or water vapour.

Spectroscopic results in the early twentieth century for Mars were also contradictory. In 1908 Very found evidence for both oxygen and water vapour, but Campbell disputed this in 1910. Walter Adams and Edward St John at Mount Wilson found evidence again for both oxygen and water vapour in 1925, but Adams and Dunham could find no evidence for oxygen when they used a new and much more accurate spectroscope eight years later. In 1952 Gerard Kuiper found a trace of carbon dioxide.

In the nineteenth century, absorption bands of unknown elements had been found in the atmospheres of Jupiter, Saturn and Uranus, and in the early twentieth century Slipher had shown that Neptune had similar lines. He also showed that the lines were darker, the further the planets were from the Sun. Rupert Wildt of Göttingen finally found the origin of these lines in 1932, when he discovered that they were produced by methane and ammonia, which was confirmed by the careful observations of Dunham in 1932 and Mecke in 1933.

Henry Norris Russell showed theoretically that, in the absence of oxygen, the atoms of carbon and hydrogen would generally combine at very low temperatures to form methane (CH_4), while those of nitrogen and hydrogen would generally form ammonia (NH_3). Slipher and Adel had observed that the methane absorption bands increase in strength in going from Jupiter to Neptune, whereas the ammonia bands are strong in Jupiter, weak in Saturn, and non-existent in Uranus and Neptune. Dunham suggested that the ammonia has been frozen out of the atmospheres of Uranus and Neptune by their lower temperatures, and Russell pointed out that this would make the upper atmospheres more transparent, thus enabling methane to be detected more readily. Hydrogen, helium and nitrogen are very difficult gases to detect in planetary atmospheres but, in 1934, Jeffreys concluded, using a mixture of theoretical and observational analysis, that the atmospheres of Jupiter and Saturn consisted of hydrogen, helium, nitrogen and methane with clouds of ammonia crystals.

Thus by the 1950s the planets seemed to be separated into two groups as far as their atmospheres were concerned. Venus and Mars appeared to have carbon dioxide as their main ingredient, whereas Jupiter, Saturn, Uranus and Neptune had atmospheres of hydrogen, helium, nitrogen and methane, with the addition of ammonia for Jupiter and Saturn.

Later work on planetary atmospheres, including the result of spacecraft observations, has been covered in the appropriate planetary chapters above.

Photometry

Early Work

• *Instrumental Developments*

Hipparchos produced the first known star catalogue in 129 BC, listing the magnitudes (i.e. brightnesses) of stars on a scale of 1 to 6, where 1 is the brightest and 6 the faintest star just visible to the naked eye. Many astronomers improved and extended his catalogue over the years, but near the end of the eighteenth century the stellar magnitudes were still estimated by eye, although by then telescopes were used to enable astronomers to see much dimmer stars.

William Herschel produced his first view of the structure of the Milky Way in 1785, assuming the number of stars per unit volume was the same in each part of the Milky Way. He was aware, however, that this assumption was only approximately valid and, as his researches progressed, it appeared less and less satisfactory. In 1796, therefore, he decided to try to measure the apparent intensities of stars, and then estimate their distances, assuming that they all had the same absolute intensities, at least to a first approximation. His method of measuring intensities was disarmingly simple. He pointed two identical telescopes at different stars, and stopped down one telescope until the images of the two stars appeared to have the same intensity.

In 1834 William Herschel's son, John, devised his "astrometer", as he called it, to measure the intensities of stars by comparing them with an image of the Moon which had been made into a point source by a small lens. This lens was then moved to and fro, to change the apparent intensity of the lunar image, until the latter and that of the star appeared to be the same. Unfortunately his results were inconsistent, which he put down to systematic errors. Many years later, however, J. K. F. Zöllner of Leipzig showed that the errors were due to the fact that the lunar intensity did not vary with lunar phase, or solar illumination angle, in the way that Herschel had expected.

Karl Steinheil produced a photometer in Germany, at about the same time as John Herschel's, that enabled the images of two stars to be compared side-by-side. He deliberately observed their defocused images, and then reduced the intensity of one until it appeared to have the same intensity as the other. Steinheil's photometer could only be used for stars of third magnitude or brighter because it used de-focused images.

In 1861 Zöllner designed a much more accurate system, where the real star was compared with an artificial star consisting of an illuminated pinhole, and where the intensities were equalised by using polarising Nicol prisms. Zöllner used this system not only to measure the brightness of stars, but also to study how the intensity of the Moon varied with phase.

Lambert's theory for the reflection from a flat surface, which John Herschel had used in analysing his "astrometer" results, implied that the brightness of the Moon should change only slightly near full Moon as the angle of the incident solar illumination changes. In fact, Zöllner found a large change of 2.5% in intensity for a 1° change in angle* from new Moon (i.e. the phase coefficient was said to be 0.025). Zöllner then realised that Lambert's theory, which had been produced for a flat surface, was inapplicable to the Moon, as the very irregular surface of the Moon causes shadows, and hence reduces the intensity faster than would be the case for a less irregular surface. The phase coefficient was thus found to be a good indication of surface roughness.

Over the next few years Zöllner also measured the brightness of the planets, but in order to calculate their albedo, or reflectivity, he needed to estimate the intensity of the sunlight illuminating them. This proved to be very difficult because the Sun was so much brighter than any other celestial object that it was difficult to estimate its relative brightness although, after many attempts, he eventually concluded that the Sun was about 620,000 times as bright as the full Moon. This led him to estimate the albedo of the Moon to be 0.17, which is a little over twice its modern value. As for the planets he found, in general, that they had much higher albedos than the Moon, but their results tended to be on the high side of their modern value.

Edward Pickering devised a system at the Harvard College Observatory in 1879 to accurately measure stellar intensities. Zöllner's system was

---

*This is the angle between the Sun and the Earth, as seen from the Moon.

difficult to use, as the artificial star never looked like a real star, so Pickering set up a specially designed meridian instrument (i.e. with fixed east–west movement) to simultaneously observe both the object star and the pole star. This Harvard Meridian Photometer, as it was called, had two 2 inch (5 cm) diameter objectives, together with a mirror and prism system that enabled the observer to see both the pole star (Polaris) and the target star, as it crossed the meridian, in the same field of view. The intensities were then made equal by means of a calibrated Nicol prism. This enabled Pickering to publish the Harvard Photometry Catalogue listing the intensities of 4,260 stars in 1884. Hertzsprung found in 1911, however, that Pickering's reference star, Polaris, was a variable star, fluctuating in brightness by several tenths of a magnitude. So the Harvard Photometry Catalogue was not as accurate as originally thought.

• *The Standard Photometric Scale*

The 1 to 6 stellar magnitude system of Hipparchos was fine for general visual purposes, but when the telescope was invented at the start of the seventeenth century stars much fainter than the sixth magnitude could be seen, and over the next 200 years or so there was no generally accepted way of classifying them. In the late eighteenth century astronomers also tried to introduce more accuracy into their intensity estimates by subdividing the existing six magnitudes, particularly to improve their recording of the intensity changes of variable stars. The final push to adopt an intensity standard came early in the nineteenth century when John Herschel and others started to introduce visual photometers to measure stellar intensities.

Karl Steinheil and Theodore Fechner independently discovered that the eye sees intensity differences logarithmically, rather than linearly, and in 1835 Steinheil showed that the average intensity ratio between two adjacent stellar magnitudes varied between 2.2 and 2.8. In 1856 Norman Pogson at Oxford proposed that a standard ratio of 2.512 be adopted between any two adjacent magnitudes, which would make a first magnitude star exactly 100 times more intense than a sixth magnitude star. This scheme was eventually adopted as an international standard, but, in doing so, it was found that some stars that had been considered as of first magnitude were, in fact, brighter than this. This led to four stars being listed as having magnitudes of less than zero, the brightest of which was Sirius at -1.46.

As far as the colour of the stars is concerned, various attempts had been made to produce a colour scale against which the stars could be measured. Julius Schmidt of Athens and Hermann Klein of Cologne used a scale varying from 0 for white, to 4 for yellow, to 10 for deep red, and in 1884 Franks used letters indicating colour, and numbers indicating saturation. Basically, however, much more attention was paid to stellar intensities and spectra, than colour, in the nineteenth century.

Planetary Photometry

In the late nineteenth century, Gustav Müller and Paul Kempf undertook an extensive investigation at Potsdam into the brightness of the planets and their phase coefficients (or change in brightness with solar aspect angle). Their results, compared with those of the Moon, were as follows:

	Albedo	Phase coefficient
Moon	0.17	0.025
Mercury	0.07	0.037
Venus	0.59	0.013
Jupiter	0.56	0.015

So Mercury was found to be rougher (higher phase coefficient) and darker (lower albedo) than the Moon, whereas Venus and Jupiter were both found to be relatively smooth and very bright. This implied that Mercury was cloudless (like the Moon), but that Venus and Jupiter were very cloudy. For Saturn the phase coefficient without the rings was the same as for Jupiter, but when the rings were included the phase coefficient increased dramatically to 0.044, indicting their particulate nature.

N. Barabashev measured the brightness of apparently flat parts of the Moon in 1923 and found that the phase coefficient was the same as for the whole Moon, indicating that the apparently flat areas were, nevertheless, still very irregular.

Stellar Photometry

Zöllner had suggested to his student Hermann Vogel in the 1860s that an accurate photometric survey should be carried out for all stars visible from his observatory in Germany, but it was not until Vogel became director of the Potsdam Observatory that the survey was undertaken. Then between 1885 and 1905, Müller and Kempf, using a Zöllner photometer, measured the magnitude of all stars seen from Potsdam down to a magnitude of 7.5. Their Potsdamer Photometrische Durchmusterung was published progressively between 1894 and 1907, giving the magnitudes of over 14,000 stars, with a mean error in their measurements for each star of only 0.07.

At about the same time, Edward Pickering of the Harvard College Observatory decided to extend his Harvard Photometry Catalogue of 1884, using an improved version of his meridian photometer with 4 inch (10 cm) diameter objectives. During the period from 1882 to 1888 he measured over 20,000 stars with magnitudes down to the faintest (i.e. magnitude 9.5) that had been measured visually by F. W. A. Argelander et al. for their BD catalogue of 1862.* It had been known for some time that Argelander's

*In this catalogue, Argelander and his assistants Eduard Schönfeld and A. Krüger had listed the magnitudes of over 300,000 stars estimated by eye, without the use of a photometer, as they passed through the field of view of their meridian telescope.

catalogue did not follow Pogson's intensity ratio of 2.5 between adjacent magnitudes, for the fainter stars. Pickering measured stars in selected regions of the sky to limit the magnitude of his task, and produced a set of correction factors which enabled all of the magnitudes in the BD catalogue to be re-estimated.

When Pickering had finished measuring the stars from Harvard, he sent Bailey to Peru with the Harvard Meridian Photometer to undertake a similar survey in the southern hemisphere. The photometer was transported to a number of sites, sometimes by mules, before Solon Bailey set up camp at Arequipa, that was to become, in 1891, the southern station of the Harvard Observatory. Bailey measured a total of 7,922 stars in selected areas of the sky, down to magnitude 10, over the period from 1889 to 1891, before returning to Harvard.

On Bailey's return, Pickering re-measured all the stars in his 1884 catalogue, and included all stars visible down to magnitude 7.5 in the process. In 1899, Bailey was again sent to Peru to complete the same task there, which he finished in 1902. As a result, in 1908 Pickering was able to publish the magnitudes of over 45,000 stars in both hemispheres, which were complete down to magnitude 7.0, with measurements in selected areas down to magnitude 10. This Revised Harvard Photometry Catalogue was the standard reference work for many years, mainly because it comprehensively covered both hemispheres.

Karl Schwarzschild, who had moved from Vienna to Göttingen, used his extra-focal photographic method (see Page 287) to measure the brightness of stars between 1904 and 1908. Unfortunately, although his work was very accurate, with mean magnitude errors per star as low as 0.02 to 0.04, he only completed the survey between declinations of 0° and 20°. In all, he determined the photographic magnitudes of just over 3,500 stars down to magnitude 7.5, which he listed in his Göttingen Actinometric published in 1910.

The accuracy of the visual photometers of Zöllner, Pickering, and their contemporaries, of the late nineteenth and early twentieth century, depended on the accuracy with which the eye could determine if two light sources were of equal intensity. This yielded errors of a few percent. A much more accurate photometer, that eliminated the subjective element, was developed in the early twentieth century by Joel Stebbins using a selenium detector.

Stebbins started a programme in 1903, at the University of Illinois Observatory, to measure the relative magnitudes of double stars. He used a visual photometer for the first few years but, in 1907, he met F. C. Brown at the university, who showed him a selenium cell that was sensitive to light. Stebbins was intrigued, and with Brown's help, developed a device that could be used to measure the intensity of bright stars. They discovered, in particular, that the signal-to-noise ratio of the device improved dramatically at low temperatures and, after further refinements, were able, in 1910, to detect the secondary minimum of Algol (see Figure 8.2), which caused a reduction of intensity of 0.06 magnitude, or 6%. The human eye had been replaced by a selenium detector, with the brightness of the image converted to an electric current.

The selenium cell worked because selenium, in a crystalline form, changes its electrical resistance when exposed to light. The effect produces a basically linear electrical response to light but, unfortunately, that response changes with time. An alternative device, the photoelectric cell, also became available at about the same time, in which electrons are emitted in a vacuum from certain surfaces when they are exposed to light. These photoelectric cells, which also produce a linear response to light, are much more stable with time. They were developed by the German physicists Julius Elster and Hans Geitel, and first used in 1912 by Paul Guthnick and Richard Prager in Berlin, and Hans Rosenberg and F. Meyer in Tübingen. Guthnick used his cell on the 12 inch (30 cm) refractor of the Berlin–Babelsberg Observatory, but Rosenberg never put a cell on his telescope, preferring, instead, to use it to measure stellar intensities from photographic plates. This enabled him to measure many stars at his leisure, but the results were no longer linear.

Stebbins heard of Guthnick's work, and in 1913 began using photoelectric cells built at the University of Illinois by Jacob Kunz, because of their greater sensitivity compared with his selenium cell. Then in 1922 Stebbins moved to the Washburn Observatory of the University of Wisconsin, and started an extensive programme to measure the variability and colour of stars, using photoelectric devices.

The electric currents generated by the early photoelectric cells were very small and difficult to measure, but in 1932 Albert Whitford, who was working with Stebbins, used electronic tube amplifiers to increase the photocell outputs. At the same time, Whitford also tested a prototype RCA photomultiplier tube, where the electrons emitted by the photosensitive surface are amplified within the detector itself. By 1937 Gerald Kron was able, using a photomultiplier tube, to measure the light curves of eclipsing binary stars at the Lick Observatory with a relative accuracy of 0.002 magnitude.

In order to measure the magnitude of one star accurately it has to be compared with a standard star. In the 1880s, Pickering had done this by using two modest-sized telescopes, one pointing at the target star, and the other pointing at the reference star, such that both images were brought into the same field of view of the eye. It may be thought that such instantaneous comparisons were now no longer necessary, as the measurements were made electrically, but this was not so. The photocell sensitivities drifted sufficiently, in the time taken to move a large telescope from the target star to the reference star, to affected the accuracy of measurement. Kron solved the problem by placing, in the telescope, a mirror which rapidly moved from the target star to a reference star in the same telescopic field of view.

The photoelectric surface used in the early photocells was generally potassium, sensitised with hydrogen. More complex surfaces of caesium and antimony were produced by P. Görlich in 1938, which had higher sensitivity than potassium. Cooling the photomultiplier tube was also found to reduce the dark current, in a similar way to that discovered by Stebbins for selenium cells. All these developments meant that, by 1960, stars of magnitude 23 could be measured to an accuracy of 0.05 magnitude, using the 200 inch (5 m) Palomar telescope, although a whole night's exposure was required to get down to such levels.

As more and more stellar magnitudes and parallaxes were being measured in the early twentieth century, it became clear that some standard was required in order to define the absolute magnitudes of the stars. Many astronomers wanted to define the absolute magnitude as the apparent magnitude that a star would show if it was 1 parsec (3.26 light years) away from the Earth (where one parsec is the distance of a star that shows a parallax of 1 arcsec) but, because no star (other than the Sun) was this close, it was decided to use 10 parsecs as the standard distance instead. So the absolute magnitude M of a star of apparent magnitude m, having a parallax of p arcsec, is given by:

$$M = m + 5 \log 10p$$

so for Sirius where $m = -1.4$, and where $p = 0.37$ arcsec:

$$M = -1.4 + 5 \log 3.7 = 1.4$$

Colour

In 1893 Wilhelm Wien discovered that the wavelength of peak radiation emitted by a black body is inversely proportional to its absolute temperature, so as bodies become hotter their colour changes progressively from red to blue. If the colour of stars could be measured or estimated accurately, therefore, their surface temperatures could be deduced.

It was found that the original photographic emulsions were more sensitive to blue light than the eye, and so a colour index could be produced by comparing the magnitudes of stars estimated visually with those estimated photographically. The first accurate photographic intensities had been produced in 1899 by Schwarzschild using his extra-focal method, so Schwarzschild suggested that the photographic minus the visual magnitude could be used to signify the colour index of a star. It was these colour indices that Ejnar Hertzsprung had used in his important colour–magnitude plots of 1911.

Colour indices were a good indication of the surface temperature of a star, but measurements of the shape of the radiation curve with wavelength should give more accurate results. In 1909 J. Wilsing, J. Scheiner and W. Münch in Potsdam, and Charles Nordmann in Paris began a detailed measurement campaign in which they visually estimated the relative brightness of different parts of a star's spectrum, but these estimates were highly subjective and limited to visible wavelengths. Five years later, Rosenberg visually measured the density variation along a photographic spectrum, and in 1922 the American W. W. Coblentz measured the spectrum directly on the telescope using a thermocouple. The most reliable of these early methods was introduced in 1923 by Ralf Sampson, the Astronomer Royal for Scotland, who used a photocell to measure the density variation along a photographic spectrum.

In all of these methods, account had to be taken of the spectral sensitivity of the eye, photographic plates, thermocouple, or whatever other detector was used, before the true wavelength of maximum energy could be determined. The determination of these correction factors was a very long and complicated process, and two more simple methods were used, which were found to give results that correlated quite well with those produced by more rigorous methods. They also allowed estimates to be made on dimmer stars.

One of these simpler methods involved putting a coarse wire grating in front of the objective of the telescope. This produced two spectra on either side of the main image in the focal plane of the telescope, with the blue ends of both spectra nearest to the central image. A photographic plate was used to record the spectra, and the distance on the plate between the maximum of both spectra was measured to estimate what was called the "effective wavelength" of the star. The other method involved estimating the colour index.

Photographic plates and photoelectric devices were manufactured with an ever increasing range of spectral sensitivities in the first half of the century, and after the Second World War it became essential to standardise on the spectral sensitivity of detectors when determining stellar magnitudes. The ISU had agreed in 1910 (see Page 288) that the visual and photographic magnitudes of class A0 stars should be the same between magnitudes 5.5 and 6.5, but this definition was too imprecise for the mid twentieth century. The system adopted in the late 1940s, and published in its definitive form by Harold Johnson and W. W. Morgan in the *Astrophysical Journal* of 1953, was the so-called UBV system. In this, ultraviolet (U), blue (B) and visible (V) stellar magnitudes were defined against scales with both clearly defined zero points and a clearly defined set of spectral response curves, with maxima at 360 nm (U), 420 nm (B), and 540 nm (V), the latter being a compromise between the eye's sensitivity in normal and dark-adapted vision. Six A0 main sequence stars, with apparent magnitudes about 6, were specified as having the same U, B and V magnitudes. Ten stars were then defined as primary standards, and 280 as secondary standards.

Following Schwarzschild's proposal (see above), the colour index of stars was generally defined as the excess of the photographic magnitude (using standard photographic plates) over the visual magnitude, and it ranged from -0.3 for bluish-white B0 stars with a surface temperature of 20,000 K, to $+1.6$ for red M0 stars with a surface temperature of 3,000 K. By convention, the zero of the colour scale corresponded to Vega (type A0, temperature 11,000 K). Colour indices based on the difference between the U and B magnitudes, and between the B and V magnitudes were also used.

Stebbins and Whitford extended the UBV system to include spectral sensitivities centred on 700 nm (Red) and 900 nm (Infrared) in the early 1950s, and later bands centred on 1.25 μm (J), 2.2 μm (K), 3.4 μm (L), 5.0 μm (M) and 10.2 μm (N) were added to correspond with the transmission windows of the atmosphere in the infrared. In the 1950s the Earth's atmospheric absorption had been measured very accurately using new photoelectric devices, and the magnitudes of stars, as they would appear outside of the Earth's atmosphere, were known to better than 0.01 magnitude.

Other Tools and Techniques

Introduction

In this section I will outline the development of a number of key techniques and devices that have contributed to our knowledge of astronomy over the last hundred years or so. Constraints of space only allows a brief summary in each case, however, focused on the most important developments; a historical review in this case being impractical. I have, therefore, deliberately chosen examples of techniques which were important at different parts of the twentieth century to retain some sense of balance.

Radiometry

• *Early Work*

Radiometry was still very much in its infancy in the nineteenth century, and the thermopiles in use were only just good enough to enable reasonable estimates to be made of the temperature of the Moon. The Fourth Earl of Rosse measured the radiation coming from the Moon in a 3 year investigation that lasted until 1872, and found that it basically consisted of reflected sunlight at wavelengths below 700 nm, and heat radiation above 1 μm. He used a sheet of water to filter out the latter and enable an estimate to be made of the two different components.

Lord Rosse concluded that the difference in temperature on the Moon between lunar noon and midnight was about 300 K, with the midnight value being not much greater than absolute zero. In 1886 O. Boeddicker, Rosse's assistant, also showed that the lunar surface cooled down very rapidly during a lunar eclipse.

As far as the Sun is concerned, it was impossible to obtain any good estimates of its temperature, because the relationship between the energy radiated by a gas and its temperature was still in the process of being clarified (see Page 5).

• *The Early Twentieth Century*

The bolometer had been invented by Samuel Langley in 1880, but in the early twentieth century much more sensitive detectors like thermopiles and

Crookes' radiometer were developed. Seth B. Nicholson and E. Pettit found in 1927, using a radiometer at Mount Wilson, that the lunar temperature fell from 340 K to 150 K during a lunar eclipse. In the same year, Donald Menzel of the Lick Observatory concluded, after analysing W. W. Coblentz and Carl Lampland's measurements at the Lowell Observatory, that the temperature of Mars varied from about 170 K at dawn to about 280 K at noon, which compared with the Moon and other planets as follows:

	Maximum (K)	Minimum (K)
Mercury	670	
Venus	330	250
Moon	340	100
Mars	280	170
Jupiter	140	
Saturn	120	
Uranus	70	

The higher minimum temperature on Mars than on the Moon, in spite of it being further from the Sun, was considered to be either the result of the atmosphere on Mars (there being no atmosphere on the Moon), or of a surface on Mars which retained heat better than the Moon, or both. The relatively small temperature range on Venus indicated that, if the solid body was rotating very slowly on its axis (i.e. once per Venusian year), the clouds must be moving much faster. The low temperatures observed for Jupiter and Saturn indicated that most of the heat must come from the Sun and not, as previously thought, from the retained heat at formation still in their interiors.

Measuring the colour index of a star was considered to be a quick way of enabling its temperature to be determined, although this index was somewhat unreliable for very red stars, as the photographic plates were insensitive to infrared light, which was an important part of these stars' spectra. For these very red stars, therefore, a heat index was devised in the 1920s, that was the difference between the visual and radiometric magnitudes of stars, with the zero of the scale (like the zero of the colour index) being for a type A0 star.

• *Cavity Radiometers*

The radiometers that were available at the start of the space age, were not accurate enough to measure small variations in the heat received even from a source as powerful as the Sun. The main problem was with the detectors, mainly thermopiles, which had hardly changed in basic design for decades, although cavity detectors, which were much more accurate and stable, started to become available in the early 1960s.

Cavity detectors consisted of an insulated black cavity that was kept in thermal equilibrium. The cavity had two apertures, one for the incident radiation of unknown energy, and one to allow radiation to leave the cavity.

The cavity was maintained at a constant temperature by a heater, and so the unknown energy of the incident radiation could be determined by subtracting the energy supplied by the heater from the energy radiated from the cavity. The latter being a function of its temperature, which was known and fixed.

Joseph Plamondon and Richard Willson of JPL decided, in 1964, to design a radiometer based on a cavity detector, to improve the heat transfer measurements made during pre-flight simulation testing of spacecraft (see Page 19). James Kendall joined the team in 1965, and by 1970 his laboratory instruments were so accurate that they were adopted as international standards. Richard Willson, in the meantime, had been adapting Kendall's instruments for use outside of the laboratory and, in 1968, he flew two of his instruments on a high altitude balloon. By 1973, he had achieved measurement accuracies of $\pm$ 0.2%, and a side-by-side comparison with a thermopile radiometer on a sounding rocket 3 years later showed the superiority of the cavity radiometer design. In future, cavity radiometers would be chosen for space flights.

Polarimetry

Barnard Lyot of the Meudon Observatory near Paris had, in 1924, improved the instrumental measurements of polarisation sufficiently to be able to analyse the change in polarisation of light with solar incident angle for the Moon and planets to an accuracy of about 0.2%. Over the next few years he compared the shape of these curves with those for materials that he measured on Earth, and concluded that the reflections from the Moon and Mercury indicated that they were both covered in volcanic ash, whereas for Venus it appeared as if some of the light was being scattered by a haze layer over the top of the clouds. The measurements for Mars were consistent with a sandy surface being disturbed from time to time to produce large dust storms.

Lyot's early work on polarimetry was carried out using various equipment arrangements but he always used the eye as a detector. In 1923, however, he attempted to measure polarisation photoelectrically, but only met with limited success. Measurement accuracies gradually improved, and in 1943 Yngve Öhman of the Stockholm Observatory was able to measure the polarisation of parts of the lunar surface, using a photoelectric polarimeter that had a photomultiplier as a detector.

The Michelson Interferometer

The French physicist, Armand Fizeau explained in 1868 how it was possible, in principle, to measure the diameter of stars or planetary moons by observing the disappearance of interference fringes seen in the focal plane of the telescope, as two apertures, placed in front of the telescope, are gradually moved apart. Experiments in the 1870s to measure the diameters of stars were unsuccessful, but in 1891 Michelson was able to measure the diameters of the four large moons of Jupiter from the Lick Observatory using what is now called a Michelson stellar interferometer, based on this principle. Michelson made no attempt to measure the diameters of stars, however, as they were presumed to subtend too small an angle at the Earth, requiring an impossibly large separation between the apertures.

At about the turn of the century, J. E. Gore estimated that Arcturus may have a diameter about the size of the orbit of Venus. In 1905 Ejnar Hertzsprung suggested that some red stars were giants (see Page 129), and over the next few years further research by Hertzsprung and Henry Norris Russell seemed to confirm this. If some stars had these sorts of dimensions, then a Michelson interferometer of feasible proportions may be able to measure them.

George Ellery Hale started to consider the possible use of an interferometer to measure the separation of double stars when the 100 inch (2.5 m) reflector was nearing completion on Mount Wilson in 1917. Michelson was interested in designing a suitable interferometer, and such a device was eventually fitted to the 100 inch reflector by Michelson and F. G. Pease in 1920. It consisted of a 6 m steel girder placed across the open end of the 100 inch telescope, with two small mirrors (instead of apertures), whose separation could be varied from 2 to 5.5 m, mounted near each end of the girder. Light from the target star was reflected off these movable mirrors, along the girder, to two fixed mirrors, which were situated a little to either side of the optical axis of the telescope. These fixed mirrors then reflected the light downwards into the telescope. On 13th December 1920, after some earlier work with the stars Algol and Bettatrix, F. G. Pease and John Anderson first observed the interference fringes disappear for Betelgeux at a mirror separation of 2.5 m, indicating a star diameter of 0.047 arcsec (indicating a stellar diameter of about 300 times the diameter of the Sun).

The Coronograph

The solar corona had been shown to be self-luminous in 1869 when the Americans Charles Young and William Harkness discovered a bright green spectral line in its spectrum (see Page 5) during a total solar eclipse, but in 1871 the corona had also been shown to scatter sunlight. The shape of the corona was also found to vary with the phase of the sunspot cycle.

The biggest problem in investigating the corona was that it could only be seen during a total solar eclipse, because it was so faint compared with the Sun. A number of astronomers tried, in the late nineteenth century, to photograph the corona in full sunshine with various coloured filters, but with no success. Hale then tried in 1893 to isolate the corona by observing it in the light of the bright green coronal line only, but this failed also.

It had been realised by these early experimenters that the scattering of sunlight in the Earth's atmosphere was a key problem in trying to observe the corona in full sunshine, so they had taken their equipment to high altitudes to reduce his effect. It was not until about 1930, however, that the various problems to be overcome were systematically assessed and solved by Bernard Lyot, of the Meudon astrophysical observatory in France.

Lyot started his investigations with an apparently top quality lens, but even this produced a halo around the Sun over a hundred times brighter than the solar corona. He then had another lens made from a specially selected disc of glass, but this also produced a halo caused, he concluded, by defects in the lens (tiny air bubbles in the glass, faint marks on the surface caused by polishing during manufacture, and variations in refractive index), diffraction by the edge of the lens, and dust on the lens and in the system.

Having established the causes of the problems, Lyot then set out to minimise them. He minimised atmospheric scattering by observing at the high altitude observatory on the Pic du Midi, at an altitude of about 2,900 m (9,400 ft), and had his telescope lenses made very carefully from bubble-free glass. He eliminated the diffraction of the edge of his objective by focusing it on a diaphragm that was slightly smaller than the image of the objective, and put grease on the inside of the telescope tube that projected beyond the objective, to trap dust particles. As a result of these and other measures, Lyot was able in 1931 to take the first photographs of the solar corona in full sunshine; the best results being achieved after a heavy snowfall had cleared the air of dust and other particles.

Charge Coupled Devices

Charge coupled devices, or CCDs, have revolutionised astronomical imaging in recent years, and are now used by amateur astronomers as well as professionals. They have a very high photon detection efficiency,* and produce digital signals that readily lend themselves to image processing by computers.

The first CCDs were developed in 1969 by Willard S. Boyle and George E. Smith at the Bell Telephone Laboratories in New Jersey, where the cosmic microwave background radiation had been detected 4 years earlier (see Page 256). Boyle and Smith announced their invention in April 1970, and within 3 years a 100 × 100 pixel array was being produced.

A CCD is basically an array of semiconductors etched onto a silicon wafer that converts incoming photons of light into electrons. These electrons are prevented from escaping from the silicon by a coating of silicon dioxide and, as each pixel is isolated from adjacent pixels, a voltage gradually builds up in each pixel that is proportional to the amount of incident light. These voltages are measured at the end of the exposure, and the image recon-

---

*The photon detection efficiency (PDE), or quantum efficiency, is the percentage of incident photons of light that are counted. For CCDs today, the PDE of a cooled detector can be as high as 80%, compared with only 2% or 3% for very high speed modern photographic emulsions. When the 200 inch (5 m) Palomar telescope went into operation in 1949 the fastest photographic emulsions had a PDE of only 0.3%!

structed on a computer screen. As the images are produced in digital form they can be stored on magnetic tapes, floppy discs or laser discs.

Between 1973 and 1977, Texas Instruments (TI) developed CCD arrays of 100 × 160 and 400 × 400 pixels, with the aim of producing an 800 × 800 pixel device. In 1973 they started working with JPL to produce a large-area CCD for space astronomy and, in 1976, the first astronomical image was taken by Bradford Smith of the University of Arizona, James Janesik of JPL, and Larry Hoveland. They imaged Jupiter, Saturn and Uranus using a prototype TI array on the 61 inch (1.5 m) telescope of Mount Biglow, near Tucson, Arizona. When Smith and his colleagues looked at the image of Uranus they were surprised to find, for the first time, evidence of particles high above the methane layer in its atmosphere. This clearly proved that CCDs were here to stay.

All the early TI devices were earmarked for organisations connected with NASA's Galileo programme to Jupiter, which caused frustration and some resentment among the remainder of the astronomical community. The company Fairchild Semiconductor started to sell a 100 × 100 pixel CCD in the mid 1970s, but this was of an inferior design, so when RCA started producing a 512 × 320 pixel CCD array in the late 1970s, astronomers breathed a big sigh of relief. Unfortunately, RCA withdrew from the market in 1985 but, in the meantime, GEC in the UK, and Thomson–CSF in France, had started to produce devices.

Today a number of other companies have started manufacturing CCDs for both the specialist and mass markets, ranging from research-grade devices to detectors for home video cameras. Not only have pixel numbers and physical array sizes increased over the last 20 years, with the largest array currently available being 4096 × 4096 pixels, but noise levels have reduced by a factor of 50. The detailed construction of the devices has also improved to extend their spectral sensitivity into both the ultraviolet and infrared spectral bands, which are of great interest to astronomers.

15 Radio Astronomy

Early Radio Astronomy

Jansky and Reber

Karl Jansky, of the Bell Telephone Laboratories, was studying the cause of interference at a frequency of 20.5 MHz (14.5 m wavelength)* on the newly-opened transatlantic radio link, when he discovered in 1932 that there was a directionality in the hiss coming from the sky. At first he thought that the signals must be coming from the Sun, but he soon realised that they peaked once every sidereal day (of 23 h 56 min), rather than once every solar day (of 24 hours), so the Sun could not be the source. Jansky spent a year observing and analysing the signals, and deduced that they were coming from the Milky Way at about 18 hours right ascension, 10° declination, the declination estimate having a large uncertainty. He assumed that the signals had a non-stellar origin in the Milky Way, as he could detect no radio waves coming from the nearest star, i.e. the Sun, but he could not be certain. This led him to suggest two possible origins, either the centre of the Milky Way, at approximately 17 h 30 min R.A., −30° Dec. (i.e. in Sagittarius), or the direction of space towards which the Sun is moving, at 18 h R.A., +30° Dec. (i.e. in Hercules).

Jansky first published his findings in the December 1932 issue of *The Proceedings of the Institute of Radio Engineers*, and presented them in April 1933 to a meeting of the International Union for Radio Science. His discovery was reported by the *New York Times* of 14th May 1933. He had

*For electromagnetic waves, whether they be visible light or radio waves, $\lambda\nu = c$, where λ is the wavelength and ν the frequency of the waves, and c is the velocity of light, i.e. 3×10^8 m/s. So if $\nu = 20.5$ MHz (i.e. 20.5×10^6/s), for example, $\lambda = 3 \times 10^8/(20.5 \times 10^6) = 14.5$ m.

another paper published by the Institute of Radio Engineers in October 1933, and he presented a paper at a meeting in Detroit in July 1935, but only about 20 people attended. His results were largely ignored by the astronomical community as they seemed of no great significance and, as astronomers were all optically-oriented at the time, they regarded radio technology with some suspicion. Jansky proposed building a dish-type radio telescope to further examine the radio signals, but he was asked by Bell Labs. to move on to other work, thus ending the first all-too-brief skirmish with radio astronomy.

Grote Reber, a radio engineer, read Jansky's papers, however, and decided to try to detect the signals at higher frequencies, where they should be more intense if they followed the black body radiation curve. His first efforts at 3,300 MHz (3.3 GHz) and 910 MHz (wavelengths of 9 and 33 cm respectively) were unsuccessful, indicating that the source was not radiating as a black body, but in October 1938 he finally detected the signals at 160 MHz (1.9 m wavelength).

Reber then spent six years mapping this sky noise at 160 and 480 MHz, clearly showing the outline of the Milky Way together with areas of higher intensity. The most intense signals were coming from the centre of the Milky Way (one of Jansky's suggested sources), but the other bright areas did not coincide with any bright stars, galaxies or nebulae. Two of Reber's sources have now been recognised as Cassiopeia A and Cygnus A (see below). Reber suggested that the radio waves could be caused by bremsstrahlung ("braking radiation") caused by the deceleration of electrons in the centre and spiral arms of the Milky Way, while the German Karl Kiepenheuer suggested, in 1950, that it could be synchrotron radiation caused by electrons spiralling in the galactic magnetic field. Albrecht Unsöld, Martin Ryle, Jan Oort and Gary Westerhout, on the other hand, favoured the theory of radio-bright, optically-dim stars.

The Sun

Thomas Edison, the American electrical engineer and inventor, had made an attempt as long ago as 1890 to detect radio signals from the Sun, but without success. The actual discovery of radio signals from the Sun was made during the Second World War in peculiar circumstances.

On 12th February 1942 the German Army had managed to jam the British coastal defence radars and allow the two battle cruisers "Scharnhorst" and "Gneisenau" to escape from Brest to Kiel. As a result, an urgent investigation was ordered in the UK of the jamming of British radars and, 2 weeks later, on 27th and 28th February, British Army radar sets picked up interference on the wavelength of 4.2 m used by their anti-aircraft radars. This was initially attributed to another case of jamming by the Germans but no aircraft attack materialised, and then James Hey of the Army Operational Research Group, who was ordered to investigate, realised that the Sun was the source. The interference had peaked on 28th February and, at the same time, a large solar flare had been seen associated with a large sunspot that was then on the Sun's meridian. Four months later George Southworth, working at the Bell laboratories, independently discovered radio emissions from the Sun, and Reber also detected them in the following year.

In October 1945 Joe Pawsey, Ruby Payne-Scott and Lindsay McCready, at Sydney, discovered that the average intensity of 1.5 m radio emissions from the Sun correlated with total sunspot area, clearly linking radio emissions with sunspots. Pawsey et al. in the following year then deduced, from radio measurements of the quiet Sun, that the solar corona had a temperature of the order of 1 million K, which was a similar order of magnitude to that deduced from coronal spectra in 1941 (see Page 10). Edward Bowen and Joe Pawsey at Sydney, and Martin Ryle at Cambridge University, developed Michelson-type radio interferometers to allow more accurate studies of the Sun. This enabled Ryle, with the help of a research student D. D. Vonberg, to show that radio bursts from the Sun came from discrete areas on the Sun's disc, and not from the disc as a whole.

Early Radio Sources

While this solar work was being undertaken, research was also beginning to unravel the sources of Jansky's and Reber's radio waves. The English radio astronomers James Hey, S. J. Parsons and J. W. Phillips, using the antenna of an anti-aircraft radar, accidentally discovered in 1946 that radio signals coming from a region in the constellation of Cygnus were fluctuating over a period of only a few seconds. Martin Ryle and F. Graham Smith at Cambridge then decided to use their radio interferometer, that had been constructed originally to examine radio burst from the Sun, to investigate this fluctuating source in Cygnus, now called Cygnus A. On their first night of observations in 1948 they picked up signals from Cygnus A but, 3 hours away in R.A., they also found another powerful source in Cassiopeia. They had re-discovered Cassiopeia A which had been originally found some years earlier by Reber.

In parallel with this work, John Bolton and G. J. Stanley of the Australian Radiophysics Laboratory at Sydney showed, using the cliff-top interferometer, that Cyg A was small, being less than 8 arcmin in size. In the process, they also discovered three other radio sources, namely Taurus A, Centaurus A and Virgo A. (Incidentally, the fluctuations from Cyg A were shown in 1950 to be caused by the Earth's ionosphere.) In the following year, Bolton, Stanley and O. B. Slee succeeded in correlating Tau A with the Crab Nebula (the remains of a supernova that exploded in 1054), Cen A with the galaxy NGC 5128, and Vir A with the elliptical galaxy M87. So the three radio sources, other than the Sun, that had been identified with optical sources turned out to be two galaxies, plus what is now called a supernova remnant. In general, however, the positional accuracies, typically 1°, estimated for all but the very brightest sources, were too large to make correlations with visible objects with any confidence. In fact, the positive identification of Cen A and Vir A was not accepted at first by a number of astronomers who were convinced that the radio sources were all optically dim radio stars in the Milky Way.

Calculations showed that if the Andromeda galaxy, the nearest spiral galaxy to the Milky Way, was emitting radio signals with the same general intensity as the Milky Way, they should just be detectable with the 218 ft (65 m) transit radio telescope at Jodrell Bank. So Robert Hanbury Brown and his research student Cyril Hazard decided to try to detect these signals, operating the telescope at a frequency of 158.5 MHz (i.e. a wavelength of 1.89 m). They found faint radio signals coming from the Andromeda galaxy, and surveyed the galaxy over a total of 90 nights in the autumn of 1950, gradually building up a contour map showing the intensity of the signals. Although the radio energy was found to be much weaker than that in visible light, these and subsequent investigations showed that it comes from a region ten times larger than the optical image of the galaxy.

The discovery that a galaxy, other than the Milky Way, was clearly emitting radio waves did not solve the problem of whether radio sources were galaxies or radio stars within the Milky Way. The radio emissions from the Andromeda galaxy were of a similar intensity to those coming generally from the Milky Way, so if radio sources were mostly outside of the Milky Way, they must be emitting very much more energy than the Milky Way and the Andromeda galaxy. So when Hanbury Brown and Ryle presented their analyses of radio sources to the Royal Astronomical Society in 1951 there was an almost unanimous agreement that they were unseen radio stars within the Milky Way.

Cen A and Vir A had, however, been identified as the galaxies NGC 5128 and M87, and then in 1951 Graham Smith at Cambridge managed to reduce the positional errors of bright radio sources to 1 arcmin, enabling Walter Baade and Rudolph Minkowski to use the 200 inch Palomar reflector to find the sources of Cas A and Cyg A. The first was identified as a faint dispersed nebula which has fast-moving gas filaments, but the second, the second strongest radio source in the sky, was attributed to an inconspicuous 16th magnitude galaxy! These and similar discoveries over the next few years finally showed that many of the radio sources were galaxies outside of the Milky Way.

The number of radio sources increased steadily in the 1940s, so that when Martin Ryle, F. Graham Smith and Bruce Elsmore produced the first Cambridge (or I C) Catalogue in 1950, they were able to list 50 northern sources, while Stanley and Slee of Sydney listed 22 southern sources in the same year. The second Cambridge Catalogue was published in 1955 listing 1,936 sources, but many of these were later found to be spurious, and when the third Cambridge Catalogue was issued in 1962 it contained only a quarter of the sources of the second catalogue.

The 21 cm Line

In 1944 Jan Oort had suggested to H. C. Van de Hulst in the Netherlands that he should examine gas spectra theoretically, to see if there was a spectral line at radio wavelengths that could be used to map the structure of the Milky Way. Van de Hulst found that the hydrogen atom has one state in which the spins of the proton and electron are parallel, and one in which they are anti-parallel. The transition from one state to the other would generate energy with a wavelength of 21.2 cm. Although the transition probability was low, about one transition per atom per 10 million years, there should be enough interstellar hydrogen in the Milky Way for it to be detectable.* Oort, C. A. Muller and van de Hulst set out to observe this emission, but Harold Ewen and Edward Purcell of Harvard were the first to detect it in 1951 when van de Hulst was spending a sabbatical at Harvard. W. H. Christiansen and J. V. Hindman confirmed the 21 cm emission just 2 months later in Australia, and Muller and Oort also succeeded in the Netherlands, following technical advice relayed by van de Hulst from Harvard.

In its early days, radio astronomy was very much separated from optical astronomy, as the techniques and vocabularies of both were so different that the two sets of astronomers found it difficult to understand each other. In the 1950s this began to change as the resolutions of radio telescopes were dramatically improved, with the result that more and more radio sources could be linked with objects visible at optical wavelengths. It is therefore at this stage that I leave this outline of radio astronomy as a separate subject. I have integrated more recent astronomical discoveries made with radio telescopes, for example the discovery of the microwave background by Arno Penzias and Robert Wilson in 1965, and of pulsars by Jocelyn Bell and Anthony Hewish in 1967, within the general text above.

Radio Telescopes

Early Dish and Wire Antennae

The first radio telescope was the 95 ft (30 m) long antenna of the American radio engineer Karl Jansky, that operated in 1932 at a frequency of 20.5 MHz and rotated horizontally every 20 minutes. Nicknamed "the merry-go-round", the array was made from brass pipes and timber, and was mounted on the wheels taken from an old model T Ford. The array was not designed specifically to look for radio sources in the universe, having detected them by accident, and the first antenna designed specifically for this was the 31 ft (10 m) diameter parabolic dish (see Figure 15.1) built by Grote Reber in 1937, with the help of two assistants, in the back garden of his house in Wheaton, Illinois. Reber's dish, which he operated at frequencies from 160 to 3,300 MHz, could be moved in declination (up and down) but not in right

*At almost the same time, Iosif Shklovskii had independently come to a similar conclusion in the USSR.

Figure 15.1 *Grote Reber and his 31 ft (10 m) diameter radio telescope, re-erected at the National Radio Astronomy Observatory. (Courtesy* NRAO/AUI.*)*

ascension. It was, therefore, used like a transit telescope, relying on the rotation of the Earth to traverse the sky in right ascension.

Optical reflecting telescopes have mirrors accurate to about 1/10th the wavelength of light. Similarly, radio telescopes operating at frequencies of 3,000 MHz, for example, which corresponds to a wavelength of 10 cm, need dishes with surfaces accurate to about 1 cm. If the frequency is 30 MHz, the wavelength is 10 m, and the dish accuracy required is only about 1 m. So, although dishes like Reber's are needed to observe sources at radio wavelengths in the centimetre range, they are not required at wavelengths of tens of metres, and wire arrays can be used instead.

Radar and radio technology improved immensely during the war, and at the end of hostilities there was an enormous amount of surplus equipment, and plenty of brilliant young scientists and engineers to use and develop it for the benefit of what was to become the science of radio astronomy. The main centres of excellence in this new field were led by Joe Pawsey in Australia, Jan Oort and van de Hulst in the Netherlands, Martin Ryle at Cambridge, and Bernard Lovell at Manchester. These eventually led, for example, to the construction and completion of the Jodrell Bank 250 ft (75 m) diameter steerable radio telescope near Manchester in 1957 (see

Figure 15.2 The 250 ft (75 m) diameter Jodrell Bank radio telescope nearing completion in early 1957. (Courtesy Aerofilms Ltd.*.)*

Figure 15.2), and the Parkes 210 ft (65 m) steerable radio telescope in Australia in 1961. The Jodrell Bank radio telescope was originally designed to operate at metre wavelengths, and so was to be built with a wire mesh dish. During construction, however, the 21 cm hydrogen line was discovered, and so the dish was built with a solid surface instead. Just after it was complete, by a fortunate accident of timing, the USSR launched their first Earth satellite, Sputnik 1, and the Jodrell Bank telescope was immediately pressed into service to track the satellite and its carrier rocket. This generated a tremendous amount of public interest, but it also showed the military, on both sides of the Atlantic, that such equipment was essential to give early warning of the launch of Soviet intercontinental ballistic missiles, giving a military dimension to future technology developments.

Early Interferometers

The problem with the early radio telescopes was their poor resolution, because radio waves are so much longer than visible light and the resolution of a telescope is inversely proportional to wavelength. In 1947 Lindsay McCready, Joe Pawsey and Ruby Payne-Scott designed an interferometer to solve the problem by placing the antenna on a high cliff near Sydney, Australia; the position of the source being determined by measuring the interference pattern between the direct beam received from the source, and that reflected from the sea. Martin Ryle and F. Graham Smith, at Cambridge, used a different type of interferometer in which they combined the signals from two 27 ft (8 m) dishes separated by about 1,000 ft (300 m). Then, in 1970, the Westerbork Synthesis Radio Telescope was completed in the Netherlands consisting of 12 equatorially-mounted 82 ft (25 m) parabolic dishes on a 1 mile (1.6 km) long baseline.

All of these interferometer systems depend on the fact that a radio telescope of large aperture can be synthesised, from a resolution point-of-view, by two* or more dishes linked together electronically, in an analogous way to the use of two mirrors of a Michelson interferometer (see Pages 306–307). The resolution of the interferometer is determined by the separation of the dishes, the larger the separation, the better the resolution along the line joining the dishes (and only along that line), and the intensity of the received signal is given by the total area of the dishes.

A similar type of device, using the technique of "unfilled apertures", was the Mills Cross built near Sydney, Australia, by Bernard Mills and A. G. Little in the mid 1950s. This radio telescope consisted of two 1,500 ft (450 m) linear arrays of wires arranged at right angles to each other. It operated at a frequency of 85 MHz, and while it produced a high resolution in two directions at once, the intensity of the received signal was low and the beam-width was narrow, being only 50 arcmin across.

Clearly interferometers and such arrays were the only practical way to resolve radio sources, but large dishes were still required to detect the signals from faint sources and to provide radio spectroscopy.

Later Dish Antennae

The American National Radio Astronomy Observatory (NRAO) was founded in 1954 at Green Bank, West Virginia, and was provided, in 1959, with its first operational radio telescope, funded by the National Science Foundation. This was the 85 ft (25 m) Howard Tatel telescope, that could be used at wavelengths as short as 2 cm. In 1962, a 300 ft (90 m) transit telescope was commissioned, and in 1965 a 140 ft (45 m) fully steerable dish was completed at a cost of $14 million. The 300 ft transit telescope dish was covered with wire mesh, making it simpler and cheaper to build than the 140 ft dish, which had a continuous surface, but this meant that the 300 ft dish could only operate at wavelengths above 20 cm, whereas the fully steerable dish could operate down to 2 cm. The NRAO also built a 36 ft (10 m) millimetre-wave telescope on Kitt Peak, because high altitude is important for wavelengths below about 1 cm, where the atmosphere is no longer transparent.

Deer Creek Valley, near Green Bank was chosen as the main site for the NRAO because it is surrounded by mountains, which help to shield the radio telescopes from man-made radio interference, and because there are

---

*In the cliff interferometer, the sea reflection plays the same rôle as the second dish.

few inhabitants and little industry in the area to cause electrical interference. The American Government also purchased an area of 4 square miles (10 square km) around the antennae, and put into effect special measures to keep the area as free from electrical interference as possible in the future.

Recently, on the evening of 14th November 1988, disaster struck at Green Bank, when the 300 ft transit telescope collapsed, shortly after a group of astronomers had posed beneath it for a photograph, depriving the observatory of its largest dish, and the northern hemisphere of one of its major radio telescopes. Fortunately, the United States Congress rapidly authorised funding for a replacement.

In the 1950s a group of American astronomers decided to try to measure radar signals reflected from electrons in the ionosphere, but were disappointed to find that they would need a 1,000 ft (300 m) dish to stand any chance of picking up the very faint signals. The DOD (Department of Defense) discovered, however, that radar could be used to detect signals from the ionisation trails left by artificial satellites, and so they agreed to pay the cost of a large dish facility. A site was found at Arecibo in Puerto Rico, where a fixed dish could be built in a natural depression, and Cornell University were contracted in 1960 by the DOD to design and build the facility. As the dish was nearing completion in 1962, it was discovered theoretically that the radar reflections that the astronomers wanted to detect in the ionosphere should be detectable with a dish 1/10th the diameter of the dish then under construction. The 1,000 ft (300 m) dish was completed the following year, nevertheless, and has been a resounding success ever since as a high-quality radio telescope.

Originally the Arecibo dish was made of wire mesh, because it was designed to operate at wavelengths longer than 50 cm, but this mesh was replaced in the early 1970s by 38,000 adjustable panels, enabling it to be used at wavelengths down to 6 cm.

The first plans for carrying out radio observations at Bonn Observatory had been produced in 1951, leading 4 years later to the commissioning of an 82 ft (25 m) diameter radio telescope, and the founding of the Max Planck Institute for Radio Astronomy. Then, in 1968, construction was started at Effelsberg, near Bonn, of an enormous 330 ft (100 m) diameter dish, which is still the largest fully-steerable dish available anywhere in the world. The structural engineers realised that it was impossible to design a rigid dish that remained parabolic within the 3 mm required by the specification for the edge of the dish, to enable it to operate at wavelengths as short as 8 mm. So they came up with a design that allowed the antenna to distort by up to 75 mm, provided the surface remained a paraboloid with deviations of less than 3 mm. The 75 mm distortion was then compensated by moving the receiving antenna at the focus of the dish.

Large Interferometers

Construction of the 27–dish Very Large Array (VLA) located near Socorro, New Mexico, was started in 1974 and, within 3 years, the first scientific results were being produced when the first six dishes were operational. The complete array, consisting of twenty-seven 82 ft (25 m) parabolic dishes, was arranged in a Y-shaped pattern, with each of the three arms of the Y being about 20 km long. The array was completed in 1981, operating at wavelengths between 1.3 and 22 cm, and achieving a resolution equal to a radio telescope dish of 27 km (17 miles) diameter. This was the same resolution as that of the best ground-based optical telescopes of the time.

Experiments with long baseline interferometers, where the signals are combined from radio telescopes many kilometres apart, were first carried out in the 1950s and 1960s. In 1965 radio telescopes at Jodrell Bank and Defford, which are separated by 130 km, were joined by a microwave link, providing a maximum resolution of 0.02 arcsec at a wavelength of 1.3 cm. It was found difficult to preserve the phases of the signals, however, using coaxial cable or microwave links over much larger distances, and in 1967 a new technique was developed using atomic clocks. The signals from each radio telescope were recorded at the same time on their own wide-band magnetic tape recorder, alongside the signals from the atomic clock, and were then combined later, using the atomic clocks to provide precise phasing information.

This new interferometer technique was developed in parallel by a team of radio astronomers in both Canada and the United States. The Canadians used radio telescopes 3,000 km apart, operating at a wavelength of 67 cm, and giving a maximum resolution of 0.06 arcsec. The Americans used radio telescopes at Haystack Hill, in Massachusetts, and at the NRAO in West Virginia, which are 850 km apart, working at a wavelength of 18 cm, and giving a maximum resolution of 0.05 arcsec. Today, using telescopes separated by almost the diameter of the Earth, and using very short wavelengths, resolutions have been achieved of better than one thousandth of an arc second.

Thus in less than half a century, radio telescopes have gone from resolutions of the order of a degree (in 1948) to resolutions of the order of one thousandth of an arc second today, an improvement of more than a million times.

16 Space Research

Introduction

Sounding rockets and space satellites are, in principle, the same as optical and radio telescopes, being simply tools that enable astronomers to understand the universe. In a book on astronomy the results are more important than the tools, and so, in this chapter, I have outlined only some of the sounding rocket and spacecraft missions, to give a flavour of their contributions to astronomy. Their more detailed contributions are included in the general text above.

Results from Early Sounding Rockets

The father of modern rocketry was Konstantin Tsiolkovskii, a physics teacher at a school in Kaluga, Russia, who started to develop his ideas in the 1890s. In 1902 he published his first paper, in which he explained the advantages of using fuels like hydrogen and kerosene in rockets, instead of gunpowder. He also explained, in one of his later papers, the advantage of using multistage rockets.

Tsiolkovskii was a theoretician who was virtually unknown in the West, and the first practical step to make a rocket was taken by Robert Goddard, an American born in Worcester, Massachusetts, in 1882. Goddard was appointed a professor of physics in 1914 at Clark University in his home town, where he became interested in the concept of using rockets for upper atmospheric research. His initial work which was published in 1919 concentrated on solid-fuel rockets, but he then turned his attention to liquid fuel designs. In 1926 Goddard launched the first liquid-fuelled rocket to usher in the modern era of rocketry.

The third independent researcher into rocketry was Hermann Oberth, a

Romanian-born schoolteacher working in Germany, who published a small book on rocketry in 1923. The first edition sold out rapidly, as did a reprint, in spite of criticism from a number of professional scientists and engineers. Undeterred by this professional scepticism, however, Oberth founded the German Society for Space Travel, one of whose members was Wernher von Braun, who began firing their own liquid fuel rockets in 1930. The German army rapidly became interested in developments, and von Braun was made a consultant to their new rocket development team under Walter Dornberger in 1932, which was originally located at Kummersdorf, outside Berlin. The facilities were moved to Peenemünde on the Baltic in 1936, and there, over the next few years, the V2 rocket was developed, using alcohol and liquid oxygen as propellants. It was 15 m (48 ft) long and 1.7 m (5.4 ft) in diameter, had a top speed of 5,500 km/h (3,500 miles/h), and reached a height of over 100 km (60 miles) before it started to descend towards its target. Over 4,000 V2s were launched against England, towards the end of the Second World War, by a progressively more desperate Hitler.

Just before the end of the war, the American and British governments decided to launch "Operation Paperclip", to remove from Peenemünde and the underground V2 factory at Nordhausen as much hardware as possible before the Russians occupied the area. Most of the technical staff from Peenemünde, including von Braun, had fled to Oberjoch, in the Austrian Alps, to escape from the SS and Russians, and at the end of the war 130 Germans were brought back to the USA in order to assemble and test the captured V2s.

The captured V2s were tested at the White Sands Missile Range in New Mexico, with various scientific instruments replacing the normal 900 kg payload of explosives. Their normal maximum altitude as a sounding rocket was 150 km, although 180 km was reached on one occasion. The early work was very much involved in getting used to handling the missile and in developing stabilisation and payload ejection systems. Nevertheless, Richard Tousey, who was head of the Optics Branch of the Naval Research Laboratory (NRL), obtained the first ultraviolet spectrogram of the Sun below 310 nm, where the atmosphere is opaque, on 10th October 1946. Unfortunately, the spectrograms cut-off at 210 nm and, with their limited resolution of 0.2 nm, they provided little useful information, but it was a start, even making the front page of the Washington Post. In late 1947, James Van Allen, who was later to become famous for his discovery of the radiation belts around the Earth, became chairman of the V-2 scientific panel, and then in February 1949 Tousey detected solar X-rays at wavelengths below 1.0 nm.

Tousey relied on retrieving the instruments on their return to Earth to obtain his results, as they were in hard copy. Herbert Friedman, on the other hand, who was head of the Electron Optics Branch at NRL, decided to use electronic methods of detection (photon counters), and the vehicle's telemetry system to relay data back to Earth. Friedman flew his first such experiment on V-2 number 49 on 29th September 1949, and detected solar radiation in the Lyman-α region of hydrogen (wavelength 121.6 nm) above a height of 75 km, and soft (0.8 nm) X-rays above 87 km. So the Lyman-α radiation was identified as the source that produced ionisation in the D-layer of the Earth's ionosphere, and the soft X-rays were similarly identified as the source of the E-layer. Although the role of Lyman-α radiation in forming the D-layer had been largely predicted, the role of soft X-rays in producing the E-layer was less expected, even though it had been predicted independently in the 1930s by Edward Hulburt of the NRL and by the Norwegian Lars Vegard. A more sophisticated experiment failed in the following year when the rocket barely lifted off its launch pad and fell over, engulfing the block-house, in which Friedman and his colleagues were ensconced, in flames. Fortunately no-one was injured.

In 1952, Friedman's group made the first measurements of the absolute intensity of the Lyman-α radiation from the Sun, and in 1955 detected discrete ultraviolet sources in the night sky. William Rensé, of the University of Colorado, produced the first spectrogram to reach down to the solar Lyman-α line on 12th December 1952, using a spectrograph fitted to an Aerobee sounding rocket, and four years later they obtained the first Lyman-α photograph of the Sun.

The V2 was a single stage rocket, as was the WAC Corporal, a small* sounding rocket whose development had been started in 1944 at the Jet Propulsion Laboratory (JPL) of Caltech in Pasadena, California. In February 1949 the first multistage rocket was launched with a V2 as first stage, and WAC Corporal as second stage, reaching an altitude of 390 km and allowing the first in-situ measurements of the electron density of the Earth's F-layer.

The biggest problem with the V2 was that the graphite deflectors in the rocket exhaust eroded unevenly during firing, causing problems with the rocket's stability. The US Navy decided, therefore, to design their own rocket, the Viking, with a gimballed engine which did away with the need for any deflectors. The Viking, which in its initial configuration was of the same length as the V2, but just half of its diameter, became the basis for the first stage of the Vanguard which was intended to launch America's first Earth-orbiting satellite during the International Geophysical Year (IGY), which was to last for 18 months from 1st July 1957 to 31st December 1958. The Free World was, however, about to get a nasty shock.

Sputniks and the Formation of NASA

The world's first artificial satellite, Sputnik 1, was launched by Sergei Korolev and his team on 4th October 1957, as part of the Soviet Union's contribution to the IGY. Although both the Soviet Union and the United States had promised to launch satellites during the IGY, the fact that the Soviets had beaten the technological might of the United States to launch a satellite first was an enormous shock to the West.

---

*It was 4.8 m (16 ft) long, but only 0.3 m (1 ft) in diameter.

The military significance of the launch of Sputnik 1 was obvious, as the Soviet Union had now shown that it could target anywhere in the world with its intercontinental ballistic missiles, but what was the scientific significance of Sputnik 1? When could a satellite be sent to the Moon, and how soon could a man be launched into space?

The USSR had occupied the German testing range at Peenemünde at the end of the war, and had continued to use it to develop and test military rockets. Because the USSR had no allies in the late 1940s that were geographically close to the USA, they had to use intercontinental ballistic missiles to threaten the USA with nuclear warheads, whereas the Americans, having air bases in Europe, could use bombers. So, immediately after the war, the USSR introduced an accelerated programme to develop intercontinental ballistic missiles, while the Americans felt that they could take these developments much more slowly.

In the early 1950s, when the Americans decided to accelerate their military rocket programme because of the perceived Soviet threat, they had much more sophisticated, and lighter, warheads than the USSR, so their long-range missiles did not need to be as large. There was also an organisational difference between the two superpowers, as the Americans kept their military rocket developments, under Wernher von Braun, separate from their civil rocket programme, whereas the USSR civil and military programmes were closely integrated. So the American rockets were smaller than those of the USSR, and the American civil rocket programme was isolated from its military programme, making matters worse.

The Soviet Union had foreseen the propaganda value that the launch of Sputnik would have showing, as it did, that the Soviet Union was now a superpower. It could over-fly the USA at will, and probably launch missiles at American cities. The Americans, on the other hand, saw satellites for the IGY as a purely scientific enterprise, and did not realise their propaganda value until faced with the Soviet *fait accompli*.

The scientific world were generally agreed that it would not be too difficult for the Soviet Union to send a satellite to the Moon, now that they had launched an earth-orbiting satellite, but the possibility of launching men to the Moon and planets was a very different matter. Harold Spencer Jones, for example, who had just retired as Astronomer Royal in England, was convinced that generations would pass before man landed on the Moon, and that man could never visit the planets. The distinguished British physicist, George Thomson, on the other hand, foresaw men landing on the Moon within a few years' time.

If the USSR had surprised the world with Sputnik 1, they astonished it a month later with Sputnik 2 which weighed half a ton (500 kg), six times as much as Sputnik 1, and carried a dog as payload. The Americans had already announced that their satellite, when launched, would weigh about 20 to 30 lbs (9 to 14 kg). So it was not just a question of the Americans launching their satellite, and all would be well, but they were also considerably behind the Soviet Union in the size of the satellites that they could put into orbit.

Stung into action by the success of Sputniks 1 and 2 and taunted by the USSR,* President Eisenhower set up the President's Science Advisory Committee in November 1957. In the following month things got even worse, when the first American attempt to launch a satellite, using a Vanguard launch vehicle, blew up on the launch pad. The President now authorised the use of American military rockets for launching satellites, and on 31st January 1958, Wernher von Braun and his Army team of engineers successfully launched Explorer 1, the first American satellite, with a four stage rocket based on the Jupiter-C intermediate range ballistic missile. Explorer 1, which was designed and built by JPL and weighed just 31 lbs (14 kg), found that the Earth was surrounded by intense radiation belts, now called the Van Allen belts, after the leader of the JPL scientific team that found them. The first successful Vanguard launching took place just over 6 weeks later on 17th March.

The Science Advisory Committee recommended to the President, in the summer of 1958, that a civilian space agency be set up, independent of the military, to manage the American civil programme and, despite his initial resistance to the idea, continuing problems with the American space effort made Eisenhower change his mind. So on 1st October 1958 this new civilian space agency, called the National Aeronautics and Space Administration, or NASA, started work, just before the first anniversary of the launch of Sputnik 1.

The Race to the Moon

America had been humiliated by the Soviet Union with their successful Sputnik launches, and it was beginning to look as though the USSR would easily put the first man into space, with the superior lift-capability of their launchers. The Americans desperately needed to do something to upstage the USSR, so they decided to try to use their more sophisticated technology to send the first spacecraft to the Moon. As a result, Pioneer 1 was launched to the Moon by the USA on 11th October 1958, before any attempt from the USSR, but it only travelled about 30% of the way there, before returning to Earth and burning up in the atmosphere. In the next 2 months, the USA launched two more unsuccessful spacecraft to the Moon, Pioneers 2 and 3, before the Soviet Union humiliated them again by sending Lunik 1 within 6,000 km of the Moon on 4th January 1959.

In the midst of adversity the USA upped the stakes, and announced on 17th December 1958, the 55th anniversary of the first manned flight by the Wright brothers, that they were to undertake a programme called "Project Mercury" to put a man in space. There would be a series of manned sub-orbital test flights using the Army's Redstone short-range ballistic missile,

---

*The Soviet delegation to the United Nations offered to include the USA in the USSR's programme of technical assistance to developing nations.

followed by orbital flights using the Air Force's Atlas intercontinental ballistic missile.

The first successful American spacecraft to the Moon, Pioneer 4, was launched on 3rd March 1959, but it passed the Moon at ten times the distance of Lunik 1, so it appeared to be only partially successful. Lunik 2, the first spacecraft to hit the Moon, was launched by the USSR in September 1959 and, in the following month, Lunik 3 sent back to Earth the first photograph of the far side of the Moon. Although crude by today's standards, this photograph had an enormous impact on the public imagination, as it was the first photograph taken from another world, and was of a part of the Moon that had never been seen before.

The next three moon-shots were all American, but none of them reached the Moon. Then, on 12th April 1961, the USSR put a man in space, with Yuri Gagarin making one orbit before returning safely to Earth. Although a Soviet manned flight had been anticipated in the West as a natural consequence of their space programme, the Soviets, as was their habit, made no pre-launch announcement. So the sudden announcement of the successful mission was another blow to American prestige. The United States, anxious to preserve some of their credibility, launched Alan Shepard on a sub-orbital flight from Cape Canaveral in the following month but, rather than being a triumph, it only went to show how far behind the Americans were. Twenty days later, President Kennedy upped the stakes once more, when he committed the United States to land a man on the Moon, and return him safely to Earth, before the end of the decade.

On 7th August 1961, Gherman Titov completed 17 orbits of the Earth, causing NASA to immediately cancel its sub-orbital flight programme after just two flights, and concentrate on orbiting a man in space to avoid further humiliation. Then, after a number of frustrating delays, John Glenn made three orbits of the Earth on 20th February 1962, helping to restore national pride. Many Mercury, Gemini and Apollo flights followed before, just over 7 years later, on 20th July 1969 Neil Armstrong and Buzz Aldrin landed on the Moon, and returned safely to Earth a few days later. Today, 25 years later, the Russians have still not landed a man on the Moon, nor even put one in lunar orbit. The Americans had clearly won the race and there was nothing to be gained by being second.

President Kennedy had announced the Apollo manned lunar programme in 1961, at the height of the Cold War. That year the East Germans had built the Berlin wall and, in the following year, the world teetered on the brink of nuclear war during the Cuban missile crisis. It was imperative to American politicians at that time that the United States proved to the world that it was still its technological leader and military superpower. The reason for putting a man on the Moon in 1969 was to salvage American pride, and to prove to the world that America was still the most powerful country on Earth and could do anything that it set its mind to. It had nothing to do with the requirements of astronomical research, and yet, not for the first time, astronomy benefited from military and political imperatives. For example, samples of lunar rock were brought back by the astronauts, and seismometers (to measure lunar quakes) and a laser reflector (to assist in distance measurements) were installed on the Moon.

Interest in the Apollo moonshots quickly waned after the first manned lunar landing, except for a brief period of interest and concern when Apollo 13 got into difficulties in space. The United States had proved their point, and the political requirements had changed by the end of the 1960s, so they decided to cancel the last seven moonshots after Apollo 17 and put the unused hardware into storage.

The main scientific results of the various manned and unmanned missions to the Moon are included in Chapter 2 and will not be repeated here, nor is there space to outline all these lunar missions. Instead, I will now consider the spacecraft that have been sent to other members of the solar system.

Early Solar Plasma Research

The environment of interplanetary space was examined by some of the earliest spacecraft. In particular in May 1958 Sputnik 3 carried ion detectors into space to measure the plasma of charged particles above the Earth's atmosphere, and the following year the Soviet Union launched their three probes (Luniks 1, 2 and 3) to the Moon. Lunik 2 carried four ion detectors on its mission to the Moon in September, and was the first spacecraft to measure ions in the solar wind outside of the influence of the Earth's magnetic field (the so-called interplanetary space). The results indicated that the interplanetary plasma contained high-speed ions, thus supporting Ludwig Biermann's theory of a dynamic environment (see Pages 15–16). The measured plasma density, however, was two orders of magnitude lower than that predicted by Biermann.

The first American probe to measure the solar wind in detail near interplanetary space was Explorer 10 launched in March 1961. It had an apogee at 240,000 km, and at this height it was expected that it would be well outside the influence of the Earth's magnetic field. Unfortunately, it became gradually clear when analysing the results that the probe had not got completely free of the Earth's magnetic field. Nevertheless, Herbert Bridge, who was head of the group that designed the plasma detector, was able to broadly confirm the Lunik results, although the measured plasma density was intermediate between those measured by Lunik and Biermann's prediction. Explorer 10 was able to measure the direction of flow of the solar wind, which Lunik 2 could not, and confirm Biermann's prediction that it was flowing away from the Sun.

The plasma environments measured by Lunik 2 and Explorer 10 were similar to each other, although there were differences, and there was concern because Explorer had not escaped completely from the influence of the Earth's magnetic field. A new measurement campaign was required to produce definitive results.

Conway Snyder and Marcia Neugebauer, of the Jet Propulsion Laboratory, designed a more sensitive ion detector to fly on the Ranger series of spacecraft, which were to fly beyond the influence of the Earth's magnetic field. Unfortunately, the first two Ranger spacecraft, launched in 1961, failed to get beyond their low parking orbits but, in October of that year, NASA decided to launch a modified Ranger spacecraft called Mariner 1 towards Venus. The launch of Mariner 1 in July 1962 was a failure, but the launch of Mariner 2 on 26th August 1962 was a resounding success. It broadly confirmed the results of Explorer 10, but it also found an excellent correlation between the velocity of the solar wind and magnetic disturbances on the Earth, proving that magnetic storms on Earth are caused by high-speed ions from the Sun.

The results of Mariner 10 and subsequent solar exploration spacecraft are outlined in more detail in Chapter 1.

Missions to the Terrestrial Planets

Venus

A satellite could, theoretically, be launched towards Venus at any time, but there are alignments of the Earth and Venus that occur about every 19 months when the propulsive energy that needs to be given to a spacecraft is a minimum. There is a period, called the launch window, of about 2 or 3 weeks on either side of this optimum alignment, when the spacecraft energy requirements are still acceptable. Similar considerations apply to launches to all the planets, the launch window for Mars, for example, has about the same width and occurs at intervals of about 26 months.

The first admitted attempt to reach any planet was made by the USSR on 12th February 1961, when Sputnik 8 was placed in a parking orbit around the Earth en route to Venus. A 640 kg probe called Venera 1 was then ejected from Sputnik 8 towards Venus, but contact was lost a few days later. Bernard Lovell and his team at Jodrell Bank had made a name for themselves using the 250 ft (75 m) diameter radio telescope to track early Soviet spacecraft, and two Soviet scientists spent several weeks there trying to contact Venera 1, but to no avail. A similar spacecraft, Sputnik 7, weighing over 6 tons, had been placed in Earth orbit a week earlier but, although no probe had been ejected towards Venus, it was generally considered to have been the first Soviet attempt to reach the planet.

The next launch window for Venus was in the late summer of 1962, when the USSR made three unsuccessful attempts to reach the planet, with none of the spacecraft leaving their Earth parking orbits.

The first United States attempt to reach Venus was that of Mariner 1 on 22nd July 1962, but the launch safety officer was forced to blow up the Atlas-Agena B launch vehicle shortly after take-off. A failure analysis was quickly undertaken by NASA, because the sister ship, Mariner 2, was due to be launched to Venus in the same launch window. Fortunately, however, the problem was found to be due to faulty guidance parameters, so these were quickly corrected and the Agena B upper stage, with the 200 kg Mariner 2 spacecraft still attached, was successfully launched into its parking orbit on 27th August 1962. Shortly afterwards, the Agena B's engines were re-ignited to send the spacecraft on its way to Venus, which it flew by at a distance of only 34,800 km on 14th December 1962. It was the first successful spacecraft to reach the planet. Mariner 2 showed that the surface temperature of Venus was about 700 K, confirming earlier estimates based on its radio emissions, and that it had no magnetic field. Radio contact with the spacecraft was lost on 3rd January 1963, when it was 87 million km from Earth, after a highly successful mission.

The Soviet Union was now becoming much more cautious about announcing the launch of satellites, because of the failure of their three 1962 attempts to leave Earth orbit. Cosmos 27, launched on 27th March 1964, appears to have been their next attempt but, if so, it failed. Zond 1, which was launched the following month, at least managed to get into the correct transfer orbit to Venus, but communications were lost with it shortly afterwards. It was during the 1965 launch window, however, that the USSR had the first of a series of successful missions to Venus.

Veneras 2 and 3, which were launched within 4 days of each other in November 1965, were almost twins. The scientific instrument package on Venera 2, which was to fly-by Venus was, however, replaced on Venera 3 by a 90 cm (35 inch) diameter instrument capsule designed to land on the planet. Communications were lost with Venera 2 just before it reached the planet, but the probe from Venera 3 was the first spacecraft to land on Venus, on 1st March 1966. It was shown later, however, that the capsule had been crushed during its descent to the surface by the very high atmospheric pressure. Capsules from Veneras 4, 5 and 6 all landed on Venus in the late 1960s, providing data on the atmosphere as they descended, but it was a capsule from Venera 7 that was the first to transmit data from the surface. The Venera 7 capsule worked on the surface for 23 minutes in 1970 showing that, not only is the temperature in excess of 700 K (high enough to melt lead), but the atmospheric pressure is 90 times that on Earth. It was amazing that any satellite could have survived this very hostile environment, even for 23 minutes.

Four years later, the American Mariner 10 sent back photographs of the cloud-tops of Venus as it passed by on its way to three successful encounters with Mercury. In fact, Mariner 10 was the first spacecraft to use the gravity of one planet to assist it to reach another in a swing-by or sling-shot encounter. The Mariner 10 photographs of Venus, which were taken in the ultraviolet to exaggerate the cloud markings seen from Earth, proved that the clouds circle Venus once every 4 days at a velocity of 400 km/h.

The Soviet Veneras 9 and 10 spacecraft landed probes on Venus in 1975 that photographed the surface. Both probes carried searchlights, as it was thought that the dense cloud layer would let through very little sunlight, but the illumination level, likened to that on Earth during a thunderstorm, rendered them unnecessary. The landing sites, which were widely separated, were both near to Beta Regio in the northern hemisphere. One was seen to be on a hillside strewn with boulders, whereas the other was on a fairly flat rocky plain.

The American Pioneer-Venus 1, that went into orbit around Venus in December 1978, had a radar altimeter in its payload, in order to make the first surface maps of the planet. It mapped 93% of the surface with a resolution of 100 km, and showed that Venus had rolling plains on 70% of its surface, with 10% highland and 20% lowland areas. Pioneer-Venus 2, which had a similar orbiter design to Pioneer-Venus 1, released four landing probes in November 1978, one of which survived the landing on 9th December and continued transmitting data from the surface for just over an hour.

Veneras 13 and 14 both landed probes in March 1982 to take colour photographs and analyse soil samples. The sky was found to be bright orange, and the soil was found to be basalt with, in some samples, an unusually high percentage of potassium. In 1984, Veneras 15 and 16 mapped 30% of the surface, using a radar system that provided a surface resolution of 2.5 km, and found a great many impact craters and small volcanoes.

The American spacecraft Magellan was launched to Venus from the space shuttle Atlantis in May 1989, sporting the most sophisticated radar payload yet launched into interplanetary space and, after a 15 month journey, it was put into orbit around the planet. After a number of glitches with its control system, Magellan completed its third radar mapping cycle in September 1992, having imaged about 99% of the surface at least once with a resolution of about 150 m. It found that the surface is largely volcanic, but also discovered signs of tectonic activity.

Mercury

So far only one spacecraft has visited Mercury, the American Mariner 10, that flew past Venus in February 1974, and reached Mercury less than 2 months later, on 29th March. Because it had to travel so close to the Sun during the Mercury encounters, Mariner 10 would receive about six times as much solar heat as a spacecraft in Earth orbit, and this required it to have a specially-designed thermal subsystem to keep it cool. The solution consisted of an umbrella-type sunshade, multilayer thermal blankets and louvred side panels.

Mariner was put into an orbit around the Sun, after the first Mercury encounter, that enabled it to intercept Mercury on consecutive Mariner orbits; Mercury having completed just over two orbits of the Sun in between each encounter. As Mercury rotates on its axis three times in two Mercurian years, this meant that the spacecraft saw the same part of Mercury at each encounter with the planet. Encounters two and three took place in September 1974 and March 1975 and, on the last visit, the spacecraft passed within 327 km of the surface.

Mariner 10 showed Mercury to have a surface that looked superficially very much like that of the Moon, with a large number of impact craters, although the lava flows on Mercury are not nearly so extensive. Mercury also appeared to have a large iron core and a small magnetic field.

Mars

The Soviet Union had three unpublicised attempts to send a spacecraft to Mars, before in November 1962 they were able to send Mars 1 on the correct trajectory to the red planet. Just 10 weeks before encounter, however, communication with Mars 1 was lost when it was 105 million km from Earth. The first American attempt was also a failure, but, exactly as with Venus, the second American spacecraft to Mars, Mariner 4, was a success.

Mariner 4 passed within 9,800 km of the surface of Mars during its fly-by in 1965, returning 21 images that revolutionised our view of the planet. The surface was found to be pock-marked with a large number of impact craters, much to the surprise of most astronomers (see Page 56), and the dark areas visible from Earth were found to be low-albedo areas, not water-filled seas or large tracts of vegetation. The atmosphere was also more tenuous than expected, with a surface pressure less than 1% that on Earth.

The successes that the Soviet Union had had launching satellites to Venus in the late 1960s seemed to desert them in the case of Mars. Zond 2, their second publicised spacecraft to Mars, was a failure in late 1964 and, although Mars 2 and Mars 3 both dropped a capsule onto Mars in 1971, no useful data were returned.

In the meantime, the Americans had repeated their successful Mariner 4 fly-by, with the two further fly-bys of Mariners 6 and 7 in 1969, but they showed nothing fundamentally new. Mariner 9, on the other hand, was a different matter. Put into orbit around Mars on 13th November 1971, it was the first spacecraft to allow a view of the whole planet. It had arrived at Mars during an extensive dust storm but, when the dust settled, Mariner 9 showed a surface with high volcanoes, deep valleys and features that looked like dried-up river beds. These new features were in parts of the planet that had not been visible to the previous Mariners.

The next four spacecraft sent to Mars, the Soviet Union's Mars 4 to 7, all failed, but Vikings 1 and 2, launched by the United States in 1975, were highly successful. The two Viking spacecraft were identical in design, each having an orbiting spacecraft and a sophisticated lander. The lander was attached to the orbiter and put into orbit around Mars, until the cameras and sensors on the orbiter had found a safe landing site on the planet.

The Viking 1 lander touched down in the Chryse Planitia (Plain of Gold) region of Mars 23° north of the equator on 20th July 1976, and the Viking 2 lander touched down 6 weeks later in the Utopia Planitia region, also north of the equator but on the other side of the planet. After the probes had landed, the two orbiters continued to operate their cameras and scientific instruments, also acting as relay stations, relaying data and images from the landers.

Each of the landers had a mass spectrometer, to sample the atmosphere during the descent to the surface, a robotic arm for scooping up surface material to test for living organisms, meteorological sensors, a seismometer, and a camera system. The experiment searching for evidence of life was generally concluded to have found no such evidence, although some scientists thought that the results were inconclusive. The landers took the first photographs from the surface of Mars, both landers showing a boulder-strewn surface under a pinkish sky.

The Soviet Union sent two spacecraft to Phobos, the larger of the two moons of Mars, in 1988. Contact was lost with Phobos 1 en route to Mars in August 1988 when the wrong command was sent to the spacecraft, and contact was lost with Phobos 2 on 27th March 1989, 2 months after it had entered into orbit around the planet, and just before it was due to encounter Phobos.

The Americans, unfortunately, have had no more success with their latest probe to Mars, called the Mars Observer, as contact was lost with it on 21st August 1993, just as it was about to be injected into orbit around Mars. So the age of spacecraft exploration of Mars has come full circle with spacecraft failures from both the major space powers.

Pioneers 10 and 11

Spacecraft launched towards Venus, and particularly Mercury, have to be designed to withstand the extra heat as they get closer to the Sun, but at least they can supply enough electrical power from their solar cells. Spacecraft heading towards Jupiter and beyond, on the other hand, have to be designed to survive low temperatures and work with less electrical power from their solar cells, or take their own electrical generators. They also have to fly through the asteroid belt, where collisions with gram-sized particles could write off the whole mission. Unfortunately, this risk could not be quantified before the launch of Pioneers 10 and 11, the first spacecraft to visit Jupiter, because the density of small particles in the belt was completely unknown. Pioneer 11, which used Jupiter's gravity to swing it on to Saturn, also faced another unknown risk of collision as it had to pass through the plane of Saturn's rings.

Jupiter was known to possess a very active magnetosphere prior to the Pioneer missions, but knowledge of its configuration and the density of the high energy particles within it had to await the arrival of the Pioneers. It was known that solid state devices carried by all spacecraft could be badly affected by such high energy particles in space, so here was yet another risk of unknown magnitude. Happily both spacecraft came through relatively unscathed.

Pioneers 10 and 11, which had been launched from Cape Kennedy in March 1972 and April 1973, respectively, each carried two radioisotope thermoelectric generators to provide electrical power. The generators were mounted on booms deployed from the spacecraft to minimise interference with the on-board experiments. The whole Pioneer spacecraft operated on a power budget of a just little over 100 W, and their 8 W signals when received by the NASA Deep Space Network (DSN) antennae had a power level of only 10^{-18} W.

To put a spacecraft into Earth orbit requires it to be given a velocity of 18,000 miles/h (29,000 km/h), but to escape from the Earth's gravity requires a velocity of at least 25,000 miles/h (40,000 km/h). Visiting the outer planets of Jupiter, Saturn, Uranus and Neptune, another consideration enters the calculation, and that is the time taken to reach the target. So Pioneer 10 was launched with a velocity of 32,600 miles/h (52,100 km/h), enabling it to reach Jupiter on 4th December 1973, after a journey of just 21 months, 6 months of which was in the asteroid belt, which fortunately caused no problems. After its fly-by of Jupiter, Pioneer 10 carried on its journey out of the solar system, but it was not targeted at any other planets as they were not suitably aligned at that time. Pioneer 11, on the other hand, which was launched one year after Pioneer 10, was due to fly-by Saturn after its visit to Jupiter.

Pioneer 11 reached Jupiter after a journey of less than 2 years, but it took almost another 5 years to reach Saturn where it arrived in good condition on 1st September 1979, flying through the ring plane completely unscathed. It passed just 29,000 km from the outer edge of the outer ring, and approaching to within 21,000 km of the surface of the planet.

The Pioneers found that the magnetosphere of Jupiter is 50 to 100 times as large as that of the Earth in linear extent. Io, the innermost of the Galilean satellites, was found to lie well within Jupiter's inner magnetosphere, and have a significant modifying effect by absorbing electrons and protons. Similar but less pronounced effects were observed with the other Galilean satellites, and so, when the number of energetic particles was observed to decrease within about 120,000 km of the planet, Mario Acuña and Norman Ness suggested that Jupiter may have a satellite or a ring system there. Neither could be found by the Pioneers.

The spacecraft showed intriguing glimpses of the complex cloud features on Jupiter with a resolution much greater than that possible from Earth but, as the Pioneer encounters were comparatively brief, few cloud motions could be measured.

Jupiter had been known to possess a very active magnetosphere before launch, because it emitted radio waves, but it was not even known if Saturn possessed a magnetic field, let alone radiation belts. In the event, Pioneer 11 detected a magnetosphere around Saturn intermediate in size between those of the Earth and Jupiter.

The images of Saturn and Titan, its largest satellite, were disappointing, as they showed Saturn had no Jupiter-type spots and Titan was seen to be completely featureless. Saturn's atmosphere was found to be warmer than expected, and Iapetus, Rhea and Titan, its three largest satellites, were found to be of low density, indicating that they were mainly composed of ice.

A new ring, designated the F ring, was discovered just outside of the outermost of Saturn's bright rings, the A ring. As in the case of Jupiter, evidence for the existence of satellites, or rings, between this new outer ring, and Saturn's inner satellite (Mimas), was found from measurements of unexplained minima in the charged particle fluxes. These additional satellites, or rings, could not be found by the Pioneers, however.

Voyagers 1 and 2

The Background

A report published in 1965, and sponsored by the National Academy of Sciences, was the first clear attempt to set a policy for the exploration of the outer planets. It suggested that NASA could either launch separate spacecraft to fly-by each of the more distant planets of Saturn, Uranus, Neptune and Pluto, or send a number of orbiters and atmospheric entry probes to Jupiter only. In the same year, Gary Flandro who was working at JPL showed that Jupiter, Saturn, Uranus and Neptune would soon be positioned so that a single spacecraft could visit them sequentially, using the gravity of one planet to swing the spacecraft on to the next. This Grand Tour spacecraft, as it became known, had to be launched between 1976 and 1980, as a suitable alignment of planets would not occur for another 175 years.

Over the next few years, a number of studies were undertaken looking at both the Grand Tour and alternative exploration missions for the outer planets. One question that received considerable attention was whether a Grand Tour spacecraft should fly outside the rings of Saturn, adding 3 years to the flight time to Neptune, or whether it should be sent between the inner ring and the surface of the planet, hoping that it would not be hit by any small orbiting particles.

Dick Wallace, a JPL mission analyst, suggested in 1968 that the Saturn problem could be solved by sending two Grand Tour probes, one to Jupiter, Uranus and Neptune, and another to Jupiter, Saturn and Pluto which would avoid Saturn's rings. In the following year NASA's Outer Planets Working Group agreed with Wallace's suggestion, and proposed an extensive programme that included a test flight of the new Grand Tour spacecraft. They suggested that this test spacecraft should be launched in 1974 towards Jupiter, and that it should use Jupiter's gravity to swing it out of the plane of the ecliptic and over the poles of the Sun. The Working Group also included Jupiter and Saturn orbiters and entry probes in their proposed programme.

In 1970, funding constraints forced NASA to drop the out-of-ecliptic test flight and, in 1971, further funding constraints forced NASA to abandon their plans for a newly-designed spacecraft for the two spacecraft Grand Tour. In January 1972 it was agreed, instead, to send two Mariner-class spacecraft to Jupiter and Saturn, which were eventually to become Voyagers 1 and 2 (see Figure 16.1). So NASA's first attempt to produce a broad-based, co-ordinated programme of exploration of the outer planets

Figure 16.1 *The Voyager Spacecraft. (Courtesy* National Space Science Data Center, World Data Center-A for Rockets and Satellites, NASA.*)*

had failed, as the programme was too ambitious. The scientific community had not helped matters either, as it could not agree on what it wanted.

In 1976, a Jupiter Orbiter and Probe mission (now called the Galileo mission) was approved and, in 1977, the out-of-ecliptic mission, called the International Solar Polar Mission (ISPM), was also approved as a joint mission with the European Space Agency (ESA). NASA and ESA were each to send a spacecraft over the pole of the Sun, one spacecraft going over one pole, while the other spacecraft went over the other pole. Budget constraints forced NASA to cancel their ISPM spacecraft in 1981, but the ESA spacecraft survived a difficult few years, and was eventually launched as Ulysses in October 1990, swinging by Jupiter in February 1992. It will observe the south pole of the Sun from June to October 1994, and the north pole from June to September 1995.

The Galileo spacecraft, which consisted of a Jupiter orbiter and atmospheric descent probe, was launched in October 1989 aboard the space shuttle Atlantis, after a $3\frac{1}{2}$ year hibernation on the ground following the Challenger Space Shuttle disaster of 1986. The latter had forced NASA to replace the Centaur liquid hydrogen–liquid oxygen upper stage that was to launch the Galileo spacecraft towards Jupiter from an Earth-orbiting shuttle, as the Centaur was considered too hazardous. As a result, Galileo had to use the less-powerful Boeing Interim Upper Stage, followed by a series of gravity-assist manoeuvres to get to Jupiter, via Venus (in February 1990), and the Earth (in December 1990 and December 1992). En route* it

*The high gain antenna failed to fully unfurl when commanded to do so on 11th April 1991, and repeated attempts to release it over the next 3 years failed. It was initially thought that this would severely restrict the results from the mission, but in 1992 NASA announced that 70% of the images could be successfully received on Earth, using sophisticated data-compression techniques and improvements to the Earth-based receiving dishes.

visited the asteroids Gaspra (in October 1991) and Ida (in August 1993), and is due at Jupiter in December 1995, over 6 years after launch. A far cry from the 18 months flight time of Voyager 1, to which we shall now return.

The Mission

Voyager 1 was launched on 5th September 1977 to fly by Jupiter and Saturn, while Voyager 2, which was launched two weeks earlier, was to fly by Jupiter and Saturn, and possibly Uranus and Neptune. Both spacecraft were also expected to observe a number of Jupiter's and Saturn's satellites, of which Titan was considered by far the most important, because a study of its atmosphere should yield important clues about the evolution of the early atmosphere on Earth. Voyager 1 was planned to fly by Titan but, if it failed, Voyager 2 would be re-programmed to fly by Titan, even though it would then have been unable to visit Uranus or Neptune.

The Voyagers, each weighing over 800 kg, were over three times as heavy as the Pioneer spacecraft and, as they were designed 5 years later, they could use more complex and more sophisticated equipment than their illustrious predecessors. The main operational difference was the use of a much more powerful on-board computer system that allowed all spacecraft functions, except for trajectory changes, to be controlled automatically on-board. This was essential as a round trip for radio waves from Neptune to Earth would take over 8 hours.

All planetary spacecraft are sophisticated systems that are expected to work remotely in extreme environments for periods of many months, if not years. Not only are the g-levels experienced during launch severe, but so are the temperatures on the Sun-facing and space-facing panels, and the high-energy particles in the solar wind can cause electrical discharges (arcing) on-board the spacecraft and destroy integrated circuits. Oil evaporates in space and can ruin lenses and mirrors, so dry lubricants have to be developed. All these problems, and many more, have to be overcome, and in the case of Voyager 2 the spacecraft would have to operate satisfactorily when it was at Neptune some 4,430 million km away from Earth 12 years after launch (assuming that Voyager 1 had made a successful fly by of Titan).

Inevitably there are problems with all sophisticated spacecraft, and the Voyagers were no exception, but they produced excellent results during the encounters, to the relief and admiration of everyone.

Jupiter

Voyager 1, which overtook Voyager 2 on the journey to Jupiter, began its Jupiter approach phase 2 months before encounter, with the 1,500 mm *f*/8.5 telephoto-lens camera taking colour images of the planet every 2 hours. A month before encounter, and still over 15 million km from the planet, the images exceeded the highest spatial resolution achieved by Pioneer. As the spacecraft got nearer and nearer, details began to be discernible on the

Galilean satellites of Callisto, Ganymede, Europa and Io, and the image of the planet became larger than the camera's field of view even using the 200 mm wide-angle lens. About 17 hours before encounter, an image was taken looking to the side of the planet, at the request of Tobias Owen and Candy Hansen who were looking for the ring suspected as a result of the Pioneer mission, and, against all odds, a faint ring was discovered.

Voyager 1 imaged the four Galilean satellites and Amalthea, the small satellite discovered by Edward Barnard in 1892. Three new satellites were discovered from the Voyager 1 images, but the most dramatic discovery of the Voyager 1 mission at Jupiter was made by Linda Morabito of JPL on 9th March 1979, 4 days after closest approach. Linda, a member of the Voyager navigation team, found a bright volcanic plume rising hundreds of kilometres above the limb of Io, on an image which had been made to check on the spacecraft's trajectory. She and her colleagues then analysed all the images of Io taken during the Voyager 1 encounter, and eventually discovered a total of eight active volcanoes.

Voyager 1 was now on its way to Saturn, as Voyager 2 closed in on Jupiter for its closest approach 4 months later, on 9th July 1979. Europa, which had not been seen very well by Voyager 1, was found to be the smoothest world in the solar system with the maximum surface relief of a few hundred metres, but all eyes were on Io. Would the active volcanoes be the same as a few months earlier?

Of the eight active volcanoes discovered on Io by Voyager 1, one was found to be inactive at the Voyager 2 encounter, one was out of view, but the other six were still active. Voyager 2 also found two new volcanic vents, although they were inactive during the fly-by.

Saturn

There was more than the usual apprehension as Voyager 1 approached Saturn in the autumn of 1980, because of the relatively featureless images returned by Pioneer 11 the previous year. Fortunately this concern turned out to be unnecessary, as distinct cloud features were seen on the planet during the final few weeks of approach. The features all had low contrast, however, and computer-enhancement was needed to show them clearly.

On 6th October, less than 6 weeks from closest approach, Richard Terrile was surprised to discover spoke-like markings across ring B (see Figure 5.5 earlier), with moving images showing them up very well. Three small satellites were also discovered during the Saturn approach phase, two of them on either side of the F ring, apparently stabilising or "shepherding" the ring, and the third just outside the outer edge of ring A, apparently stopping the A ring from expanding outwards. As Voyager 1 got closer and closer to Saturn, the ring structure became more and more complex and, at the limit of resolution of the Voyager 1 cameras, the rings were found to be subdivided into more than 1,000 narrow rings. In addition, the F ring discovered by Pioneer 11 was found to consist of three narrow rings that appeared to be intertwined one with another.

Titan was an important target for the Voyager 1 spacecraft, which it successfully flew by at a distance of only 6,500 km. It could not detect any surface features beneath its thick atmosphere, however, and only the very faintest markings were apparent at the top of the atmosphere, even after the most sophisticated image processing routines had been used.

On 19th December 1980, the two cameras on Voyager 1 were turned off, at the completion of a fabulous mission of discovery, meanwhile Voyager 2 was getting ready for its fly-by of Saturn in the following August. Voyager 2, as the second visiting spacecraft of the pair, was inevitably not destined to make the same level of discoveries as its predecessor. It was used to look in more detail at some of the features observed by Voyager 1, namely the markings in Saturn's atmosphere, and the spokes in the rings, and also observed four satellites that had been poorly placed for observation by Voyager 1, namely Iapetus, Hyperion, Tethys and Enceladus. Voyager 2's camera platform jammed near the end of the encounter, however, causing the best images of Tethys and Enceladus to be lost.

As Voyager 1 had successfully encountered Titan, Voyager 2 was targeted to fly by Uranus and Neptune in 1986 and 1989, respectively. Fortunately, the problem with the camera platform was overcome prior to those encounters.

Uranus

As 1985 drew to a close, Voyager 2 was prepared for its brief meeting with Uranus and its satellites. In December 1985, it was realised that Uranus' satellites were not exactly where they were expected to be in their orbits, because Uranus was slightly heavier than previously thought, and this required the spacecraft's trajectory to be slightly modified. Then on the last day of the year Voyager discovered its first new satellite of Uranus, now called Puck, and on 21st January 1986, 3 days before closest approach, it discovered two small shepherding satellites on either side of the epsilon ring.

A few days before closest approach, a strong burst of polarised radio signals indicated the presence of a planetary magnetic field, and then $10\frac{1}{2}$ hours before closest approach this was confirmed when the spacecraft crossed the bow shock. A day later it became evident that the magnetic axis was tilted at an amazing 60° to Uranus' rotational axis.

The images of Uranus were disappointing, showing a bland disc with only slight markings, even after computer enhancement. The satellites, however, were a different matter, and here Miranda stole the show with its strange chevron-shaped feature, where pronounced surface grooves intersected at right angles (see Figure 5.9 earlier). It was pure good fortune that Voyager had had to pass just 29,000 km above this, the most interesting of Uranus' satellites, in order to get the required gravitational boost from Uranus to take Voyager to its encounter with Neptune.

Voyager imaged each of the five satellites known before launch, and discovered a total of ten more. It also found more very narrow rings and provided key data on the Uranian atmosphere and magnetosphere. Because the orbits of the satellites are almost perpendicular to the orbit of Uranus around the Sun, the whole encounter phase with Uranus and its satellite system had taken only five hours, with Voyager passing at almost 900 through the orbital plane of the satellites.

Voyager now left Uranus behind on its way to its final encounter due $3\frac{1}{2}$ years later with Neptune.

Neptune

One of the key trade-offs that had to be made before the Neptune encounter, due on 25th August 1989, was how close should Voyager fly to Neptune and Triton, by far its largest satellite. In 1985, the Voyager Science Steering Group had recommended a Triton fly-by distance of 10,000 km, which meant that the spacecraft would fly just 3,400 km above Neptune's clouds. Then, later that year scientists announced that Neptune's radius was 1,000 km larger and its mass 1.5% smaller than previously thought, and Triton was 8,000 km away from its previously-estimated position. So the previously-agreed Voyager trajectory needed to be re-examined, as a 10,000 km closest approach of Triton would now have taken Voyager only 1,250 km above Neptune's clouds. Eventually a compromise trajectory was agreed that took Voyager to within 38,000 km of Triton, and 4,850 km above Neptune's clouds.

On 7th July 1989, Voyager 2 made its first discovery as it closed in on Neptune when it found a new satellite, now called Proteus, orbiting well within the orbits of the only two satellites, Triton and Nereid, known previously. Voyager was destined to find five more satellites of Neptune over the next 3 months.

During the Neptune encounter, Voyager found that Neptune's magnetic axis was tilted at 47° to the planet's rotational axis, and that it was offset from the centre of Neptune by over half of the planet's radius. Before this was known, scientists had thought that the similar arrangement with Uranus' magnetic field was unique, but now they had two almost identical cases to explain.

The cloud-tops of Neptune showed much more structure than those of Uranus. A Great Dark Spot was found, very reminiscent of the Great Red Spot of Jupiter, and winds of up to 2,000 km/h were measured in the upper atmosphere, which are the fastest in the solar system.

Voyager confirmed that Triton is one of only three satellites in the solar system to have an atmosphere (the others being Titan and Io). It was found to consist mainly of nitrogen, with a surface pressure of 10^{-5} times that of the Earth. Long black streaks were also found on the icy surface of Triton that had a temperature of only 38 degrees above absolute zero. Later analysis of the Voyager images showed that these streaks were caused by geyser-like eruptions of nitrogen from beneath the surface.

So ended the remarkable set of Voyager encounters, with both Pioneers and both Voyagers now following trajectories that would take them away from the solar system and into interstellar space. As a farewell, Voyager 2 took one final set of images; that of the solar system from outside.

The Halley Intercepts

An armada of five spacecraft flew by Halley's comet in 1986, to observe that most famous of all comets from close quarters. In addition, the Americans re-targeted one of their spacecraft to study the comet from a somewhat greater distance.

Halley's comet had been chosen because it is the most active of the periodic comets that have well-known orbits, and it could also be observed from Earth at the time of spacecraft intercept, allowing simultaneous space and ground-based observations. Halley's perihelion was due on 9th February 1986, when it would be 0.59 AU from the Sun.

The orbit of Halley's comet is inclined at 180 to the ecliptic, and the ideal intercept points for a spacecraft, requiring the least launch energy, is at either of the two places where the comet's orbit crosses the ecliptic, one before and one after perihelion. The first such intersection occurred on 9th November 1985, but at this time Halley's comet was still 1.85 AU away from the Sun, whereas at the time of the second intersection on 10th March 1986 it was only 0.89 AU away. Comets are more active the nearer they are to the Sun, so this second date was chosen as the target fly-by date for the fleet of spacecraft.

It was important to coordinate the operations of all the spacecraft due to intercept the comet, so that the data from the earlier intercepts were available to help the later ones. In 1981 the four Space Agencies involved, representing the Soviet Union, Europe, Japan and the United States agreed, therefore, to set up an Inter-Agency Consultative Group (IACG) to coordinate the planning and implementation of the intercepts. This arrangement, which was informal, was based on a similar group (CGMS) that had been set up some years earlier between the same four Agencies, to coordinate the planning and operation of geosynchronous meteorological satellites.

The six spacecraft which were due to observe Halley's comet were:

Spacecraft	Agency	Date in 1986	Fly-by distance from nucleus (km)
Vega 1	Intercosmos (USSR)	6th March	8,890
Suisei	ISAS (Japan)	8th March	151,000
Vega 2	Intercosmos (USSR)	9th March	8,030
Sakigake	ISAS (Japan)	11th March	6,990,000
Giotto	ESA (Europe)	14th March	596
ICE	NASA (USA)	25th March	28,100,000

Halley's comet, unfortunately, orbits the Sun in a retrograde orbit, and this meant that the spacecraft would meet it head on, resulting in a velocity relative to the comet of an incredible 250,000 km/h. As the Giotto spacecraft was going to approach very close to the nucleus, the spacecraft engineers had to design a spacecraft that would survive the impact of a large number of dust particles travelling at this speed, and provide a camera that was able to pan the nucleus and produce sharp images when the angle of the target was changing very quickly. Furthermore, as the spacecraft was about 150 million km from Earth at the time of intercept, the round-trip time for radio signals was 16 minutes, so all the intercept commands had to be pre-programmed on-board the spacecraft.

Giotto solved the dust impact problem by having a two-layer bumper shield designed on the Cobham armour principle, where the first layer vaporises the incident particles and the second one protects the spacecraft from the resultant products. The shield was designed to withstand the impact of a 0.1 g dust particle travelling at 250,000 km/h, which would have the same kinetic energy as a 600 kg vehicle travelling at 100 km/h. Unfortunately, by their very nature, some of the spacecraft experiments had to have an unrestricted view to the comet, so protection of those could not be 100% complete.

Giotto was designed to get very close to the nucleus, but it was recognised that it may not survive the intercept because a very small dust particle could still destroy it. It was basically a kamikaze mission, so any data that were received as Giotto was leaving the comet would be a bonus.

NASA had hoped to fly a dedicated spacecraft to Halley's comet, like the European Space Agency, the USSR, and Japan, but funding constraints meant that they had to re-target one of their existing orbital spacecraft, ISEE-3, to Halley. This spacecraft, renamed the International Cometary Explorer (ICE), intercepted the comet Giacobini-Zinner on 11th September 1985, passing through the comet's tail at a distance of 7,800 km from the nucleus, before continuing on its way to Halley. Unlike the Vega and Giotto spacecraft, however, it had no camera on board.

Vega 1 was the first spacecraft to reach Halley's comet. It, and its twin spacecraft Vega 2, were both launched in December 1984 towards Venus, where they dropped probes onto the surface in June 1985, before continuing their journey to Halley. Vega 1 crossed the comet's bow shock 1.1 million km from the nucleus about 4 hours before closest approach, and showed an indistinct nucleus of about 4 km in diameter when it was closest to the target. Three days later, Vega 2 showed a larger, more elongated, nucleus that had clearly rotated since the first encounter.

The Japanese spacecraft Suisei and Sakigake were designed mainly to examine the huge neutral hydrogen corona which was expected to extend some millions of kilometres from the nucleus. Suisei showed that this corona extended to 10 million km from the nucleus, and that its brightness was varying with an average period of 53 hours.

The position of the nucleus could be determined from the ground to only within about 400 km, which was not accurate enough if Giotto was to approach within less than 1,000 km of the nucleus and automatically photograph it while travelling past it at high speed. Fortunately, the USSR agreed to provide information on the position from their Vega spacecraft, which were due to arrive at Halley a few days before Giotto, but the location of these spacecraft themselves was only known to within about 400 km. NASA solved this problem, however, by agreeing to track the Vega spacecraft with their Deep Space Network, which resulted in the targeting error of Giotto being reduced by a factor of ten, enabling a fly-by distance of 540 $\pm$ 40 km to be selected.

Giotto crossed the bow shock 1.3 million km from the nucleus, and recorded its first dust impact at 290,000 km out. About a minute before closest approach, the spacecraft crossed the ionopause 4,700 km from the nucleus, the only spacecraft to do so, where the interplanetary field fell to zero. Thirty-three seconds before closest approach one instrument was damaged by a dust impact, at 21 seconds two more instruments failed, and at 9 seconds the camera failed. The last complete image was transmitted from a distance of 1,930 km. Two seconds later, all communications with the spacecraft were lost, and it looked as if high velocity dust had finally destroyed the spacecraft.

Confusion reigned at the ESA ground station where 600 news reporters were gathered. Most people thought the spacecraft was dead and quickly wound up their live television coverage, but an intermittent signal was picked up from the spacecraft after a few seconds, indicting that it was nutating, or wobbling, owing to an off-axis impact by a dust particle. Half an hour later, the on-board nutation damper had stabilised the spacecraft, and it was sending back full telemetry as it receded from its target, battered but not silenced.

One week later, Giotto was retargeted to fly by the Earth on 2nd July 1990, enabling it to successfully intercept the comet Grigg-Skjellerup on 10th July 1992, 7 years after launch.

Orbital Observatories

Most of the spacecraft discussed so far in this chapter visited the object of their investigations and made in-situ measurements. In this section I will introduce some of the orbital observatories which have been put into space to avoid problems posed by the Earth's atmosphere. I will briefly mention some of their findings, although some of their more detailed results have already been mentioned in previous chapters.

Until the 1940s, our knowledge of the universe had been gained by looking at the sky with the eye or by taking photographs, but our eyes and photographic plates are only sensitive to a limited range of wavelengths of electromagnetic radiation. The colour receptors in the eye are sensitive from 400 to 700 nm, and the monochromatic receptors from 390 to 610 nm. Photographic plates available in the mid 1930s were sensitive from the near ultraviolet to the near infrared, but our knowledge of the universe was non-existent at gamma (γ) ray, X-ray, far ultraviolet, far infrared and radio wavelengths.

The problem is that the Earth's atmosphere is basically transparent only from 310 to 900 nm (i.e. the visible spectrum, plus a very small part of the ultraviolet and infrared), and between 1 mm and 30 m (radio wavelengths). So, observations outside of these ranges have to take place outside of the atmosphere.

Atmospheric absorption in the infrared is somewhat complicated, however, with wavelength "windows" where the absorption is not too great and where observations can be made from the surface of the Earth. The best windows are at 2.2 and 4 μm but, even observing at these wavelengths, it is important to be above as much of the atmosphere as possible by, for example, locating observatories on high mountains or carrying telescopes aloft with sounding rockets or balloons.

Astronomers thought in the 1940s that they would be able to improve their knowledge of the universe if ultraviolet and infrared observations could be achieved, but virtually no-one expected that radio, X-ray and γ-ray astronomy would yield much of interest (apart from data on the Sun). In that they were, happily, dramatically wrong.

Infrared (Wavelengths from 0.7 to 300 μm)

In the 1960s, the American astronomers Gerry Neugebauer and Bob Leighton carried out the first infrared (IR) sky survey using a simply-constructed, Earth-based 1.5 m telescope, fitted with a lead sulphide cell cooled* with liquid nitrogen. They worked in the 2.2 μm near-infrared window for 6 years, and produced an Infrared Catalogue containing 5,612 stars. This Two Micron Survey, as it became known, was followed in the early 1970s by one carried out by the United States Air Force Geophysical Laboratory which used a 6 inch (16 cm) telescope operating at 4, 11, 20 and 27 μm on 11 rocket flights. The total observing time was less than 1 hour, but in this time 90% of the sky was surveyed and 2,000 stars and nebulae were detected. At 2.2 μm very cool stars were seen with temperatures of about 1,400 K, but at wavelengths of 27 μm, objects and dust with temperatures of only about 100 K were detected as, according to Wien's law, the wavelength of the peak emission of a black body is inversely proportional to its temperature.

By the 1980s, satellites had come into their own, with the launch of the Infrared Astronomical Satellite (IRAS), which was a joint venture between the USA, the UK and the Netherlands. The satellite, which was launched on 25th January 1983, carried a 24 inch (55 cm) diameter telescope, with detectors cooled by liquid helium to 2 K. In 9 months IRAS surveyed 96% of the sky twice, at 12, 25, 60 and 100 μm, and it was three quarters of the way

---

*Infrared energy is emitted by objects at all temperatures, but the lower the temperature, the less energy is emitted. In infrared telescopes, therefore, it is essential to cool the detectors to as low a temperature as possible to reduce the detector noise levels.

through the third survey when the liquid helium supply finally ran out. The IRAS catalogue, issued in 1986, is a mine of information containing data on 245,000 point sources and 16,700 extended sources, plus 5,400 infrared spectra.

IRAS examined galaxies that appeared to be producing large numbers of stars, and found evidence that these "star-burst" galaxies are often the result of galaxies colliding. They appeared to be emitting about 99% of their energy in the infrared, possibly generated by their colliding gas clouds. One particular galaxy, IRAS 01003–2238, which is about 2 billion light years away, showed evidence of a large number of Wolf–Rayet stars, but a small number of O-type stars. As O-type stars take about 10 million years to turn into Wolf–Rayet stars, this indicated that star formation occurred in a sudden burst a little over 10 million years ago.

In 1986, Dana Backman and Frederick Gillett of the Kitt Peak National Observatory found particle clouds around six nearby stars using IRAS data, and by 1988 they had found such clouds around 25 out of the 134 stars that they had studied. Using a combination of IRAS data and ground-based observations, they studied three stars in detail, namely Fomalhaut, Vega and Beta Pictoris and discovered, for each of them, a dust-cleared region between the cloud and the star which, they suggested, could contain planets that have cleared the region of dust during and after their formation. In 1992, Melissa Nischan and Dana Backman of Lancaster, Pennsylvania and Ben Zuckerman of the University of California, examined the dust cloud around Beta Pictoris at infrared wavelengths, using the 1.5 m telescope at Cerro Tololo. They concluded that the cloud consists of dirty water, methane and ammonia ices, which are similar to the materials that make up the satellites of Jupiter and Saturn, adding further to planetary speculation.

A new 2.2 μm survey was started in 1988 by Mike Selby and Francisco Garzon, using the 1.5 m Carlos Sanchez Telescope at the Teide Observatory on Tenerife. The survey, which was to map and catalogue that part of the Galactic plane and bulge visible from Tenerife had, by mid 1993, catalogued 470,000 sources in just 255 square degrees of the sky. This Two Micron Galactic Survey (TMGS) will be followed shortly by the DEep Near Infrared Survey (DENIS), using the European Southern Observatory's 1 m telescope at La Silla, Chile. This survey of the whole of the southern sky at 0.9, 1.2 and 2.1 μm will be an invaluable aid to the Infrared Space Observatory (ISO) spacecraft, due to be launched in 1995.

Ultraviolet (9 to 310 nm)

According to Wien's law, objects have the peak of their radiation curves at shorter and shorter wavelengths the hotter they are. Thus cool objects, with temperatures of 4,000 K or less, have their peak intensities in the infrared, whereas hot objects with temperatures of between 10,000 and 300,000 K have their peak intensities in the ultraviolet (UV). Objects at the top end of this temperature range are not ordinary stars, but can be, for example, old stars that have suddenly blown away their outer envelopes to reveal their hot core, or streams of very hot gas flowing from one star to another. The most abundant atoms in the universe (i.e. hydrogen, helium, carbon, nitrogen, oxygen and silicon) also have their most prominent spectral lines in the UV.

OSO-1 (Orbital Solar Observatory-1), which was launched in 1962, was the first spacecraft to observe the Sun in the ultraviolet, but the first dedicated UV orbiting observatory was the American OAO-2* (Orbital Astronomical Observatory-2), launched in December 1968 and having 11 telescopes with apertures ranging from 8 inches (20 cm) to 16 inches (40 cm). OAO-2 was designed to measure the ultraviolet brightness of as many stars as possible, and obtain the spectra of selected sources in the range from 100 to 400 nm. Shortly after launch it discovered a large spherical hydrogen cloud surrounding the nucleus of the comet Tago-Sato-Kosaka, and similar clouds have since been found around other comets.

OAO-3, renamed Copernicus after launch, carried a 32 inch (80 cm) telescope into space in August 1972, to take UV spectra from 70 to 320 nm. It also had three small X-ray telescopes operating at 7 nm, and a proportional counter operating at 0.2 nm. The Copernicus satellite enabled Princeton astronomers to detect deuterium absorption lines in the spectrum of the star β Centauri. This was the first time that deuterium had been detected anywhere in the universe, except on Earth, and its existence in some quantity was a key prediction of the Big Bang theory. The density of deuterium measured indicated that the universe was "open" and would continue to expand for ever, unless there was at least 90% of dark matter in the universe to eventually stop the expansion (see Pages 257–258).

Copernicus operated for 9 years, and was followed during this period by the joint USA/UK/ESA International Ultraviolet Explorer (IUE) launched on 26th January 1978. IUE had an 18 inch (45 cm) reflecting telescope that took UV spectra from 115 to 320 nm, and a small camera that showed the optical image at which the UV telescope was pointing. IUE returned key spectroscopic data on cool, dark, active regions on stars (star spots), showing that the emission lines in the UV are enhanced whenever the dark spotted side of the stars face the Earth. Star spots have been found to cover up to 50% of a star's surface, compared with sunspots, which cover less than 1% of the surface of the Sun. IUE also assisted the international effort to observe and understand the supernova 1987A in the Large Magellanic Cloud (see Page 195).

Just over a decade after the launch of IUE, the Astro-1 Mission was flown on the space shuttle Columbus from 2nd to 11th December 1990. The payload of Astro-1 consisted of an Ultraviolet Imaging Telescope (UIT) for imaging stars and galaxies, an ultraviolet telescope for studying the spectra of quasars and active galaxies, a photometer for measuring sources in polarised ultraviolet light, and a broad-band X-ray telescope to examine

---

*OAO-1 was a failure.

active galaxies. The UIT had a 15 inch (40 cm) mirror, and recorded images on film from 120 to 320 nm. By early 1991 the Astro astronomers were publishing their first results, including an improved estimate of the distance to the Perseus cluster of 350 million light years.

Finally, on 7th June 1992, NASA launched their Extreme Ultraviolet Explorer (EUVE) to study the extreme ultraviolet (variously designated EUV or XUV) between 7 and 76 nm. It was the first observatory designed to map the entire sky at all EUV wavelengths, although Rosat had surveyed the sky at up to 30 nm 2 years earlier. The EUVE completed its all-sky survey in 1993, and 356 objects were included in the Bright Source List published in the same year. EUVE also found the first unambiguous evidence for helium in the Martian atmosphere.

X-Rays (0.01 to 9 nm; photon energies 100 to 0.1 keV)

In 1949, Richard Tousey of the United States Naval Research Laboratory (NRL) discovered that the Sun was a source of X-rays, by using phosphor-coated metal strips flown on a V2 rocket. The Sun's photosphere is far too cool to emit X-rays, however, and only prominences and the corona are hot enough (> 300,000 K) to be seen at these wavelengths. The spacecraft Vanguard 3, which was launched in 1959, was the first spacecraft to study solar X-rays and, in the following year, the first X-ray photograph of the Sun was obtained by Herbert Friedman and his team using a pinhole camera on board an Aerobee sounding rocket.

President Kennedy decided shortly after becoming President of the United States in 1961 to fund research into detecting high energy particles emitted by nuclear weapons, and the AS & E company of Cambridge, Massachusetts, received a contract to produce sounding rocket payloads to measure these particles. AS & E was also funded, using similar technology, to investigate whether the Moon produced X-rays caused by the impact of high-speed particles emitted by the Sun.

A few astronomers thought that there may be some optically-dim, X-ray-bright objects in the sky, and hoped that these would be found accidentally by the Aerobee rocket carrying Bruno Rossi and Riccardo Giacconi's X-ray detector in October 1961 to detect X-rays from the Moon. Unfortunately, the doors on the rocket payload module failed to open, but on the next launch attempt in June 1962 the doors opened correctly at an altitude of 225 km. The research team thought, at first, that they had detected X-rays from the Moon but, on further analysis, they realised that they had detected an X-ray source, now called Scorpius X-1, in the constellation of Scorpius. The same rocket flight also discovered a diffuse background of X-rays.

That part of the diffuse X-ray background in the 4.4 to 7.0 nm range was found, in 1966, to have a minimum intensity in the plane of the Milky Way, and a maximum towards the galactic poles. This implied that it originates outside the Milky Way, which is attenuating the signals. (At optical wavelengths a similar situation prevails, where dust in the plane of the Milky Way attenuates the light from distant galaxies; see Page 225.) It was suggested that this diffuse X-ray background could be due to unresolved quasars at a density of 5 per square degree, or possibly intergalactic hydrogen gas at a temperature of about 1 million K.

The detector used by Giacconi in 1962 was not collimated, and so it was difficult to determine the position and size of Scorpius X-1. In the following year, however, Friedman's group at the NRL flew an X-ray detector with an restricted field of view to obtain a reasonably accurate estimate of the position of Scorpius X-1, and that of any other source it may discover. Two sources of X-rays were detected. Scorpius X-1 was found to have an angular diameter of less than 10 arcmin, but its position did not seem to correlate with that of any radio or unusual optical objects. The other source was the Crab nebula.

Over the next 4 years, more than 30 discrete sources of X-rays were found using various rocket experiments, but the optical counterpart of Scorpius X-1 was still a mystery. Then, on 8th March 1966, the AS and E group and visiting scientist Minoru Oda flew an experiment on an Aerobee rocket that enabled the angular size of the Scorpius X-1 source to be determined as no larger than 20 arcsec. This led to a new search for its optical counterpart, which Japanese astronomers identified as a blue 13th magnitude star. Historical records showed that this star had exhibited fluctuations in its optical brightness, and it was concluded that it was the remains of an old nova. In 1975 measurements of the Doppler shift of its spectral lines showed that it was part of a binary pair with a period of 0.79 days, and simultaneous X-ray and optical variations that were found in 1981 indicated that there was an accretion disc around the X-ray star.

The American SAS-1 (Small Astronomical Satellite-1), was the first satellite to carry out an all sky survey of X-ray sources. It was launched from Kenya on 12th December 1970, the anniversary of Kenya's independence, and was called Uhuru, the Swahili word for freedom. Uhuru observed a total of 339 X-ray sources during its two year lifetime, the most interesting of which was probably Cygnus X-1, an X-ray binary consisting of a heavy O-type star and an invisible companion whose mass appeared to be too high (at least 6 solar masses) for it to be a neutron star. So this companion was thought to be a black hole (see Pages 204–205).

Uhuru's first surprise was, however, the discovery that Centaurus X-3, which had been discovered in 1967, was a pulsar of period 4.84 seconds with a secondary period of 2.09 days, indicating that the pulsar was part of a binary pair. The optical counterpart of one of these stars was found to be a blue supergiant. Uhuru then discovered that Hercules X-1 was also a binary pulsar with periods of 1.24 seconds and 1.7 days. Again the optical counterpart was found to be a blue star, HZ Herculis, having the same 1.7 day period, but there was more to it than that as the light curve of HZ Herculis was far more complex than that of a simple binary. The strange optical variability of HZ Herculis had been known for some time, and the discovery that its companion was a pulsar (a pulsating neutron star) explained its behaviour. X-rays from the neutron star Hercules X-1 are apparently heating up one side of HZ Herculis to 20,000 K, compared with the temperature of the un-radiated side of 7,000 K. So as the system revolves once every 1.7 days, we see alternately the hot and cool sides of HZ Herculis, hence the strange variability.

The Doppler shifts of the X-ray pulses from Hercules X-1, and of the visible light from HZ Herculis, enabled the masses of the two stars to be calculated. The neutron star Hercules X-1 and the variable star HZ Herculis were found to have masses of 1.3 and 2 solar masses respectively.

It was later found that the 1.24 second pulse rate of the Hercules X-1 pulsar was not slowing down, like the Crab pulsar and others, but speeding up. This was attributed to gas being pulled from HZ Herculis onto an accretion disc which surrounds the neutron star. This gas, in turn, falls from the accretion disc onto the neutron star, thus causing it to speed up.

Uhuru could only determine the location of X-ray sources to about 1°. The first satellite with an imaging X-ray capability was HEAO-2 (High Energy Astronomical Observatory-2) that was launched on 13th November 1978, with a thousandfold increase in sensitivity over Uhuru. HEAO-2 was renamed the Einstein Observatory after launch, to recognise the centenary, in 1979, of Einstein's birth. The satellite was 19 ft (6 m) long, weighed 3.5 tons, and had a maximum resolution of 2 arcsec, which is within an order of magnitude of the best resolution achieved at ground level with optical telescopes of the period.

The Einstein Observatory had four detectors, two of which were imaging cameras (the HRI and IPC), and two of which were spectrometers. The HRI had a high angular resolution (of 2 arcsec), but no energy resolution, whereas the IPC has a low angular resolution (of 1 arcmin), but a high sensitivity over a broad energy range of from 0.2 to 4 keV (i.e. wavelengths of 5 to 0.3 nm).

The Einstein Observatory surprised astronomers by showing that faint, red, cool dwarf stars are strong X-ray emitters, emitting about 10% of their total energy in the X-ray band. Prior to Einstein, two red dwarf stars were known to be flare stars, often emitting X-rays during violent flares (which are like large solar flares), but Einstein found that these stars were also X-ray emitters in their quiescent state, possibly due to hot coronas like that of the Sun.

Einstein also found that very young massive stars, that have recently been formed in nebulae like the Orion and Carina nebulae, are X-ray emitters. This was another surprise, as such young stars were not expected to have high-temperature coronas, and so should not emit X-rays. It now appears that these nebulae have within them some very massive stars which, although young, are well into their hydrogen burning phase and, in the case of η Carinae (see Pages 171–172), getting near to the end of their lifetime.

A number of supernova remnants in the Milky Way and in the Magellanic Clouds were examined by the Einstein Observatory. It also measured a number of X-ray bursters, which are objects that have sudden increases in their X-ray brightness of a factor of 10 or 100 over periods as short as one second. It was thought that these X-ray bursters are probably neutron stars sucking in gas from a low-mass companion star.

Powerful X-ray emissions were also found by Einstein to be coming from the centres of active galaxies which, in some cases, showed intensity fluctuations in their X-ray emissions over a period of only a few hours. These fluctuations indicated that the sources of the emissions were small, being only a few light hours across. Potential sources were thought to be gas clouds surrounding supermassive black holes in the centres of these active galaxies.

The Einstein Observatory failed in 1981, after producing more than 7,000 X-ray images, and was succeeded by Exosat that was launched by the European Space Agency (ESA) in May 1983.

The European Space Research Organisation (ESRO), a forerunner of ESA, had investigated in 1968 the possibility of producing a joint X-ray and γ-ray satellite called Cos-A. This concept was simplified in the following year, because of mass and cost considerations, by the elimination of the X-ray element and the decision to build a γ-ray only satellite, Cos-B. The idea of a complementary X-ray only mission was not forgotten, however, and after the launch of Cos-B work was started on the X-ray satellite called Exosat.

Exosat was a much smaller satellite than the Einstein Observatory, with a diameter of 7 ft (2.1 m), height of $4\frac{1}{2}$ ft (1.4 m), and mass of 500 kg, but its performance belied its size. Exosat's payload consisted of two identical imaging telescopes, each equipped with a channel multiplier array (CMA) and a proportional counter (PSD) operating in the range 0.1 to 2 keV, a medium energy detector operating in the 1 to 50 keV range, and a gas scintillation spectrometer. The CMA gave a maximum resolution of 8 arcsec, but with no energy resolution, whereas the PSD, which had a low angular resolution of 50 arcsec, had a high energy discrimination which allowed both the multicolour mapping of diffuse X-ray sources and more precise spectroscopy than had been possible with the Einstein Observatory.

Exosat's main mission was to provide more information on known X-ray sources, rather than to find new ones. Early in its mission, for example, it observed the source V0332 + 53, which was thought to be a black hole candidate because its X-ray intensity was known to flicker rapidly. Exosat found that the source was a binary pulsar with periods of 4 seconds and 34 days, but its orbit was not circular, however, and the rapid flickering of the X-ray output was observed to correlate with the closest approach of the compact object to its companion. This proved that not only is V0332 + 53 not a black hole, but rapidly-flickering X-rays can be associated with neutron stars, and not just black holes as previously thought.

The Röntgen X-ray satellite, more commonly known as Rosat, was launched on 1st June 1990 with an 83 cm diameter, grazing-incidence X-ray telescope on board. It was a collaborative project between Germany, the UK and the USA, and produced images with a sensitivity 100 times better than the Einstein Observatory. Rosat completed its all-sky survey in February 1991, discovering a number of previously unknown supernova remnants, and showing that a great deal of the sky-glow, previously discovered in soft

X-rays, is coming from a vast number of very distant quasars. Prior to Rosat, only about 10,000 X-ray objects were known, but during its all-sky survey Rosat detected about 100,000.

Rosat had three detectors, two of which were almost identical proportional counters with a field of view (FoV) of 2° and a resolution of 30 arcsec, and a high resolution imager with a FoV of 30 arcmin and a resolution of 5 arcsec sensitive to soft X-rays in the 0.1 to 2 keV range. In addition, Rosat carried a wide field camera operating in the extreme ultraviolet (XUV) band from 6 to 30 nm.

Prior to Rosat, only a handful of XUV sources were known as the hydrogen clouds between the stars absorb energy at these wavelengths, generally limiting observations to relatively near objects a few hundred light years away. Rosat detected over 700 of these objects, however, of which 384 were published in the first Rosat catalogue. The majority proved to be either white dwarfs, with surface temperatures in excess of 25,000 K, or the coronas, with temperatures of a few million K, surrounding F to M type stars.

Gamma-Rays ($\lambda < 0.01$ nm; $E > 0.1$ MeV)

The American SAS-2 that was launched in 1972 was the first satellite to have a gamma (γ)-ray telescope, but it only operated for 7 months. It was followed by the ESA γ-ray satellite, Cos-B, which operated for 7 years from its launch in 1975.

Cos-B while only having a resolution of 2°, was able to show that two of the most intense γ-ray sources are the Vela and Crab pulsars, the former being the brightest source of γ-rays in the sky. It also showed that γ-rays originate not only from energetic sources like pulsars, supernova explosions and gas surrounding possible supermassive black holes but, surprisingly, also from cold gas clouds. The latter do not propagate γ-rays as primary sources, however, but they are generated when high energy cosmic rays collide with atoms in the cold gas clouds.

Some γ-ray sources could be correlated with sources seen at other wavelengths, but there were a number of Unidentified γ-ray Objects (UGOs) that could not be associated with such sources. One particular object, called Geminga, which had been discovered by SAS-2 in the constellation of Gemini, was located by Cos-B to within $\frac{1}{2}$°. It is the second most intense source of γ-rays and, although an X-ray source was found close by in 1979, there were no optical or radio sources that appeared to be likely counterparts. It thus joined the list of UGOs.

Gamma-rays are very rare and, unlike visible light which consists of a continuous stream of photons, with γ-rays the photons are so scarce that they are counted individually like high energy charged particles. During its 7 month lifetime, for example, SAS-2 recorded a total of only 8,000 individual γ-rays (or photons), and Cos-B recorded only 100,000 during its 7 year lifetime, or less than two per hour.

The discovery of a new astronomical object, called a γ-ray burster, was announced in 1973, but this was 6 years after the first such object had been discovered. This delay was because the γ-ray bursts had been discovered by the American military Vega spacecraft, which was on the lookout for thermonuclear explosions on Earth violating the Nuclear Test Ban Treaty. Four Vega spacecraft recorded more than a dozen short, intense γ-ray bursts over a 3 year period.

Over the past 20 years, numerous γ-ray bursts have been observed by orbital spacecraft, and the positions of about 200 sources have been determined with sufficient accuracy for their optical counterpart to be found. Unfortunately, so far nothing unusual has been discovered in the optical photographs, with one exception. The position of a burst, which was observed by nine spacecraft on 5th March 1979, appears to coincide with the supernova remnant N 49 in the Large Magellanic Cloud. This burster is one of only a handful that have been observed more than once, and so it may not be a typical burster.

On 5th April 1991, a new era of γ-ray observations was ushered in with the launch of the Compton Gamma Ray Observatory (GRO) from the space shuttle Atlantis. The GRO was the heaviest unmanned spacecraft ever launched by NASA weighing almost 16 tons, of which 5.4 tons was instruments, (by comparison, the Hubble Space Telescope weighed a total of 9.1 tons). The GRO was 9 m (30 ft) long, 4.5 m (15 ft) in diameter and orbited at the relatively low altitude of 450 km to avoid the Van Allen radiation belts.

There were four instruments on board the GRO, each an order-of-magnitude more sensitive to γ-rays than any previous orbiting instruments. The Energetic Gamma Ray Experiment Telescope (EGRET) operated in the range 20 MeV to 30,000 MeV, and was designed to measure the position of a source to within 10 arcmin. The Compton Telescope operated in the range 1 to 30 MeV, and the Oriented Scintillation Spectrometer Experiment operated from 100 keV to 10 MeV. The fourth experiment, the Burst and Transient Source Experiment, was specifically designed to study the mysterious γ-ray bursts.

The Compton GRO was expected to help in understanding γ-ray bursters. Prior to the GRO, the detection rate of γ-ray bursts was about 60 to 80 per year, whereas the GRO has been detecting about 300 per year in its first 2 years of operation. Interestingly, the sources of these bursts seem to be spread uniformly across the sky indicating that, as they are not concentrated in the plane of the Milky Way, they are either very close to us or at galactic distances.

The brightest burster found so far by the GRO was seen on January 31st 1993, the day of Super Bowl football game in the USA. This superbowl burster, as it became known, shone at its peak 100 times brighter than the brightest steady-state source of γ-rays in the Milky Way. The Compton GRO also discovered a powerful X-ray pulsar, now designated GRO J1008–57, in the constellation of Carina, on 14th July 1993, which briefly, five days later, became one of the brightest X-ray sources in the sky. It emitted an intense pulse of X-rays every 93.6 seconds, but the intensity started to fade after

2 weeks. Investigations are still underway to locate the optical counterpart of this rare transient X-ray pulsar.

In 1992, Jules Halpern of Columbia University and Stephen Holt of NASA had found, using the Rosat spacecraft, that the X-ray source adjacent to Geminga is a pulsar with a period of 237 milliseconds. Then David Bertsch of NASA's Goddard Space Flight Center found, using the GRO spacecraft, that the γ-rays of Geminga varied with the same period, showing that the X-ray and γ-ray sources were the same. So Geminga is a pulsar and, from the change in its pulsation rate, it has been estimated to have an age of about 300,000 years. Finally, in late 1992, the Italian astronomers Giovanni Bignami, Patrizia Caraveo and Sandro Mereghetti found that Geminga had moved 1.5 arcsec in 8 years, indicating that it is very close to us. Geminga appears, after all, to be a normal neutron star whose extreme brightness is due solely to its relative proximity to Earth.

The Hubble Space Telescope

Many astronomers had idly dreamed of the advantages of having a large optical telescope operating above the Earth's atmosphere, but Lyman Spitzer of Yale University was the first person to detail the advantages of placing a large optical telescope in space. This was in 1946 when the United States had just begun testing their captured V2 rockets, and over 10 years before the launch of Sputnik 1. It was not until 1962, however, that the Space Science Board of the American National Academy of Sciences organised a major review of the scientific research to be undertaken in space and considered, in the process, the possibility of following up the OAO programme with a large optical telescope in space. The Working Group on Astronomy then set out to consider a series of fundamental questions, namely, should the telescope be located in space or on the Moon, what size should be its primary mirror, and should it rely on astronauts for regular servicing?

It is amazing to look back now and realise the scope of the programme that was being discussed in the year that America had launched its first astronaut into Earth orbit in a tiny Mercury capsule. But America had committed itself to putting a man on the Moon by the end of the 1960s, and if they could do that, they could certainly launch a large space telescope in the following decade. This "can do" attitude of NASA was backed by the President of the USA with a seemingly unlimited amount of money, following the humiliations of the early American space programme.

The first detailed design study for what was then called the Large Space Telescope, or LST, was carried out by Boeing in the mid 1960s. It was for a telescope with a 3 m (120 inch) diameter primary mirror operated alongside a manned space station in low Earth orbit, the telescope size being limited by the capability of the Saturn V launch vehicle. But could a 3 m mirror be made light enough to survive the launch and achieve maximum ("diffraction limited") performance? As the largest spacecraft mirror then being designed was an 80 cm (32 inch) to be carried by OAO-3, the jump in size looked too large to make in one step, so an intermediate size 1.5 m (60 inch) telescope was examined. Such an intermediate instrument would reduce the risk associated with the jump straight to a 3 m telescope, and would provide some astronomical results sooner, but the total cost of a 1.5 and 3 m programme would be larger, and the final LST would be later.

NASA had used an alternative strategy to the conservative approach that an intermediate LST implied when they developed the massive Saturn V rocket. Instead of building an intermediate size rocket and flying that, NASA had undertaken extensive ground testing of the Saturn V and its associated components before launching the real thing. The Saturn V programme was a resounding success, and so NASA's Astronomy Planning Panel suggested a similar approach for the LST; cut out any intermediate design and go for the big LST straight away, but undertake an extensive programme of ground testing at all stages during development. After much deliberation, this strategy was agreed but, as money became more and more scarce in the 1970s and early 1980s, the scope of the extensive testing programme was gradually reduced, along with the size of the mirror which was reduced from 3.0 to 2.4 m (95 inches).

The launch of the Hubble Space Telescope, or HST, as the LST was now called, was delayed by a number of technical problems but, with less than a year to go to its rescheduled launch in the autumn of 1986, it was further delayed by the Challenger Shuttle disaster in common with many other spacecraft. The Challenger delay, as it turned out, was over 3 years, and it was not until 10th April 1990 that the launch countdown was started. A faulty power unit in the shuttle caused a launch abort 4 minutes before lift-off, and it was not until 2 weeks later that the HST (see Figure 16.2) was finally placed in orbit.

The HST carried four main instruments each contained in 2.2 m (7 ft) long units behind the primary mirror; namely the Wide Field/Planetary Camera (WF/PC), the Goddard High Resolution Spectrometer, the High Speed Photometer and a European instrument called the Faint Object Camera. Incidentally, Europe also made the large retractable solar arrays to bring ESA's financial contribution to the HST programme to 15%, consequently allowing European observers 15% of the observing time on the telescope. The key requirement for the HST optics was to have 70% of the light from a star contained within a circle of 0.1 arcsec diameter, which at the time of its design was way in advance of the performance of any other telescope. Astronomers eagerly awaited the first images from space to see if this had been achieved.

The first image was taken with the WF/PC on 20th May 1990. To most people this looked as sharp as could be expected before the focus of the telescope was fine tuned, but to Roger Lynds, of the National Optical Astronomy Observatories and a member of the WF/PC team, there was something seriously wrong. At a meeting on the following day he suggested that the optics were suffering from spherical aberration, a claim that was made independently by Chris Burrows of ESA just 24 hours later. Over the next 2 or 3 weeks, astronomers tried to focus the telescope, but

Figure 16.2 *The Hubble Space Telescope during testing. (Courtesy* Lockheed Missiles and Space Company, Inc.*)*

gradually the dreadful truth dawned on them, the primary mirror was suffering from chronic spherical aberration causing star images to be about seven times larger in diameter than expected. The discovery of the spherical aberration was made public on 27th June.

At the end of June, NASA set up a committee of enquiry under Lew Allen, the director of JPL, to detail the fault and examine its cause. The committee rapidly concluded that the spherical aberration had not been found in ground testing because of a fault in the test set-up for the primary mirror. Normally such an error would have been picked up when the optical system, including both the primary and secondary mirrors, had been tested as a complete system but, unfortunately, financial constraints had prevented such a test taking place. But whatever the cause of the problem, it was there, so what was to be done with the HST?

The spectrometer and photometer experiments were least affected by the aberration, but it caused major problems for the two imaging packages, namely the WF/PC and the Faint Object Camera. A computer program was quickly produced to "clean up" the stellar images, but the result could not be perfect, and the solution was clearly only a stop-gap measure pending a more long-term solution. After much agonising, it was decided to send up a repair mission which would replace the High Speed Photometer package with a corrective optics system, and this was achieved by a series of exhaustive spacewalks from the space shuttle Endeavour in December 1993. Subsequent astronomical images have shown that the repair was a resounding success (see Figure 16.3).

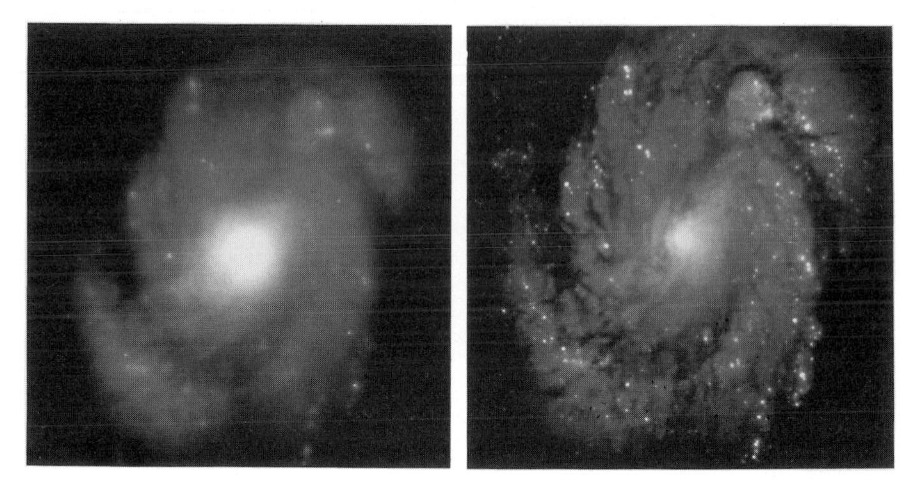

Figure 16.3 *Images of the centre of galaxy M100 before (left) and after (right) the repair of the Hubble Space Telescope. (Courtesy* Space Telescope Science Institute.*)*

17 Modern Astronomy in Context

Introduction

Astronomy over the last hundred years has gone through a number of phases, which have been described in some detail above. I will now briefly examine how external events have affected this astronomical research.

1890–1914

During the period from 1890 up to the First World War, major new observatories were established in North America, the first international star catalogue, the Carte du Ciel, was begun, and great strides were taken in the categorisation and understanding of stellar spectra. In physics, X-rays, gamma rays and the electron were discovered, and the quantum theory of radiation was developed. Einstein introduced his theory of relativity, and the first steps were taken in understanding how stars generate their energy.

In this period, the most advanced countries usually had one large government-sponsored observatory, i.e. Greenwich in England, Paris in France, Pulkovo in Russia and the United States Naval Observatory in Washington, and the main universities usually had a small astronomy department. In addition, in the USA, a number of wealthy individuals or trusts either bought instruments for existing observatories, or set up their own observatories. In England wealthy amateurs also played a key role.

This was a period of industrial growth and relative stability in the developed countries, although serious political tensions were appearing in Europe as the second decade of the twentieth century opened. This eventually resulted in the First World War of 1914–18.

1914–1939

The First World War was a tragedy for the millions of people who lost their lives. Its impact on astronomy in Europe was also devastating, lasting well beyond the end of the war, as so many young men had been killed. In Russia, with the Communist take-over of power in 1917, matters were even worse, as the country lurched from civil war to starvation and mass purges, and, in the early 1920s, letters were received in the United States from starving Russian astronomers asking for food. As a result, Edwin Frost, who was then Director of the Yerkes Observatory, organised an informal committee, and several hundred American astronomers made monthly contributions to enable food parcels to be sent to all the main Soviet observatories. The Soviet astronomers were full of gratitude, and named an asteroid Ara in recognition of the aid that they had received from the American Relief Administration.

The Allies' insistence that Germany paid reparations after the First World War, to help defray its cost, and the loss of the German Empire ultimately led to Hitler's rise to power in 1932, pledging to restore national pride and control the Jewish population which he thought was siphoning away Germany's wealth. This introduced a new and desperate phase for Germany's Jews, and many eminent Jewish scientists fled the country over the next few years, to live in the USA or elsewhere in Europe.

Meanwhile, on the other side of the Atlantic, the United States was relatively unaffected by the First World War. Wealthy individuals continued to support astronomy as patrons in the 1920s, but the Wall Street stock market crash of 1929, and the world-wide economic slump of the 1930s had its effect. So funding for astronomy was kept under strict control in both the Old and New Worlds in the 1930s. New observatories were at a premium, and astronomers had generally to "make do" with ageing equipment.

During this inter-war period, major advances were made in understanding the scale of the Milky Way and the universe as a whole. Galaxies were finally proved to be distant collections of stars receding from us at ever faster rates, the further away they were. Nuclear processes in stars were beginning to be understood, as were the mechanisms for the transfer of heat from the interior of stars. The fundamental importance of hydrogen in the universe gradually became clear, and theories were developed for white dwarfs, neutron stars and black holes, but the real significance of some of these theories only became evident much later.

As the storm clouds gathered over Europe in the 1930s, the first tentative steps were taken in a new astronomy, Radio Astronomy, and the first experimental rockets were launched in the United States and Germany. In the United States the early rocket flights were largely ignored, but in Germany the army became actively involved in rocket developments.

1939–1970

The Second World War from 1939 to 1945 had an even more profound effect on astronomy than the First. Europe was devastated once again, and this time the United States was much more involved, drafting its young people into the armed forces to fight in both Europe and the Pacific. Millions of Russians and Jews met their deaths, and among their numbers were innumerable astronomers and physicists. Amid all this misery, however, the Second World War had a positive impact on astronomy as, unlike the First, it was a much more technically-oriented conflict. Developments in rocketry, radar and nuclear physics, which had a profound effect on the conduct of the war, were also used afterwards in space research, radio astronomy, and the understanding of stellar processes.

The period after the Second World War saw the USSR take over Eastern Europe, and Communism take a foothold in the Far East, particularly in China and Korea. For the first time in its history, the United States was the only country able to resist disturbing developments of this magnitude, and, although it had allies, most of these had been decimated by the war that had just finished.

The Soviet Union, in the years immediately after the war, did not have any allies geographically close to the United States and so, in order to pose a threat to the USA directly, the Soviets had to develop intercontinental ballistic missiles (ICBMs). The Americans, on the other hand, had bases in Western Europe and, consequently, they developed ICBMs at a much slower pace. It was only when the Soviet Union exploded its first atomic bomb in 1949 that the USA began to take the Soviet ICBM threat seriously, and start to accelerate its own ICBM programme.

In the 1950s the Soviet Union wanted to demonstrate its new-found technological power to the United States and her allies, as well as to countries in the Third World to persuade them to join the Communist camp. The USSR decided that the most visible way of demonstrating its technology and gain international recognition, without divulging military information, was to enter the space race, and beat the Americans to put the first man in space, and then on the Moon. At the start, the Soviet policy worked, and they had some astounding successes; being first to put a satellite into orbit (in 1957), and first to put a man in space (in 1961). The American effort suffered in the mid 1950s, however, because they wished to avoid using military rockets for civilian applications, but, stung into action, President Eisenhower gave the necessary authorisation in 1958 to use these rockets, and the space race had begun. It lasted for a brief but highly exciting 12 years, and ended when the USA landed the first men on the Moon in 1969.

So during the period from 1945 to 1970, major funding was poured into rocket and space research by both the USA and the Soviet Union. Western Europe and Japan also participated, but at a much more modest level; Japan, in particular, being hampered by the constraints put on its military establishment after the war.

International collaboration has been the key to successful major astronomical research programmes since 1945. In Western Europe, at the end of the war, there was a strong political will to develop European institutions, in order to get the different nationalities working together, and so reduce the possibility of another European war breaking out. The EEC and the EFTA trade areas were developed and, on the scientific side, CERN (the European centre for particle physics research), ESA (the European Space Agency) and ESO (the European Southern Observatory) were established.

In parallel, in the United States national astronomical facilities were established for the first time. The National Observatory was set up at Kitt Peak, Arizona and the National Radio Astronomy Observatory was built at Green Bank, West Virginia.

Major advances in radio technology during the Second World War gave radio astronomy a kick start when the conflict ended. Not only was there a great deal of surplus communication equipment available, but there were many trained engineers to use and develop it. The large radio telescope at Jodrell Bank was completed in the late 1950s, followed shortly by the Parkes antenna in Australia, and the American Arecibo dish in Puerto Rico. Large radio interferometers were also built in Australia, the UK, the Netherlands and the USA, to improve on the resolution of radio sources.

In the 1940s there had been problems with the Big Bang theory of the universe, as the magnitudes of the galactic red shifts and galactic distances seemed to indicate an age for the universe considerably less than the age of the Earth. This discrepancy led to the birth of the Steady State theory, in which atoms are produced at just a fast-enough rate in the universe to balance its expansion, and keep its density constant. The difficulty with the Big Bang theory was largely overcome, however, when the distance scale was dramatically increased in the early 1950s, leading to a much greater age for the universe, which was now comfortably greater than that of the Earth. The discovery of the cosmic background radiation in 1965, that had been predicted by the Big Bang theory, virtually finished off the Steady State theory.

The 1960s also saw major developments in observational astronomy with the discovery of quasars, pulsars and BL Lac sources. The first X-ray source (other than the Sun) and the first gamma ray source were found, and neutrinos were discovered in cosmic rays.

1970 to The Present

Public support for the space programme rapidly faded in the United States at the end of the 1960s, once the Americans had put a man on the Moon, and the last seven Apollo missions were cancelled to save money. The Cold War was also beginning to thaw, and, in recent years, the United States has welcomed collaboration from other countries to help provide major new facilities. In particular, the USA worked with Canada and France in building a large telescope on Mauna Kea, and agreed on a 15% European contribution to the Hubble Space Telescope.

Teamwork is now very much the order of the day in observational astronomy. This is especially so in space research, where the astronomer is dependent on the work of hundreds or, sometimes, thousands of people who have designed, manufactured, launched and operated his observational platform. Major telescope facilities are also very sophisticated, using both state-of-the-art electronic devices to record the images, and complex computers to analyse them. Even theoretical work is often dependent on the use of large computers. Because of these interdependencies, it is less common today for a discovery to be attributed to one person, and it is more difficult for people to shine as individuals, as is evident from the large number of co-authors often seen for astronomical papers. This may, however, be partly due to astronomers being more ready to give credit to their team than they were before.

What has this teamwork brought us in the last 20 years?

The 1970s was a difficult period for the world economy, with the oil crisis fuelling rising inflation. The space budget in the USA was heavily cut, although many of the satellites then in orbit, under construction, or in the design phase, were still supported, albeit at a reduced level. Many new observatories and major new telescopes had also been planned by the developed countries in the 1960s, and these were generally completed, as their financial requirements tended to be relatively modest. Eight of the 12 largest telescopes available today saw first light in the 1970s, and major observatories were developed at Mauna Kea (Hawaii), La Silla and Cerro Tololo (Chile), Siding Spring and Mount Stromlo (Australia), La Palma (Canaries) and Zelenchukskaya (Russia).

Even though funding for the various space programmes has been tight since Apollo, there have been a number of astonishing successes, particularly in the area of planetary exploration. The Voyager spacecraft that visited the outer planets were probably the most conspicuous success, but the Magellan spacecraft that imaged Venus, and the armada of spacecraft that intercepted Halley's comet were also important. Pride of place must, however, go to the Voyagers and their incredible images of the sulphur volcanoes on Io, the chevron feature on Miranda, the nitrogen geysers of Triton, and the rings of Saturn which were resolved into a thousand individual rings with "spokes".

A number of highly successful orbital observatories have also been launched in the last 20 years or so to study objects in the visible, infrared, ultraviolet, X-ray and γ-ray wave bands. These have enabled active galaxies to be studied, and evidence to be sought for dark matter and black holes. In parallel, the Big Bang theory has been developed, with the addition of the inflationary theory of the early universe, and the expected small variations in the microwave cosmic background radiation have been found.

Today, it is as true as ever, however, to admit that "The more we know, the more we know we don't know". The 9.8 m (390 inch) Keck telescope has now been commissioned, the Hubble Space Telescope has been repaired, and many large Earth-based telescopes are being built and planned. So astronomers should have the tools available to start the new century, and new millennium, full of hope and anticipation, and ready to take our knowledge of the universe further into the unknown. It remains to be seen what external constraints will be placed on them, as mankind struggles to live in this small corner of the cosmos that we call "Home".

References and Further Reading

References

1. Ledger, Edmund, *The Sun: Its Planets and Their Satellites,* Edward Stanford, London, 1882.
2. Berry, Arthur, *A Short History of Astronomy*, John Murray, 1898. (Dover reprint 1961.)
3. Pannekoek, A., *A History of Astronomy*, Allen and Unwin, 1961. (Dover reprint 1989.)
4. Ball, Robert S., *The Story of the Heavens,* Cassell, Revised Ed., 1897.
5. Clerke, Agnes M., *A Popular History of Astronomy During the Nineteenth Century*, Black, 4th Ed., 1908.
6. Newcomb, Simon, *Popular Astronomy*, Macmillan, 2nd Ed., Revised, 1898.

Further Reading

The following is a selection of books which give more details of the topics covered in this book. The publishers mentioned are generally those for copies available in the UK. In other countries the publishers may differ. Earlier and/or later editions are sometimes available.

General Astronomy

Audouze, Jean and Israël, Guy (eds), *The Cambridge Atlas of Astronomy*, Cambridge UP, 1985.

De la Cotardière, Philippe (ed.), *Larousse Astronomy*, Hamlyn, 1987.

Meyers, Robert A. (ed.), *Encyclopedia of Astronomy and Astrophysics*, Academic Press, 1989.

Historical*

Berendzen, Richard, Hart, Richard and Seeley, Daniel, *Man Discovers the Galaxies*, Neale Watson, New York, 1976.
Bok, Bart J. and Bok, Priscilla F., *The Milky Way*, Harvard UP, 3rd Ed., 1957.
Brown, R. Hanbury, *Boffin. A Personal Story of the Early Days of Radar, Radio Astronomy and Quantum Optics*, Adam Hilger, 1991.
Friedman, Herbert, *The Astronomer's Universe*, Norton, 1990.
Gingerich, Owen (ed.), *Astrophysics and Twentieth Century Astronomy to 1950: Part A*, Cambridge UP, 1984.
Gingerich, Owen, *The Great Copernicus Chase*, Cambridge UP, 1992.
Goldberg, Leo and Aller, Lawrence H., *Atoms, Stars and Nebulae*, Harvard Books on Astronomy, Churchill, London, 1950.
Hearnshaw, John B., *The Analysis of Starlight*, Cambridge UP, Pbk Ed., 1990.
Hermann, Dieter B. (Krisciunas, K., trans.), *The History of Astronomy from Herschel to Hertzsprung*, Cambridge UP, 1984.
Hirsh, Richard F., *Glimpsing an Invisible Universe: the Emergence of X-ray Astronomy*, Cambridge UP, 1983.
Hoyle, Fred, *Home is Where the Wind Blows: Chapters form a Cosmologist's Life*, University Science Books, California, 1994.
Hubble, Edwin, *The Realm of the Nebulae*, Oxford UP, 1936.
Hufbauer, Karl, *Exploring the Sun*, Johns Hopkins UP, 1991.
Jeans, James, *The Universe Around Us*, Cambridge UP, Pbk Ed., 1960.
King, Henry C., *The History of the Telescope*, Charles Griffin, 1955. (Dover reprint 1979.)
Littmann, Mark, *Planets Beyond: Discovering the Outer Solar System*, Wiley, Revised Ed., 1990.
Lovell, Bernard, *Astronomer by Chance*, Macmillan, 1991.
Rowan-Robinson, Michael, *Ripples in the Cosmos*, W. H. Freeman, 1993.
Sheehan, William, *Worlds in the Sky*, University of Arizona Press, 1992.
Smith, F. Graham, *Radio Astronomy*, Penguin, 1960.
Struve, Otto and Zebergs, Velta, *Astronomy of the 20th Century*, Macmillan, 1962.
Thackeray, A. D., *Astronomical Spectroscopy*, Eyre and Spottiswoode, 1961.
Waterfield, Reginald, *A Hundred Years of Astronomy*, Duckworth, 1938.

Biography

Ashbrook, Joseph, *The Astronomical Scrapbook*, Cambridge UP, 1984.
Heramundanis, Katherine (ed.), *Cecilia Payne-Gaposchkin*, Cambridge UP, 1984.
Lightman, Alan and Brawer, Roberta, *Origins; the Lives and Worlds of Modern Cosmologists*, Harvard UP, 1990.

---

*This list includes both historical survey books and old texts explaining astronomical knowledge of the time.

Macpherson, Hector, *Makers of Astronomy*, Oxford UP, 1933.
Porter, Roy (consultant ed.), *The Hutchinson Dictionary of Scientific Biography*, Helicon, 2nd Ed., 1994.
Sharov, Alexander S. and Novikov, Igor D., *Edwin Hubble, the Discoverer of the Big Bang Universe*, Cambridge UP, 1993.

Cosmology

Ferris, Timothy, *Coming of Age in the Milky Way*, Bodley Head, 1989.
Gribbin, John, *The Omega Point*, Heinemann, 1987.
Hawking, Stephen W., *A Brief History of Time*, Bantam, 1988.
Will, Clifford M., *Was Einstein Right?*, Oxford UP, 1986.

The Solar System

Beatty, J. Kelly and Chaikin, Andrew (eds), *The New Solar System*, Cambridge UP, 3rd Ed., 1990.
Briggs, Geoffrey and Taylor, Fredrick, *The Cambridge Photographic Atlas of the Planets*, Cambridge UP, 1982.
Greeley, Ronald, *Planetary Landscapes*, Allen and Unwin, Revised Ed., 1987.
Moore, Patrick and Hunt, Garry, *The Atlas of the Solar System*, Mitchell Beazley, 1983.

Stars and Galaxies

Arp, Halton, *Quasars, Redshifts and Controversies*, Interstellar Media, Berkeley, California, 1987.
Burnham, Robert, Jr., *Burnham's Celestial Handbook*, 3 vols, Dover, 1978.
Goldsmith, Donald, *Supernova, the Violent Death of a Star*, Oxford UP, 1990.
Henbest, Nigel and Marten, Michael, *The New Astronomy*, Cambridge UP, 1983.
Murdin, Paul and Murdin, Lesley, *Supernovae*, Cambridge UP, 1985.
Osterbrock, Donald E. (ed.), *Stars and Galaxies, Citizens of the Universe (Readings from Scientific American)*, W. H. Freeman, 1990.

Space, Sounding Rockets and Observatories

Calder, Nigel, *Giotto to the Comets*, Presswork, London, 1992.
Cornell, James and Gorenstein, Paul (ed.), *Astronomy from Space*, MIT Press, 1985.
DeVorkin, David H., *Science With a Vengeance. How the Military Created the US Space Sciences After World War II*, Springer-Verlag, 1993.
Krisciunas, Kevin, *Astronomical Centres of the World*, Cambridge UP, 1988.
Smith, Robert W., *The Space Telescope*, Cambridge UP, Pbk Ed., 1993.

Units

I have tried to use a set of units which would be acceptable to both amateur and professional astronomers. Some people normally use SI units, but some feel more familiar with cgs or imperial units, so my selection has been a compromise.

The units used are as follows:

Quantity	Unit	Abbreviation	Equivalent
Distance	megaparsec	Mpc	
	parsec	pc	3.26 light years or 3.1×10^{16} m
	light year		6.3×10^4 AU or 9.5×10^{15} m
	Astronomical Unit	AU	150 million km
	kilometre	km	0.62 miles (1 mile ≡ 1.6 km)
	metre	m	39.4 inches
	centimetre	cm	10^{-2} m (1 inch ≡ 2.54 cm)
	millimetre	mm	10^{-3} m
	micron	μm	10^{-6} m
	nanometre	nm	10^{-9} m or 10 Ångströms
Angle	arc minute	arcmin	1/60 degree
	arc second	arcsec	1/3600 degree
Mass	kilograms	kg	2.2 lb (1,000 kg ≡ 1 metric ton)
	grams	g	
Density	grams per cubic centimetre	g/cm^3	10^3 kg/m^3

Pressure	bar		1.01 atmosphere or 1.0×10^5 Pa
Temperature	kelvin	K	°C + 273
Energy	electron volts	eV	1.6×10^{-19} J
Frequency	gigahertz	GHz	10^9 Hz
	megahertz	MHz	10^6 Hz
Magnetic field	gauss		10^{-4} tesla

General Abbreviations Used

AAT	Anglo-Australian Telescope
AU	Astronomical Unit
BCD	Barbier, Chalonge, Divan (stellar classification system)
Caltech	California Institute of Technology
CCD	Charge Coupled Device
COBE	Cosmic Background Explorer (spacecraft)
CPD	Cape Photographic Durchmusterung (star catalogue)
ESA	European Space Agency
ESO	European Southern Observatory
HST	Hubble Space Telescope
INT	Isaac Newton Telescope
IR	Infrared
IRAS	Infrared Astronomical Satellite
IUE	International Ultraviolet Explorer (satellite)
JPL	Jet Propulsion Laboratory
LMC	Large Magellanic Cloud
MIT	Massachusetts Institute of Technology
MKK	Morgan, Keenan, Kellman (stellar classification system)
MMT	Multi-Mirror Telescope (in Arizona)
NASA	National Aeronautics and Space Administration
NRAO	National Radio Astronomy Observatory (in the USA)
NRL	Naval Research Laboratory (in the USA)
NTT	New Technology Telescope
SNR	Supernova Remnant
UBV	Ultraviolet, Blue, Visible (stellar classification system)
UC	University of California
UV	Ultraviolet
VLA	Very Large Array (radio telescope)
WHT	William Herschel Telescope

The Greek Alphabet

α	alpha	ι	iota	ρ	rho
β	beta	κ	kappa	σ	sigma
γ	gamma	λ	lambda	τ	tau
δ	delta	μ	mu	υ	upsilon
ε	epsilon	ν	nu	φ	phi
ζ	zeta	ξ	xi	χ	chi
η	eta	ο	omicron	ψ	psi
θ	theta	π	pi	ω	omega

Subject Index

Main references are in bold type

Name Index